HAZARDOUS WASTE TREATMENT PROCESSES

WEF Manual of Practice No. FD-18
Second Edition

Prepared by **Hazardous Waste Treatment Processes Task Force** of the **Water Environment Federation**
William J. Librizzi, Jr., *Chair*

Michael L. Bradford
Edward J. Cooney
Alexander H. Danzberger
Gomes Ganapathi
Yuyang Gong
David J. Gworek
Donald J. Joffe
Steven Kennedy
James A. Kent
Robert M. Lantz
Jeffrey L. Pintenich
David L. Russell
Enos Stover
John T. Sudnick
Daniel J. Watts

Under the Direction of the **Industrial and Hazardous Wastes Subcommittee of the Technical Practice Committee**

1999

Water Environment Federation
601 Wythe Street
Alexandria, VA 22314-1994 USA

IMPORTANT NOTICE

The material presented in this publication has been prepared in accordance with generally recognized engineering principles and practices and is for general information only. This information should not be used without first securing competent advice with respect to its suitability for any general or specific application.

The contents of this publication are not intended to be a standard of the Water Environment Federation (WEF) and are not intended for use as a reference in purchase specifications, contracts, regulations, statutes, or any other legal document.

No reference made in this publication to any specific method, product, process, or service constitutes or implies an endorsement, recommendation, or warranty thereof by WEF.

WEF makes no representation or warranty of any kind, whether expressed or implied, concerning the accuracy, product, or process discussed in this publication and assumes no liability.

Anyone using this information assumes all liability arising from such use, including but not limited to infringement of any patent or patents.

Library of Congress Cataloging-in-Publication Data

Hazardous waste treatment processes / prepared by Hazardous Waste Treatment Processes Task Force of the Water Environment Federation ; under the direction of the Industrial and Hazardous Wastes Subcommittee of the Technical Practice Committee.

p. cm. — (WEF manual of practice ; no. FD-18)

New ed. of: Hazardous waste treatment processes including environmental audits and waste reduction, 1990.

Includes bibliographical references and index.

ISBN 1-57278-1-53-X

1. Hazardous wastes—Purification. 2. Hazardous wastes—Management. I. Water Environment Federation. Hazardous Waste Treatment Processes Task Force. II. Water Environment Federation. Industrial and Hazardous Wastes Subcommittee.

TD1060.H37 1999

628.4'2 21—dc21 99-039985

ISBN 1-57278-153-X

Printed in the USA 1999

Water Environment Federation

Founded in 1928, the Water Environment Federation is a not-for-profit technical and educational organization. Its goal is to preserve and enhance the global water environment. Federation members number more than 42 000 water quality professionals and specialists from around the world, including engineers, scientists, government officials, utility and industrial managers and operators, academics, educators and students, equipment manufacturers and distributors, and other environmental specialists.

For information on membership, publications, and conferences, contact

Water Environment Federation
601 Wythe Street
Alexandria, VA 22314-1994 USA
(703) 684-2400
http://www.wef.org

Manuals of Practice

(As developed by the Water Environment Federation)

The Water Environment Federation (WEF) Technical Practice Committee (formerly the Committee on Sewage and Industrial Wastes Practice of the Federation of Sewage and Industrial Wastes Associations) was created by the Federation Board of Control (now Board of Directors) on October 11, 1941. The primary function of the committee is to originate and produce, through appropriate subcommittees, special publications dealing with technical aspects of the broad interests of the Federation. These manuals are intended to provide background information through a review of technical practices and detailed procedures that research and experience have shown to be functional and practical.

A.B. Pincince
T. Sadick

Authorized for Publication by the Board of Directors
Water Environment Federation

Quincalee Brown, *Executive Director*

Contents

List of Tables

List of Figures

Preface

The purpose of this manual is to provide a description of the technical and regulatory approaches to hazardous waste treatment. It offers solutions to hazardous waste control problems and will be a useful tool to industrial environmental engineers and managers, practicing environmental engineers, and municipal agencies charged with the operation of a wastewater treatment facility or a solid waste landfill.

This manual should also be used by those responsible for hazardous waste management at private facilities or a public works department. It provides descriptions of the most widely applied technologies, including biological, chemical, and thermal processes. This manual also includes guidelines for conducting environmental audits and approaches to waste reduction. It is intended for use and reference in planning, executing, and continuing a program of hazardous waste treatment.

This second edition of this manual was produced under the direction of William J. Librizzi, *Chair.*

Principal authors and the chapters for which they are responsible are

Edward J. Cooney (7)
Alexander H. Danzberger (9)
David J. Gworek (5)
William J. Librizzi, Jr. (1)
Jeffrey L. Pintenich (2)
David L. Russell (4, 6)
John T. Sudnick (8)
Daniel J. Watts (3)

Additional review was provided by Kauser Jahan.

Authors' and reviewers' efforts were supported by the following organizations:

Bechtel Jacobs, Oak Ridge, Tennessee
ETTL Engineers & Consultants, Tyler, Texas
Global Environmental Operation, Lilburn, Georgia
Jacobs Engineering Group, Inc., Houston, Texas
Jordan Jones & Goulding, Atlanta, Georgia
New Jersey Institute of Technology, Newark, New Jersey
RMT, Inc., Greenville, South Carolina

Chapter 1
Introduction

BACKGROUND

Since World War II, the United States has experienced an industrial growth that has yielded perhaps the highest standard of living in the world. However, this way of life has created environmental concerns as to how to manage waste produced by industries and consumers. Recent estimates from the U.S. Environmental Protection Agency (U.S. EPA) indicate that the U.S. generates approximately 5.4×10^{12} kg (6 billion tons) of industrial, agricultural, commercial, and domestic waste each year. Of this figure, approximately 2.3×10^{11} kg (250 million tons) are considered hazardous waste, and more than 90% of this waste is liquid.

The management of hazardous waste has become a critical issue to all sectors of society. In the engineering field, among government, in the company board rooms, at the plant, and throughout the public sector, there is recognition that answers to managing hazardous waste must be found. This challenge is complex, multidisciplinary, and multimedia. The demand for technically sound, implementable, and cost-effective approaches for treating hazardous waste is of immediate and growing concern. Types of waste that require proper management include solids, liquids, and vapors. The wastes come from many sources, including manufacturing processes, wastewater stream spillage, discarded products, or construction materials.

A historical perspective shows that improper hazardous waste disposal has resulted in the contamination of the air, land, surface water, and groundwater, presenting potential risks of human health and environment. A number of laws enacted by the U.S. Congress have been directed toward a more effective hazardous waste management approach. In 1976, Congress enacted the Resource Conservation and Recovery Act (RCRA) authorizing the U.S. EPA to regulate current and future waste management and disposal practices. The primary goals of RCRA are to:

- Protect human health and the environment from the disposal of hazardous waste,
- Ensure that hazardous wastes are managed in an environmentally sound manner, and
- Reduce the amount of hazardous waste generated.

The definition of hazardous waste is defined by RCRA as a solid waste (that is, a solid, liquid, or contained gas) that meets certain criteria as it relates to quantity, concentration, characteristics, and an increase in mortality, serious irreversible or incapacitating reversible illness, or potential hazard to human health and the environment.

The U.S. EPA regulation establishes the process to determine whether a waste is hazardous and subject to hazardous waste regulations. The process determines whether the waste exhibits ignitability, corrosiveness, reactivity, or toxicity. The process further identifies if the waste is listed as source specific, based on specific industries (such as petroleum refining and wood preserving) or nonspecific (such as a solvent or a discarded commercial product).

In dealing with hazardous wastes, RCRA provides an important distinction in U.S. regulations for what is known as "permit by rule." The relevant portion of the reference is found in 40 CFR 261.1O, and the regulation provides a critical exemption for generators of hazardous wastes. If the waste generated is regulated under another federal permit program, such as the Clean Water Act, the waste is not hazardous. This is applicable even if the wastes generated are covered by a National Pollutant Discharge Elimination System (NPDES) permit directly or indirectly through a regulation from a publicly owned treatment works (POTW). The converse is also true, as will be illustrated by the following example.

A company in Georgia manufactured designed lighting. The company had its own wastewater treatment plant (WWTP) and combination spray field and septic tank for the treatment of its treated effluent. Because the company did not have an NPDES permit, the entire WWTP was classified as a hazardous waste treatment facility and had to obtain a RCRA Part B permit retroactively. The logic set forth for the requirement of the RCRA B permit was that the discharge was a hazardous waste, including the treated effluent, and the permit by rule provisions that would have exempted the facility from obtaining a Part B permit did not apply.

In 1984, Congress amended RCRA with passage of the Hazardous and Solid Waste Amendments (HSWA) to strengthen the approach to hazardous waste management developed in the earlier legislation. In particular, HSWA stated that the practice of land disposal was least acceptable and alternate treatment methods should be established. The direct land disposal provisions in HSWA were in response to serious problems that were consequences of Love Canal, New York, and Times Beach, Missouri, where the wastes were considered disposed because they were out of sight. The passage of the Comprehensive Environmental Response, Compensation, and Liability Act (CERCLA) and Superfund Amendments and Reauthorization Act (SARA) addressed the cleanup of inactive and abandoned hazardous waste sites.

Superfund deals with abandoned, inactive waste sites and establishes a program for private party of public funded cleanup. The federal National Contingency Plan (NCP) provides the framework for action such as the interaction between federal and state governments through the national and regional response team, policy such as "how clean is clean," operating procedures, and enforcement requirements.

The Clean Water Act (CWA) is a statute for water quality and wastewater treatment. Industries discharging into waterways or publicly owned collection systems are being required to comply with toxic reductions. The Clean Air Act (CAA) established the National Emissions Standards for Hazardous Waste Pollutants (NESHAPs) to control the release of toxic substances that may adversely affect health. U.S. EPA has established NESHAPs for asbestos, beryllium, mercury, and vinyl chloride and has proposed standards for benzene and arsenic. Other CAA program areas include New Source Performance Standards (NSPS), prevention of significant deterioration, nonattainment area regulations, and air toxic regulations. These regulatory requirements are discussed as they relate to hazardous waste treatment technologies. It is suggested that the appropriate U.S. EPA or state agency be contacted when contemplating the selection of a specific hazardous waste treatment technology.

Knowledge has been gained from the past. The policies that shape these laws present the framework for hazardous waste management for this country. These environmental mandates have three distinct focus areas:

- They address past practices through remediation. Past management of hazardous waste has created thousands of abandoned and inactive sites. U.S. EPA estimates that there are more than 217 000 sites throughout the nation that may require remediation. These sites include those identified under Superfund (NPL), RCRA Corrective Action, Underground Storage Tanks, Department of Defense, Department of Energy, other federal agencies, and the states. CERCLA and SARA provide the policy framework for dealing with these problems. The WEF Special Publication, *Hazardous Waste Site Remediation Management* (WEF, 1999), provides guidance in this particular area.

- They control and require proper management of the waste currently generated. Increased regulatory authority such as RCRA and expanded enforcement program rising costs for handling, transporting, treatment, and disposal concerns regarding future liabilities and an adverse public image have demanded that effective and efficient controls of hazardous waste be used.
- They prevent creation of new hazardous waste problems through waste minimization. The promotion and enhancement of the reduction of the amount of waste generated through one or a combination of methods, such as source reduction and recycling, has recently become a more important factor in the proper management of hazardous waste. The HSWA set national policy by declaring that, wherever feasible, the generation of hazardous waste is to be reduced or eliminated as expeditiously as possible. New legislative and regulatory initiatives at the federal level and in many states further promote the goals of waste minimization. For example, in October 1990 the U.S. Congress enacted the Pollution Prevention Act of 1990, which firmly established the national policy to prevent or reduce the generation of pollution at the source whenever possible. The act in fact codified the national hierarchy for waste management as waste prevention or reduction, recycling, treatment, and disposal. This book recognizes the importance of waste reduction and its relationship to treatment strategies by devoting attention to the subject.

Hazardous waste management along with other environmental protection programs have seen change in recent years. We have not only seen the shift from command and control to waste prevention and reduction, but we have also seen a refocus of the regulatory framework to one that is a voluntary cooperative approach. ISO 14000, an internationally developed and released standard of the International Organization for Standardization, provides industry with the framework for an overall environmental management system (EMS) that encourages a uniform program of environmental management. The standard, in this regard, allows a company to develop and operate environmental quality management programs as an integral part of its total operation with reduced effects from laws and regulation and regulatory oversight. This is particularly important for multinational corporations that deal with different environmental approaches across the globe. Recent reports indicate that companies are adopting EMS to enhance management system efficiencies, waste reduction, and proactive regulatory compliance. Specifically, the list of reasons can include:

- International trade competitiveness—meeting international guides reduces the barriers associated with national and regional standards (that is, reduced bureaucratic complexity and redundancies).
- Credibility—third-party certification and a regulatory record should improve company credibility with associates, regulatory agencies, environmental groups, and the public.

- Profit—benefits include reduced liability/risk, pollution prevention and waste reduction, better insurance potential and preference on bank loans, and improved internal management.
- Sustainable development—management standards can serve as the basis for the integration of economic growth and environmental protection.

Typically, the ISO 14000 framework that addresses short- and long-term effects on the company's products, services, and environmental credibility includes five areas: environmental management systems, environmental performance evaluations, environmental auditing, life cycle assessment, and environmental labeling. Specifically, this framework is achieved through a documented program of education and training and a sound knowledge and conformance to relevant local and federal environmental regulation. The acceptance and application of ISO 14000 through this framework can provide organizations with an effective way to:

- Establish an appropriate environmental policy that is based on environmental management among the organizations' highest priorities and includes a commitment to pollution prevention;
- Develop management and employee commitment, accountability, and responsibility to environmental protection, including recognition of legislative and regulatory requirements relative to the organizations' activities, products, and services and establishment and achievement of targeted performance levels;
- Implement an effective emergency preparedness and response program;
- Provide appropriate and sufficient resources, including training, to achieve performance targets;
- Provide an operational control and maintenance program to ensure a sustainable high level;
- Establish a management process for audit, evaluation and feedback on program effectiveness, needed improvements, and measurement of resulting environmental benefit; and
- Encourage external contractors and suppliers to establish an EMS program.

ISO 14000 as developed is similar in structure to ISO 9000, which is an accepted international standard for an environmental management system. In concept, the commonality between ISO 9000 and ISO 14000 is in the areas of management responsibility, internal auditing, and document control training. The ISO organization has established task forces to coordinate the two standards.

There are a number of documents available that can provide guidance and advice on how ISO 14000 can be planned and implemented. (For example, see *ISO 14000 Handbook* [CEEM Information Services and ASQC Quality Press, 1996].)

HAZARDOUS WASTE MANAGEMENT

CONCEPTS

There is general consensus that approaches to safe and efficient hazardous waste management are available. Extensive experience has been gained in treating industrial flows, giving us a reasonable foundation of scientific and technical understanding and knowledge. However, more needs to be learned about using these experiences to solve the complex problems faced today.

In general, hazardous waste treatment is based on chemical structure. Compounds requiring treatment can be divided into the following three categories:

- Organic, which contain carbon as derived from or found in living things or are manmade and capable of yielding electrons in oxidation reactions;
- Inorganic, which do not contain carbon as an element and are incapable of yielding electrons in oxidation reactions; and
- Mixed waste, which includes both organic and inorganic constituents and is an increased challenge for treatment.

Organic compounds as a class differ from inorganic compounds in several ways. They are combustible in air, typically have lower melting and boiling points, are less soluble in water, and are biodegradable. Reactions are molecular (that is, taking place one molecule at a time). As a result, equilibrium is very slow to develop. In addition, molecular weights can be very high, affecting properties such as solubility, volatility, and surface properties.

Organic compounds typically fall into categories based on their carbon linkage. For example, alkenes are carbon compounds linear in structure and aromatic. Carbon compounds with sulfur, nitrogen, and oxygen represent another class that includes alcohols, ethers, ketones, esters, aldehydes, and amino compounds. Compounds that contain fluorine, chlorine, bromine, or iodine along with carbon and hydrogen atoms are another class of organic compounds called halogenated organic compounds.

The influence of these properties on the treatability of organic compounds can be subtle. Several factors should be kept in mind when attempting to deal with organic compounds.

- There are more than 300 000 identified organic compounds currently cataloged by the Chemical Abstract Service.
- Approximately two-thirds of these compounds have never been observed in air or water.
- Water-soluble compounds number in the tens of thousands.
- Most situations in the field involve mixtures (that is, more than one compound).

The complexity of identifying and treating specific organic compounds can be overwhelming. For this reason, wastewater treatment practices incorporate nonspecific processes such as carbon adsorption or biological treatment. These nonspecific processes typically follow a sequence of removal of classes of compounds from the easiest to the more difficult to remove. Depending on the mixtures of compounds present, the treatment objectives of the wastewater treatment process are met incrementally with additional process units added to facilitate higher removals.

Inorganic compounds are distinguished from organic compounds by the fact that, in general, these compounds do not contain carbon. For inorganic compounds, there are two general classes that encompass the broad variety of substance of interest:

- Metals, or substances capable of giving up electrons in oxidation/reduction reactions, and
- Nonmetals, or other substances that consist of oxygen, fluorine, chlorine, or sulfur.

Metals can be distinguished from nonmetals by two distinctive properties:

- Metals are substances that form salts with oxygen-bearing anions such as sulfate or nitrate.
- Metals form at least one oxide with a reasonably strong base to the limit of solubility.

The latter property can be used to further distinguish metals. Metal oxides or hydroxides are basic in character. They react with acids, especially with strong mineral acids such as hydrochloric, nitric, or sulfuric acid to form salts. These product salts tend to be more soluble than the metal oxides or hydroxides themselves. Therefore, if a sample of a solid material exists in a water solution and this material dissolves in one of the mineral acids, the solid material is probably a metal oxide or hydroxide.

Nonmetals tend to form oxides that are acidic in character, and they tend to react with bases but not acids. If a solid residue in a water solution dissolves when caustic soda (sodium hydroxide) is added, the substance could be a nonmetal. Conversely, some insoluble hydroxides exhibit the property of being capable of being dissolved in both acids and bases. These compounds have the other characteristics of metals and are called amphoteric. Amphoteric metals include aluminum, cadmium, chromium, iron, and zinc, among others. The hydroxides of these amphoteric metals exhibit a minimum point of solubility that is pH dependent. This characteristic can be used to facilitate separation of the amphoteric metal hydroxide from other present precipitated soils.

Some metals are known as heavy metals. This characterization includes the amphoteric and other metals. Heavy metals are conveniently known by their chemical group and the relative position of the element with respect to sodium. Several heavy metals are very insoluble—a property that can be used in waste treatment to reduce the concentration of the heavy metal to very low levels.

OBJECTIVE

This MOP is a presentation of current practice in the field of hazardous waste treatment. It offers technical solutions to hazardous waste control problems and should be a useful tool to industrial environmental engineers and managers, practicing environmental engineers, municipal agencies or authorities charged with the operation of wastewater treatment facilities or solid waste landfills, and citizens concerned about hazardous waste control problems in their community. This MOP includes discussions on a number of important topics including state-of-the-art approaches for managing hazardous waste; detailed descriptions of the most widely applied technologies (these technologies include biological, physical, chemical, and thermal processes); an overview of the application and use of hazardous waste treatment technologies, including the types and characteristics of waste handled by each technology; key design parameters that must be considered in the application of a technology; and regulatory requirements and their effects on the use of each technology. Table 1.1 provides a summary of the technologies covered by this MOP, including the types of contaminants that each technology is capable of treating.

This book provides information about several important areas related to decision-making for hazardous waste management. Each technology chapter includes a discussion on waste characterization, treatability study requirements, and use of bench and pilot testing. Finally, the discussion considers safety aspects and new and innovative technologies.

The book also presents a comprehensive discussion on environmental auditing, waste reduction, and material handling and storage as an integral part of managing hazardous waste. Environmental auditing involves the identification of the points of hazardous waste generation and regulatory deficiencies in the operation. Once known, the points of waste generation can be changed or modified to improve the manufacturing and operational process to achieve better compliance with federal and state regulatory requirements. Audit elements discussed in this MOP include planning pre-audit information gathering, conducting the audit evaluation, reporting, and following up. Dealing with the issue of confidentiality is also presented. Waste reduction represents an economically beneficial response to growing regulatory requirements, higher disposal costs, and increased liability costs faced by both industry and government. Inventory controls, better housekeeping, process modifications, and materials substitution are several approaches to waste reduction. Materials control is important in waste reduction because whether they are raw materials for manufacturing, waste treatment chemicals, or wastes that

Table 1.1 Summary technologies.

Technology	Most common application[a]	Types of contaminants
Biological aerobic		
Suspended, fixed film, bioreactors, biofiltration, slurry, solid phase	Wastewater, soil, sludge, air	Volatile organics, nonvolatile organics, phenol, palate esters, polycyclic aromatic hydrocarbons, PCBs, pesticides, solvents, hydrogenated ethers
Biological anaerobic		
Fixed and fluidized bed reactors, up flow sludge blanket	Wastewater, soil, sludge, air	Chlorinated solvents, phenolic compounds, sulfonated benzene, nitrogen substituted benzene
Physical		
Coagulation, flocculation, sedimentation, filtration (granular membrane, ultrafiltration), granular activated carbon (GAC) stripping, flotation, gentrification, ion exchange, encapsulation, thermal drying	Wastewater, soil, sediment, groundwater	Volatile organics, nonvolatile organics, polynuclear aromatics, phenolic, surfactant, chlorinated aromatics, solvents, chlorinated hydrocarbons, metals
Chemical		
Neutralization, oxidation, stabilization and solidification, precipitation	Wastewater, soil, sludge, sediment	Iron, nonvolatile organics, cyanide, sulfides, phenol, benzene, metals
Thermal		
Liquid and vapor incinerators, catalytic vapor incinerators, fluidized bed, rotary kiln, multiple hearth	Wastewater, sludge, air	High-organic waste, waste oils, halogenated solvents, nonhalogenated solvents, organic liquids, pesticides, herbicides, PCBs

[a]Potential applications for all covered technologies can include wastewater, soil, sludge, sediment, groundwater, and air. The most common applications are based on generally accepted practice in the field relative to technology selection.

are being stored, poor controls result in higher costs. Changing the current operating practice for materials handling can produce short- and long-term benefits measured by less waste and more stable and safe environment.

A key aspect to hazardous waste treatment is the concept of treatment process integration. This concept is characterized by the application of segregated streams to enhance cost-effective treatment; combined treatment to effect multiple removals; and, finally, contaminants that may hinder a downstream process element. Evaluation of the myriad options possible for a particular waste stream or media will be based on a number of site-specific factors. Some of these factors are given below.

- The chemical nature of the waste. Is it organic, inorganic, or mixed?
- The waste matrix. Is it solid, liquid, or gas?
- Hazardous versus nonhazardous stream.
- Treatment requirements, including
  - Ultimate disposal reuse,
  - Delisting considerations,
  - Zero discharge possibilities,
  - Discharge limitations (POTW or receiving stream),
  - On-site and off-site disposal, and
  - Permitting procedures.
- Unit process selection. Table 1.2 presents a conceptual treatment process for a liquid mixed waste stream. Four unit processes required to adequately treat the waste flow include phase separation, metals removal, organic treatment, and polishing treatment as needed to meet stringent effluent limitations.

ORGANIZATION

This MOP has been structured so that each chapter provides information about a single topic and can be used independently. However, the MOP has both management and engineering details so that problems identified through the application of the principles in the early chapters can be solved through the use of the treatment practices described in the later chapters.

The definition of hazardous waste treatment, the reasons for its importance, and who can use the information provided in this MOP are included in this introductory chapter. Methods for analyzing the operation to determine its regulatory deficiencies are described in Chapter 2. That chapter presents the latest information about various companies and logical steps for a successful environmental audit, the elements of an environmental audit, and the regulatory policy associated with an environmental audit. Chapter 3 presents waste reduction techniques including a concise introduction to waste reduction concepts, insights into process evaluation, and a review of waste reduction such as inventory management, process modification, volume reduction,

Table 1.2 Liquid treatment processes.

Phase separations first
Oil-free aqueous streams
Water free oil streams
Remove metals next
Will poison biological treatment
Will cause scaling in physical processes
Consider various treatment combination for organics
Volatile with semivolatile
Biorefractory compounds, dioxins, pesticides
Chloride containing
Polishing treatment
Membrane
Thermal
Ion exchange

recovery, and reuse. Also presented are the components and approaches for implementing a waste reduction program. Chapter 4 presents operational techniques for hazardous materials and waste handling and storage. It includes such important topics as site location, structural considerations, compatibility, maintenance, management plans, and contingency planning.

The next four chapters describe treatment under the broad headings of biological (Chapter 5), physical (Chapter 6), chemical (Chapter 7), and thermal (Chapter 8). This MOP has recognized the potential for overlap between physical and chemical treatment; thus, chemical treatment processes covered were those that occur by chemically induced change, while physical treatment is based on physical processes. Precipitation of a hazardous waste from solution is presented as a good example of the overlap between physical and chemical treatment. Here, the addition of chemicals causes a chemically induced formation of insoluble precipitation that is ultimately removed through gravity separation, centrification, or filtration. Another overlap example is the topic of solidification and stabilization, which uses both chemical and physical change in the treated media.

Each of these chapters presents in-depth discussion of the principles of each technology, waste process descriptions, basic design parameters, equipment descriptions, applicability, waste handling, treatability studies, specific regulatory requirements, and emerging technologies. Case studies are presented where useful to describe the technology and provide useful reference data.

Chapter 9 considers the integration of technologies into treatment trains to address complex and varied waste (that is, liquid, sludges, and organic and inorganic waste). Waste character, design, design considerations, and options for ultimate disposal are covered in this chapter.

Wastes included in this definition can be properly managed by the technologies as presented in this MOP. The regulatory requirements under RCRA and other statutes should be considered an important factor in decision-making.

Table 1.3 Outline for 40 CFR Parts 260–280.

Part	Information description
Part 260	General information, definitions, petitions
Part 261	Hazardous waste identifications system, criteria for listing and characteristics, lists of wastes and off-specification chemicals
Part 262	Hazardous waste generator requirements, manifesting, pre-transport, recordkeeping
Part 263	Hazardous waste transporter requirements
Part 264	Final standards for TSD facilities, contingency planning preparedness and prevention, manifesting waste analysis closure and post closure care, groundwater monitoring, specific technical standards
Part 265	Interim standards for TSD facilities that apply until a permit is issued
Part 266	Standards for the management of specific hazardous waste and specific types of hazardous waste land disposal facilities
Part 267	Interim standards for new hazardous waste land disposal facilities
Part 268	Land disposal restrictions
Part 270	Hazardous waste permitting regulations for TSDs
Part 271	Requirements for the authorization of state hazardous waste programs
Part 280	Requirements for underground storage tanks

For example, RCRA requires that generators manifest their waste. Treatment, storage, and disposal facilities that receive hazardous waste must receive a U.S. EPA permit that ensures compliance with established designs and operating criteria for safe operation. These requirements have substantial effect on the handling, treatment, storage, and disposal of wastes using the technologies covered by this MOP. These RCRA regulations are codified in the Code of Federal Regulations (CFR), Parts 260–280. The code is updated annually and can be obtained from the U.S. Government Printing Office (Washington, D.C.). A breakdown of 40 CFR Parts 260–280 is provided in Table 1.3.

REFERENCES

CEEM Information Services and ASQC Quality Press (1996) ISO 14000 Handbook.

Water Environment Federation (1999) *Hazardous Waste Site Remediation Management.* Special Publication, 2nd Ed., Alexandria, Va.

Chapter 2 Environmental Audits: Internal Due Diligence

INTRODUCTION

There are several types of environmental audits being conducted throughout the world today. Audits basically are differentiated by their respective goals and may also use somewhat differing approaches to accomplish these goals. One of the most prevalent types of audits is the transactional audit, which is performed during the acquisition, divestiture, or refinancing of an enterprise. These enterprises would typically include manufacturing facilities, laboratories, commercial facilities, and undeveloped land. Because environmental liabilities may be transferred during a sale unless indemnities or other legal mechanisms prevail, transactional audits typically are required by the lending institutions before finalizing the deal. This is also true for refinancing because environmental liabilities can be so costly that the survival of a facility or company could be jeopardized.

Another type of audit is the regulatory or enforcement audit. These can be similar to the more routine inspections performed by regulatory personnel or can be part of a consent decree when a facility is forced to have an audit performed by an independent group. This type of audit is requested to determine the general deficiencies that might exist at a site.

In recent years, audits have been performed with an emphasis on waste reduction or elimination. These audits often focus on the identification of waste sources and result in ideas for waste minimization and toxicity reduction through recycling, reclamation, material substitution, and other solutions. It is prudent to perform an internal due diligence (IDD) audit before developing a waste reduction plan to include the complete facility and project the overall ramifications of the program. It can also be prudent to perform IDD audits subsequent to the implementation of a waste reduction program to assess its success and determine if other opportunities are available to affect further reductions.

The IDD audit is performed by company personnel or outside contractors to evaluate the status of regulatory compliance and the ability of current standard operating procedures to meet the various needs of the facility in the areas of environmental, health, and safety concerns. An IDD audit can be used to evaluate the efficiency of the current treatment technologies, or it can be used to help select various treatment options for a facility's hazardous waste. The efficiency of current waste treatment is assessed with respect to treatment performance (including maintenance and repair time), operational costs, and regulatory compliance. It is also critical to evaluate the hazardous waste treatment program in terms of capabilities versus anticipated changes (such as residual limits or permitting).

The IDD audit can be comprehensive, addressing all aspects of a facility's environmental, health, and safety operations, or it can be focused. Focused audits target specific operations or concerns of the facility and are performed after an incident, after a change in manufacturing processes, or after changes in the applicable regulations. As indicated, different audits use different approaches. Much of the information provided about IDD audits, however, is applicable to all types of audits.

PURPOSE

An IDD audit is a proactive, voluntary effort initiated by some level of management within a company. Its overall purpose is to identify any existing or potential problems so that remedies can be applied in a timely fashion.

The IDD audit typically is intended to determine regulatory compliance status, evaluate environmental management practices and facilities, or to identify risks and liabilities, including those attributable to past practices. In some cases, compliance and consistency with corporate (and even facility) directives, policies, and procedures are also addressed. The required corrective actions and recommended risk reduction measures resulting from audits can serve as the basis for long-term facility improvement plans and projected environmental control budgets.

Pertinent federal, state, and local environmental, health, and safety regulations are considered in comprehensive multimedia compliance audits. Practice and facility audits may evaluate the adequacy of operating procedures; maintenance programs; staffing (type, number, and competence); treatment, control, and monitoring devices; and laboratory testing. Risk and liability assessments are often extended to include items that are not required by law, for example, an evaluation of the potential for sudden or gradual hazardous chemical releases from underground storage tanks that in all aspects comply with regulatory requirements.

POTENTIAL APPLICATIONS

The IDD audits may be useful in obtaining a picture of existing conditions as part of evaluating hazardous waste treatment processes or selecting new candidate processes.

GENERAL. Environmental audits can take several forms, as discussed previously, ranging from comprehensive reviews of regulatory compliance and the suitability of practices and facilities to focused evaluations of waste sources or specific environmental media. Using the developing international standards on environmental management systems (EMSs), audits may also be performed to evaluate the effectiveness of a facility's or company's EMS.

WATER. Audits may focus on aqueous media. The reader is referred to the subsections entitled "Wastewater Discharges," "Water Supply," and "Injection Wells" discussed under "Conducting the Audit."

AIR. For guidance on audits pertaining to the air medium, the reader is referred to the subsection entitled "Air Emissions" discussed under "Conducting the Audit."

SOILS AND SLUDGE. The reader may find the following subsections under "Conducting the Audit" in this chapter useful with respect to soils and sludge: "Solid and Hazardous Wastes," "Chemical and Oil Storage," and "Air Emissions."

SPECIFIC CHEMICAL COMPOUNDS. Topics that pertain to the treatment, storage, and disposal of specific chemical compounds may be found in the following subsections under "Conducting the Audit" in this chapter: "Air Emissions," "Chemical and Oil Storage," "Polychlorinated Biphenyls," "Asbestos," "Pesticides," "Health and Safety," and "Radioactive Substances."

COMPONENTS OF SUCCESSFUL INTERNAL DUE DILIGENCE AUDITS

The performance of an IDD audit needs the support of top management. This is especially true when an audit is performed for the first time at a facility. Management must communicate to all involved that this is a serious undertaking and that honest answers must be provided, even if the information is negative. This support must be felt through all subsequent levels of management at the audited facility. Also, there needs to be a commitment to follow up on the findings of the audit (from top management down) to prevent a waste of time and funds. Commitment to followup basically includes revising current practices and facilities to effect regulatory compliance or providing better procedures by which given operations are performed. Additionally, management support is needed for the funds necessary to conduct an IDD audit. Significant expenditure of funds should be anticipated, especially if revisions are required.

Throughout the audit and the followup, individuals at the facility being audited must be prepared to spend additional time beyond that demanded by their routine job tasks. Time is spent collecting the necessary information and duplicating materials for distribution to the audit team, as well as reviewing the standard operating procedures, regulatory requirements, and other associated data with the audit team both during the audit and in preparation of the report.

Another important component of a successful IDD audit is the creation of the audit team. The first aspect is the selection of the team itself. The audit team leader must have previous experience in conducting environmental audits, a broad command of environmental regulations and environmental science and technology, and a demonstrated ability to lead audit teams. It is necessary to find individuals whose expertise can assist the audit team leader in developing an accurate picture of the status of the facility. In addition, these individuals must have the time available to perform their assigned tasks and a commitment to see that they are completed in a timely fashion. The effect on the daily duties of the individual, of course, varies from site to site. Each team member must not only be knowledgeable in his or her area of expertise, as previously mentioned, but must also be able to translate the requirements of the regulations and operating procedures to the real-world operations of the facility.

The team members must be impartial in their review of the status of the facility. Frequently, the most knowledgeable individual for a given area is the manager or supervisor of that area. If the audit team member is such an individual, negative or deficient areas within his or her realm of responsibility must be accurately (and impartially) reported. If this individual is too close to the situation to be objective, another audit team member or outside reviewer must be able to supply an accurate assessment. It is typically easier for an outside or independent auditor to be impartial, and such individuals may have the advantage of exposure to a wider range of situations. In any case, periodic audits by individuals completely independent of the audited facility may be advisable.

A clear definition of the purpose of the audit should be provided to the audit team and to those individuals who are required to respond to the demands of the team. Similarly, each member of the audit team must have a clear definition of his or her role in the conduct of the audit.

To expedite the audit, it is critical that the pertinent documents and personnel be readily available to the team. Availability means not only a presence in a file, but also that copies of all pertinent documents, such as laboratory results, permits, reports from internal or external consultants, relevant standard operating procedures, and training files, are available for review and distribution. If these are not available, it is necessary that facility personnel know where any missing information can be obtained. Indeed, part of the audit process is to examine the adequacy of records retention practices. As will be discussed later in this chapter, guidance in this area of information retrieval and organization is often provided by a questionnaire. It is recommended that this questionnaire be given to facility personnel well in advance of the on-site phase of the audit, with sufficient lead time to answer the questions and prepare the document files.

Throughout the audit, cross-checks need to be in place to ensure accuracy, consistency, and thoroughness. Cross-checks may be performed by asking the same question in different ways, asking the same question of facility personnel with different job functions, or reviewing and comparing records with common elements. Examples of such cross-checks include reconciliation of waste disposal records (for example, manifests or certificates of destruction) with annual reports to regulatory agencies and comparison of hazardous material purchase records with waste disposal records.

The final phase of the IDD audit is the preparation and issuance of the audit report. This report should be concise and candid. It needs to address, in great detail, those areas of deficiency rather than those areas of good performance or compliance. Areas that are in compliance can, however, serve as a reference to help remediate deficient areas.

The audit report needs to be produced in a timely fashion so that the information in its contents remains accurate. It is also critical that the report be timely so that implementation of the recommendations can begin or additional study can be undertaken. Recommendations in the audit report might include:

- No action,
- Revisions to the standard operating procedures,
- Issuance of standard operating procedures,
- Physical plant upgrades,
- Waste reduction/pollution prevention measures,
- Obtaining of permits,
- Improved sampling and monitoring, and
- Additional study on the extent and impact of questionable areas that cannot be readily determined from existing information (data, reports, and permits). Additional studies might, for example, be needed to evaluate groundwater and soil contamination.

ISO 14000 GUIDELINES

In recent years, the International Organization for Standardization (ISO) has been developing a series of standards to help organizations implement and improve their environmental management systems (EMS). They define an EMS as that part of the overall management system that includes organizational structure, planning activities, responsibilities, practices, procedures, processes, and resources for developing, implementing, achieving, reviewing, and maintaining the environmental policy of an organization. Key principles in implementing or improving an EMS include the following:

- Recognize that environmental management is among the highest corporate priorities;
- Establish and maintain communications with internal and external interested parties;
- Determine the legislative requirements and environmental aspects associated with the organization's activities, products, or services;
- Develop management and employee commitment to the protection of the environment, with clear assignment of accountability and responsibility;
- Encourage environmental planning throughout the product or process life cycle;
- Establish a process for achieving targeted performance levels;

- Provide appropriate and sufficient resources, including training, to achieve targeted performance levels on an ongoing basis;
- Evaluate environmental performance against the organization's environmental policy, objectives, and targets and seek improvement when appropriate;
- Establish a management process to audit and review the EMS and to identify opportunities for improvement of the system and resulting environmental performance; and
- Encourage contractors and suppliers to establish an EMS.

The various standards and guidelines are published as the ISO 14000 series.

The ISO series calls for periodic audits of the EMS by properly trained, impartial auditors (internal or external) to determine whether the system conforms to planned arrangements and has been properly implemented and maintained. Audit programs and procedures are suggested to ensure that the EMS audits accomplish their objectives. The audits culminate in the preparation of a report with written findings and a closing meeting to discuss the findings with facility and organization management personnel. Three standards deal with the EMS audit:

- ISO 14010—Guidelines for environmental auditing, general principles of environmental auditing;
- ISO 14011—Guidelines for environmental auditing, audit procedures, auditing of environmental management systems; and
- ISO 14012—Guidelines for environmental auditing, qualifications criteria for environmental auditors.

U.S. EPA'S POLICY ON ENVIRONMENTAL AUDITS

In July 1986, the U.S. Environmental Protection Agency (U.S. EPA) issued its first formal policy statement on environmental auditing (51 FR 25004). This policy encourages the regulated community to develop, implement, and upgrade environmental auditing programs, because audits help ensure compliance with environmental regulations and lead to integrated management of environmental hazards. U.S. EPA believes that auditing should be a voluntary activity and published a restatement and clarification of the agency's policy on environmental auditing in July 1994 (59 FR 38455). The agency continues to encourage the use of environmental auditing and is participating in the development of the ISO standard, among other guidelines.

Copies of environmental audit reports will not be routinely requested by U.S. EPA because such requests could serve to inhibit auditing activities. On a case-by-case basis, however, U.S. EPA may request all or portions of an audit report. Examples of these requests given in the policy statements are:

- Audits conducted under consent decrees or other settlement agreements;
- When a company places its management practices at issue by raising them as a defense;
- Criminal investigations in which state of mind or intent are of importance; and
- When the agency determines it is needed to accomplish a statutory mission.

The regulated community is cautioned that certain audit findings may, by law, have to be reported to government agencies (for example, the discovery of an unreported reportable quantity [RQ] release under the Comprehensive Environmental Response, Compensation, and Liability Act [CERCLA]). Communities should also be aware that audit reports cannot shield monitoring, compliance, or other information that would otherwise be reportable or accessible to U.S. EPA.

As a matter of general policy, U.S. EPA will not eliminate inspections or reduce enforcement responses in exchange for the implementation of environmental auditing. On the other hand, effective audit programs should lead to an improved compliance status. To the extent that compliance performance is factored into assigning inspection priorities, facilities with a good history of compliance may be subject to fewer inspections. Furthermore, if a facility reports violations discovered during an audit that were not otherwise required to be reported, U.S. EPA "may exercise its discretion to consider such actions as honest and genuine efforts to assure compliance" as relates to determining enforcement responses to violations. These principles have been incorporated into the agency's enforcement response and civil penalties. In certain cases, U.S. EPA may stipulate environmental audits in consent decrees or other settlements in which auditing could produce a remedy for identified problems and reduce the possibility of recurrence. These could take the form of a strict regulatory compliance audit and/or a management audit of environmental compliance policies, practices, and controls.

Federal facilities subject to the various environmental statues and regulations are also encouraged to institute environmental auditing programs. U.S. EPA will provide technical assistance to other federal agencies in designing their audit programs. State and local regulatory agencies are encouraged to adopt U.S. EPA's auditing policies or similar policies.

STATE POLICIES ON AUDIT REPORTS

Several states have statutes, regulations, or policies that pertain to environmental compliance audits. These vary in the way they address important issues such as evidentiary privilege, immunity from administrative and civil penalties, and notifications and corrective actions for noncompliance matters. Note, however, that U.S. EPA may not agree with immunity provisions of state audit laws and regulations. This is a rapidly evolving area of environmental law and environmental auditors are encouraged to consult with legal counsel before proceeding with environmental audits of any type.

TYPES OF INTERNAL DUE DILIGENCE AUDITS

There are basically two types of IDD audits. One is comprehensive; the other is focused. The comprehensive audit addresses all aspects of environmental, health, and safety regulations and practices relevant to a given facility. This type of audit is performed to evaluate overall compliance with regulatory requirements and best management practices (BMPs). The duration of such an audit will vary according to the physical size of the facility, the complexity of the facility, and the number of members on the audit team. A comprehensive audit may go on simultaneously in different areas of the facility if the auditing team is composed of independently functioning members. For example, one group might be addressing aspects relating to air emissions, while another group is working with those individuals responsible for wastewater treatment or hazardous waste management at the facility. Because frequently only one individual has responsibility for environmental practices at a given facility, this approach may not be feasible.

It is not uncommon for a comprehensive audit to take between 1 and 5 days and be staffed by a team of up to 10 professionals. If it is a complex facility, it is helpful to convene a meeting at the end of each day to review what, if anything, significant has been found. If the audit involves simultaneous assessments of the various aspects of facility operations, it is important that the team members interact while on site to help ensure the accuracy, thoroughness, and consistency of the audit being performed.

The focused audit is, as indicated by its name, more targeted and involves less time and fewer individuals on the audit team than the comprehensive IDD audit. Focused audits can be used after a facility incident, a change in process, a change in regulation, or as a followup to check areas that were targeted for revisions after a comprehensive audit. Because of its nature (that is, less manpower, time, and cost), the focused audit can be performed more frequently and is often preferable to repeating a comprehensive audit. The results of the focused audit, however, must be included in the support documentation provided to audit teams performing future comprehensive audits.

RELATIONSHIP TO PHASE I ENVIRONMENTAL SITE ASSESSMENTS

In 1993, the American Society for Testing and Materials (ASTM) published a written practice to help participants in commercial real estate transactions satisfy due diligence responsibilities. ASTM E1527, Standard Practice for Environmental Site Assessments: Phase 1 Environmental Site Assessment Process, which was revised in April 1994, has the purpose of defining adequate commercial and customary practice in the U.S. for conducting an environmental site assessment (ESA) of a parcel of commercial real estate with respect to the range of hazardous substances within the scope of CERCLA and petroleum products. The practice is intended to permit a user to satisfy one of the requirements to qualify for the innocent landowner defense to CERCLA liability. An environmental compliance audit may be performed in conjunction with an ESA, although they are not the same.

The goal of the processes established by ASTM E1527 is to identify recognized environmental conditions (RECs), if any, associated with the target property. These RECs are defined as the presence or likely presence of any hazardous substances or petroleum products on a property in conditions that indicate an existing release, a past release, or a material threat of a release of any hazardous substances or petroleum products into structures on the property or into the ground, groundwater, or surface water of the property. The existence of RECs associated with hazardous substances may prevent the user from qualifying for an innocent landowner defense under CERCLA.

The Phase 1 ESA has four components according to ASTM E 1527: (1) records review, (2) site reconnaissance, (3) interviews, and (4) evaluation and report preparation. The types of records that an environmental professional is to review about the site and nearby properties include standard environmental record sources, physical setting records, and historical use information about the target property and surrounding area. The site reconnaissance or inspection is a visual observation to identify RECs, if any. Both owners and occupants are interviewed to obtain further information about the target property, as are one or more local governmental officials. The ESA report summarizes the work that was done and expresses an opinion regarding the existence of RECs.

STAFFING OPTIONS FOR AUDIT TEAMS

As has been discussed previously, a successful IDD audit team must comprise qualified, seasoned professionals capable of providing an impartial review of compliance and management practice situations. Company-internal personnel may form the audit team, as long as they can function independently from the entity being audited. Corporate staff often conduct the audit, or the team is formed from a combination of corporate professionals and those stationed at other facilities in the organization.

Outside consultants and specialists may serve as audit team members as long as they are fully indoctrinated with regard to the audit's purpose and protocol. Consultants may be requested to conduct the audit in its entirety. External personnel, however, may not be well suited to evaluate compliance with company policies and procedures. If a sampling and testing program is to be implemented as a part of the audit (typically of a limited nature), certified laboratories and testing companies should be selected well in advance (see Table 2.1 for reasons to use an outside audit team).

Professionals from a variety of disciplines may be effective audit team members. Most commonly, audit professionals include the following disciplines:

- Environmental engineers,
- Environmental chemists,
- Chemical engineers,
- Civil engineers,
- Hydrogeologists,
- Industrial hygienists,
- Geotechnical engineers,
- Safety engineers,
- Atmospheric scientists,
- Health physicists,
- Nuclear engineers,
- Environmental attorneys, and
- Environmental control managers.

Frequently, the experience and expertise of the auditor is more important than the educational background. For example, an environmental engineer experienced only in industrial wastewater treatment would be an inappropriate auditor for the air pollution control topic. Hydrogeologists are essential for evaluating the adequacy of groundwater monitoring programs or corrective actions at land disposal facilities, but in most cases they are not trained to assess spill prevention control facilities and plans. It is absolutely imperative that auditors possess a strong background in the federal, state, and local regulations they will be dealing with in the audit.

It is highly desirable for audit team members to have dual capabilities (a primary and secondary expertise). This allows for audit interaction and, typically, a more conclusive end product. Also, such an audit team can divide work assignments if unanticipated problems develop at the site.

Table 2.1 Reasons to use outside audit teams or team members.

Facility too small to staff a team
Facility personnel too busy to perform the audit
Facility personnel lack expertise
Management prefers an independent assessment
More objective assessment
Perceived anonymity of facility employees to independent auditors
Multiple facilities can be compared for management consistency on corporate-wide issues

AUDIT ELEMENTS

PLANNING. Because environmental IDD audits, particularly comprehensive audits, represent a major commitment of time, personnel, and funding, planning is an essential element. Planning involves communication before, during, and after the audits. This communication is necessary for scheduling the site visit, determining the availability of facility personnel, and other aspects that will be discussed later in this chapter. Communication must be complemented by organization. Typically, there is little time to waste. Examples of difficulties that could arise because an audit has been poorly organized are that people do not know where to go, that data or reports are not available, and that requirements for entry (such as physical or personnel protective equipment) have not been met. Communication is also extremely important in delivering the message to facility personnel of what is needed and why.

Other aspects in the auditing planning phase include obtaining appropriate security clearances for team members touring the facility. Another concern is adequate working space for the audit team members. Areas must be set aside where team members can write notes, review notes, review documents, and, if necessary, have confidential discussions with various site personnel and have access to support personnel for rapid, on-site duplication of documents.

The logistics of an audit, especially in a large, complex facility, can be very involved. There must be planning relating to intrasite travel. Some facilities have millions of square feet under roof. The use of in-house travel can be essential for conducting an efficient audit in a large facility. These conveyances must be available, charged, and dedicated for the period of the audit. Another important aspect is having the proper safety equipment available should areas require hard hats, glasses, respirators, or special personal protective gear. Team members need to have equipment that fits properly so that the possibility of accidental exposure is reduced; this includes fit testing respirators. Audit team personnel must have appropriate medical monitoring. Additionally, if the audit is occurring on a Superfund site, certain training requirements must be met and documented before audit team entry.

Another aspect of planning for an audit is "contingency." Some contingencies are more manageable than others. A prime example is weather. This includes weather that could delay the arrival of the audit team from off site, as well as weather that can affect the review of external areas of the facility. If sampling is involved during the audit, weather might preclude implementation of the sampling program (for example, there might be surface water or frozen ground).

Another contingency encountered by the audit team is an unanticipated manufacturing downtime at the time of the IDD audit. It is frequently critical to have the processes up and operating in a normal mode during an audit to assess ambient air sources and the resulting internal and external air quality, wastewater discharges, sources of waste generation, and so on. Therefore, audits should not be scheduled during operational downtime or during periods of nonroutine processing. (Note: Job shop facilities need to be evaluated on both an integrated and worst-case basis.) A final aspect in the planning phase is legal considerations. These are addressed in detail later in this chapter.

ETIQUETTE. An adversarial relationship should not be generated during an audit. Etiquette, or, more appropriately, the approach to facility personnel, should be friendly but serious. The auditor needs the assistance of facility personnel. An individual might have some trepidation about reporting deficiencies in his or her area. Also, it is possible that the audit may not be taken seriously. This is especially true when there have been no previous difficulties or needs for an audit. Frequently, individuals interviewed during an audit may be hostile or unable to understand exactly what is needed. Therefore, a question needs to be asked several times, in different ways, until the interviewee understands what is needed. Also, questions often need to be asked several times if the answer seems deficient or off target.

The approach should be in a friendly, nonaggressive manner. Audits are sometimes made more difficult in that inexperienced auditors may "attack" the various personnel as though they were on a witness stand. This typically is not the best way to obtain information, although in some instances it may be the only way. The auditor needs to go to a high level of supervision or management if answers and information are not forthcoming.

Another area of "audit etiquette" relates to how and when outside (regulatory) agencies should be contacted to obtain information without prematurely raising flags that could cause unnecessary problems for the facility. Auditors must understand the need for confidentiality before, during, and after the audit. It is not appropriate to share the audit results with others, including facility staff, without team leader and client approval.

As a courtesy, the auditors should, if at all possible, stick to the schedule so that the time set aside by the interviewees is not exceeded and the overall timetable for touring the facility is maintained. This, however, is frequently not possible. Facility personnel should be amenable to schedule changes when additional information is needed or delays are incurred. If possible, additional time should always be left at the end of each audit day so that timetables can be maintained and those interviews that run long can be completed.

PRE-AUDIT OPTIONS FOR INFORMATION GATHERING. The most common and useful mechanism for obtaining information about a site before the visit is a targeted questionnaire. A preprinted questionnaire with space available for answers, comments, and attachments should be forwarded to the appropriate site personnel with a request to complete and forward by a certain date so the audit team can review them before going on site. This approach is frequently used on IDD audits. It may not be a viable option for other types of audits such as enforcement audits or audits implemented before acquisitions, refinancing, and so on, which are typically performed in a very short time. The questionnaires should be as simple as possible but should address all aspects of the audit. Additionally, photographs, drawings, maps, and process flow diagrams should be provided to the audit team before going on site.

Requests to carry a camera on site and take pictures should be made as early as possible. Photographs and videos are quite useful not only for jogging the memories of the audit team members during report preparation but are also helpful for reviewing a given situation with people who have not been on site.

Receipt of documents requested is integral to developing a quality audit. These documents include permits, correspondence with regulatory agencies, analytical results, manifests, and copies of standard operating procedures relating to environmental, health, and safety programs at the facility. Also included with this information are copies of previous audit reports, both comprehensive and focused.

The time-saving aspects of pre-audit information gathering can be exemplified by the review of manifests. The review of manifests indicates where a facility's wastes were taken. It is frequently important to identify these sites to determine the status of the site regarding RCRA compliance and CERCLA or Superfund designation. Significant costs can be incurred if a facility is identified as a contributor to a Superfund site that is targeted by federal or state regulatory agencies for cleanup. This information about disposal site status, however, frequently takes some time to be received from the state and federal government. Requests often must be in writing. Any head start on this procedure would be beneficial toward the timely completion of the audit report.

CONDUCTING THE AUDIT. Site Features and Operations. Early in the audit procedures, it is essential that the audit team develop a solid understanding of the nature of the facility being audited. This should include general location and layout information, an overview of environmental conditions, and a thorough description of site operations. Table 2.2 provides a summary of the types of data that have proven useful during an IDD audit. Without this profile, the team will not be able to conduct a thorough audit of the site.

Major Topic Areas. Practitioners are well aware of the ever-increasing number of environmental statutes and regulations. In Table 2.3, the principal environmental statutes and regulations are grouped according to major topic areas for comprehensive, multimedia environmental audits. The components of typical audit evaluations for these topics are described in the following paragraphs. Undoubtedly, the first comprehensive audit at a facility will be the most time consuming. Subsequent audits can simply build on the historical data gathered during the first audit.

Table 2.2 Site features and operations profile.

General information
Location map
Site topographic and property boundary map
Site plan (buildings and environmental facilities)
Fencing and security
Building (area, type, and age)
Surrounding land usage
Closest residential area
History of property ownership
Environmental conditions
Proximity of floodplain and surface waters
Vegetative cover
Known threatened or endangered species
Proximity of wetlands, coastal zones, or other sensitive areas
Closest public water supply
Geology and hydrogeology
Nearby surface impoundments (pits and lagoons), landfills, dumps, and treatment sites
Subsurface drainage (number and locations of drains, routing, and discharge)
Previous and ongoing environmental investigations of target or nearby facilities
Community relations
Utility suppliers (electricity, water, gas, wastewater, and refuse)
Zoning and land use restrictions
Air quality maintenance area (AQMA)
Attainment and nonattainment status
Previous audit reports and findings
Facility operations
Previous owners, operators, and land use
Types and dates of significant facility changes or additions
Process description
Process flow diagrams
Applicable Standard Industrial Category (SIC) codes
Process changes over time
Primary raw materials and method of delivery
History of fires, explosions, and significant releases
Comprehensive chemical inventory
Material storage (drums, tanks, bins, etc.)
Employees (number, shifts, and unions)
Operations schedule (number of days per week, hours of operation, and so on)
Product or sales brochures
Expansion and modification plans
Environmental impairment liability (EIL) insurance coverage or denial

Table 2.3 Principal environmental regulations.

Topic	Key statutes and regulations
Air emissions	Clean Air Act (CAA)
	New Source Performance Standards (NSPS)
	State Implementation Plan (SIP)
	National Emissions Standards for Hazardous Air Pollutants (NESHAPs)
	State and local permitting and emissions regulations
	Emergency episode plan requirements
Wastewater discharges	Clean Water Act (CWA)
	Effluent limitations guidelines
	National Pollutant Discharge Elimination System (NPDES) regulations
	Pretreatment regulations
	Local sewer use ordinance
	State water quality standards
Water supply	Safe Drinking Water Act (SWDA)
	Drinking water standards and testing requirements
	State withdrawal permits
Solid and hazardous waste	Resource Conservation and Recovery Act (RCRA)
	Hazardous waste regulations (federal and state)
	State solid waste regulations
	Hazardous Materials Transportation Act (HMTA)
Chemical and oil storage	Comprehensive Environmental Response, Compensation, and Liability Act (CERCLA)
	Hazardous substance release reporting requirements
	Emergency Planning and Community Right-to-Know Act (EPCRA)
	Oil-spill prevention control and countermeasures (SPCC) plans
	Best management practices under the NPDES program
	Underground storage tank (UST) regulations
	Occupational Safety and Health Administration (OSHA) regulations for storage
	Local fire codes
	State requirements

(Continued)

Table 2.3 Principal environmental regulations (continued).

Topic	Key statutes and regulations
Polychlorinated biphenyls	Toxic Substances Control Act (TSCA)
	Usage, storage, and disposal regulations under TSCA
	Hazardous waste land disposal restrictions
Asbestos	NESHAPs
	Effluent limitations guidelines
	TSCA regulations
	OSHA and MSHA regulations for worker protection
	U.S. Department of Transportation (U.S. DOT) regulations for transport
	State regulations
Pesticides	Federal Insecticide, Fungicide, and Rodenticide Act (FIFRA)
	FIFRA registration, certification, and disposal regulations
Injection wells	Underground injection control (UIC) permit system
	Hazardous waste land disposal restrictions
Health and safety	OSHA regulations
	OSHA permissible exposure limits (PELs)
	Hazard Communication Standard (HCS)
	RCRA hazardous waste management training
	State regulations
Radioactive substances	NESHAPs
	Uranium Mill Tailings Radiation Control Act
	State regulations
	Nuclear Regulatory Commission (NRC) regulations
	U.S. Department of Energy (U.S. DOE) requirements

AIR EMISSIONS. The auditor should ask the facility to identify and locate all sources of air emissions. For each source, the contaminants present, gas flow volume, emission rates, and control devices should be defined. Emission permits should be reviewed to ensure that they have been secured when required; that they have not expired, or that renewals have been requested; and that special conditions are being complied with. Permit applications, such as the Title V application, should be evaluated to see if accurate emission and control data were supplied to the regulatory agency. Ideally, the auditor will inspect each source to verify data received and assess the efficacy of control devices.

Maintenance, inspection, and monitoring records should be evaluated to check for problems. Emission test reports and monitoring instrument calibration data (for example, for continuous emission monitors [CEMs]) should be secured and reviewed.

Facility expansion and modification dates should be correlated to applications for new permits or permit modifications. Reports of regulatory agency inspections and notices of violations (NOVs) should be summarized to assess historical (and possibly ongoing) compliance patterns.

Records retention practices should be reviewed to see if the required records are available. It should be determined if new source performance standards, national emission standards for hazardous air pollutants (NESHAPs), or state air emission standards apply to any of the emissions, and if compliance with the standard(s) is being achieved. The auditor should see if the facility has banked emission reduction credits that may be useful if expansions are desired in the future. State regulations should be reviewed to determine if an emergency episode plan is required at the facility. Also, the auditor should define air emission requirements under other programs, such as RCRA standards for hazardous waste management facilities. Finally, the auditor should determine if the facility is fully compliant with prevention of significant deterioration (PSD) requirements for attainment areas and emission offset requirements for nonattainment areas.

WASTEWATER DISCHARGES. The results of recent in-plant wastewater surveys should be reviewed in summary form so that the auditor has an appreciation of the sources, volumes, and pollutants in the wastewater discharges. Discharges should be categorized as sanitary, process wastewater, cooling water (contact or noncontact), stormwater, or combinations of these. Treatment or pretreatment facilities should be evaluated by obtaining and reviewing process flow diagrams and material balance data, sampling and analysis data, process design criteria, unit sizing, permit applications, engineering reports, plans and specifications, and operation and maintenance (O&M) manuals. A walk-through inspection should include discussions with operational supervisors. The physical condition of the treatment units, operational capability, competence of operators, operator licensing compliance, and so on should be assessed.

Outfalls or discharge points should be visually examined to check for obvious problems (for example, oily visible sheen, floatable solids, or highly colored or odorous discharges). Current versions of the NPDES permits or indirect discharge permits should be obtained and evaluated with respect to compliance, especially for effluent limits. Pertinent categorical effluent guidelines should be identified, along with the basis for the NPDES permit (effluent-limited or water-quality-limited basis).

The last 3 to 5 years of discharge monitoring reports (DMRs) should be reviewed to assess effluent performance. A copy of the local sewer ordinance should be obtained to verify compliance. The most recent permit applications should be checked to verify that accurate and complete data were provided. Sample collection, preservation, and analysis protocols should be evaluated with regard to appropriateness and permit conditions. The auditor should determine if certified laboratories are used for discharge analyses and if a laboratory audit program and quality assurance program are in effect.

Facility personnel should be asked if there are or were septic tanks at the site, what was the nature of the discharges, and if local permits and approvals were obtained. Expansion and modification dates should be correlated to new or modified permits and applications. Reports of regulatory agency inspections and NOVs should be summarized to assess historical compliance patterns and ongoing problems. The adequacy of records retention practices should be confirmed.

WATER SUPPLY. The sources of water used should be determined. The auditor should determine if local water supply connection authorizations (and backflow preventors) or surface water or groundwater withdrawal permits have been obtained and are current. Evaluations of the adequacy of water treatment systems may be part of the audit scope. If the facility has its own drinking water supply, the auditor should check compliance with the Safe Drinking Water Act (SDWA) standards and testing programs. Raw and finished water quality data should be summarized for review. The auditor should determine if there is reason to believe that present or future raw water quality will deteriorate. Nearby water usage points should be identified to determine if the facility being audited may affect the quality of off-site surface water intakes or groundwater wells. As with other topics, regulatory inspection, NOVs, and records retention practices should be included in the site audit.

SOLID AND HAZARDOUS WASTES. The facility should be asked to provide a listing of the names, characteristics, generation rates, classifications, and disposition of all hazardous and nonhazardous wastes generated by or received at the site. The disposition should discuss on-site and off-site storage, treatment, disposal, recycling, recovery, reuse, and reclamation. Historical practices should be defined when they differ from the current situation.

The auditor must determine if the facility has done an adequate job of identifying and characterizing its wastes (for example, hazardous waste determination). The hazardous waste notification form should be reviewed for accuracy (that is, consistency with manifests and current waste production). The hazardous waste site classifications should be defined. These include:

- Large quantity generator,
- Small quantity generator,
- Conditionally exempt small quantity generator,
- Transporter,

- Storage facility, or
- Disposal facility.

The audit team should evaluate the adequacy of written programs and actual generator practices for the hazardous waste facility standards, such as the waste analysis program, preparedness and prevention, contingency plans, inspections, training, storage practices, recordkeeping, reports, and so on. The quarterly, annual, or biennial reports of hazardous waste activity submitted to regulatory agencies should be checked for accuracy. Facility expansion and modification dates should be correlated to applications for new permits or modifications to current permits. Reports of regulatory agency inspections and NOVs should be summarized to assess historical and ongoing compliance patterns.

The names, locations, and license and identification numbers of waste transporters should be obtained. The auditor must evaluate the adequacy of waste transportation practices. For example, proper preparation of manifests, vehicle placards, loading and unloading procedures, and history of transportation spill incidents should be reviewed. For facilities that also own hazardous waste and material transport fleets, a focused audit of transportation procedures may be necessary (for example, incident response, driver qualifications, and training).

At permitted hazardous waste treatment, storage, and disposal (TSD) facilities, the auditors should obtain and review copies of the Part A and Part B permit applications and interim status or final operation permits. The audit team should assess adherence with pertinent regulations and with approved plans in the permit applications (for example, the waste analysis plan). For TSD facilities, it is suggested that the audit team use the checklists developed by U.S. EPA's National Enforcement Investigations Center (NEIC). For hazardous waste units subject to groundwater monitoring, U.S. EPA's RCRA Technical Enforcement Guidance Document (TEGD) on Groundwater Monitoring has a series of useful checklists.

Previous on-site disposal practices should be scrutinized extensively, including pits, lagoons, waste piles, landfills, storage areas, treatment units, burners, flares, incinerators, spill areas, and so on. The auditor should learn as much as possible about each past practice, such as size, waste characteristics and quantities, period of operations, and closure procedures. The auditors should determine the status of the RCRA corrective action program for continuing releases from previous or existing solid waste management units (SWMUs). Copies of the RCRA Facility Assessment (RFA), RCRA Facility Investigation (RFI), Corrective Measures Study (CMS), and interim or final corrective measures design and operations plans should be procured and evaluated.

The facility may have filed a CERCLA 103(g) form with U.S. EPA for notification of previous on-site disposal practices or otherwise have been listed in the Comprehensive Environmental Response, Compensation, and Liability Information System. If so, the auditor should determine if a preliminary assessment/site inspection (PA/SI) has been conducted and obtain a copy of the report. This information should be considered to determine if the facility has potential liability for CERCLA cleanup.

The restrictions on land disposal of certain hazardous waste (also called the "land ban") have had a profound effect on generators and TSD facilities (40 CFR Part 268). The auditor should assess the validity of generator certifications and disposal facility adherence to the regulations.

Current and previous off-site waste disposal practices should be reviewed to ensure compliance and define possible liabilities. The audit team should define the names, locations, and overall compliance status of off-site management facilities. The facility should be asked to describe the mechanisms in place, if any, to minimize company liability (for example, disposal contracts or periodic audits). The appropriate facility personnel should be asked if they have shipped wastes to Superfund sites and if they are a potential responsible party (PRP) in any waste site cleanup proceedings.

Waste oil handling practices and solid waste TSD practices should be evaluated for compliance during each audit. However, the regulations vary considerably from state to state, and the auditor must check each site on a state-specific basis.

CHEMICAL AND OIL STORAGE. The audit team should begin its assessment of chemical storage practices with the chemical inventory and material storage information developed as part of the site features and operations profile. In addition to compliance with the various regulatory programs, the audit team should determine if there is a significant potential for either sudden or nonsudden releases of hazardous substances or oil to the environment. This can be accomplished by reviewing leak detection devices, spill containment provisions, spill contingency plans, and so on.

The facility chemical inventory should be reviewed to determine which of the materials are hazardous substances under the auspices of the CERCLA. Facility personnel should be questioned to determine whether they are aware of the RQ values for the hazardous substances and to determine if reporting procedures are well understood and followed. The release history of the facility should be examined to determine if improved management practices are necessary or if contingency plans should be modified.

The Emergency Planning and Community Right-to-Know Act (EPCRA) is Title III of the Superfund Amendments and Reauthorization Act, the 1986 CERCLA amendments. The facility being audited may be subject to one or more of the emergency planning and reporting requirements contained in the EPCRA. The audit team should obtain copies of the facility's reports to responsible authorities for Sections 311, 312, and 313 of the EPCRA; these refer to material safety data sheets, emergency and hazardous chemical inventory forms, and toxic chemical release reporting, respectively. These should be evaluated for completeness and accuracy. The audit team should determine if the facility has or should prepare an emergency or contingency plan to deal with fires, explosions, or other hazardous substance incidents. Such a plan may transcend the hazardous waste contingency plan and oil spill prevention plan.

The oil storage inventory and practices at the facility should be reviewed to determine if there is a requirement for the facility to have an oil spill prevention control and countermeasure (SPCC) plan. If such a plan is required, the auditors should determine its adequacy according to the regulations (40 CFR Part 112) and whether a registered professional engineer has certified the adequacy of the plan.

As part of its NPDES permit, the facility may have been required to prepare a BMP plan to reduce the possibility of releases of toxic or hazardous substances. If this is the case, the audit team should review the BMP plan for compliance with permit conditions and the BMP guidance documents. The audit team should also determine if the facility is complying with the programs set forth in its own plan.

Underground storage tanks (USTs) have been widely used for the storage of petroleum and hazardous substances. The facility should be asked to provide an inventory of these tanks along with information about the results of previous leak test, installation details, and whether spill protection cathodic protection and leak detection are provided. The audit team should evaluate this information in the context of UST regulations promulgated under Subtitle I of the 1984 RCRA amendments and by state-specific regulations. Depending on the age of the tank system, the regulations provide for a phased compliance period for corrosion protection, leak detection, and spill and overfill control. There are also specific requirements for financial responsibility assurance. The regulations also contain requirements for UST system closure and corrective action in the event of a release. Appropriate facility personnel should be asked to describe closure plans and corrective action in the event of a release. Facility personnel should be asked to describe the program for compliance with these regulations, and the audit team should determine its adequacy.

Certain hazardous substances are regulated under the Occupational Safety and Health Administration (OSHA) regulations and local fire codes. These include flammable materials, among others. Facility personnel should be asked if they are aware of any building or fire code violations, and the audit team should obtain copies of recent fire insurance or fire marshal inspections. Certain states have promulgated regulations regarding the storage of hazardous substances. For example, New Jersey has enacted a discharge prevention control and countermeasure plan requirement and a discharge cleanup and removal plan requirement for certain hazardous materials as well as a toxic catastrophe prevention act.

POLYCHLORINATED BIPHENYLS. The use, storage, and disposal of polychlorinated biphenyl- (PCB-) containing materials is regulated under the Toxic Substances Control Act (TSCA). Facility personnel should be asked to provide an inventory of PCB-containing items such as transformers, capacitors, lighting ballast, high-temperature hydraulic oils, and PCB wastes. The TSCA regulations are formatted according to various "PCB items" (such as capacitors, transformers, containers, etc.) and "PCB wastes." Disposal requirements are tiered based on the type or origin of the waste, the physical state, and the concentration, with regulatory requirements beginning at 50 ppm and calling for the highest degree of treatment for liquid wastes with PCB concentrations of 500 ppm or more. Annual reports and inspection data should be correlated to manifests for off-site disposal. The hazardous waste land ban regulations also contain certain restrictions for the disposal of PCB wastes. If the facility has received a permit from U.S. EPA for PCB waste management, it will be necessary for the audit team to fully evaluate compliance with these regulations in the same manner as for the hazardous waste management regulations.

ASBESTOS. The manufacture, use, and disposal of asbestos and asbestos-containing materials are regulated principally under the NESHAPs, TSCA regulations, and OSHA regulations. The audit team should evaluate the facility for the presence of friable asbestos materials. It may be that a followup asbestos survey is in order if one has not already been completed. Asbestos-containing insulation may be present around boilers, steam pipes, and other heated vessels and in ceiling insulation and floor tiles. The auditors must determine if manufacturing and demolition or renovation work are in compliance with these regulations.

PESTICIDES. Pesticide use in the U.S. is regulated under the Federal Insecticide, Fungicide, and Rodenticide Act (FIFRA), which contains certain requirements for the registration, certification, and disposal of pesticides. Some state-specific laws also apply. These regulations are most important to manufacturers and applicators of pesticides, and audits of these types of operations typically are considered to be product liability audits rather than environmental audits. Nonetheless, facilities should be questioned regarding their use of pesticides to ensure that proper application and disposal practices are used.

INJECTION WELLS. Underground injection control is regulated by a permit system as part of the SDWA regulations. Various classes have been established for different types of injection wells. The audit team should determine if injection wells are used at the facility and if operations are in compliance with the regulations and permit conditions. It should be noted that the land ban regulations have restricted the injection of various hazardous wastes.

HEALTH AND SAFETY. Audits of this topic may be conducted separately or as part of a facility environmental audit. Health and safety audits typically concentrate on compliance with the OSHA regulations, although building and fire code topics are often included. The auditors should determine if the company requires a preemployment physical, periodic medical monitoring, and so on. Especially important are the permissible exposure limits (PELs) for various chemicals and whether the company has performed chemical exposure monitoring in the past. If PELs have been exceeded, the auditors should determine if the company has taken appropriate remedial action. The results of any noise monitoring surveys at the facility should be obtained and reviewed to determine if the facility is required to have a hearing conservation program with OSHA regulations. Hearing conservation techniques and protective equipment should be described by facility personnel, as well as the results of any audiometric testing programs being conducted. The need for and adequacy of personnel protective equipment (PPE) should be assessed. Depending on the extent of the health and safety audit, other topic areas, for example, confined space entry, lock-out/tag-out, and bloodborne pathogens, may be reviewed.

All OSHA citations, building and fire code citations, and alleged unsafe working condition grievances during the past 3 to 5 years should be reviewed to check for patterns and trends. Record retention policies should be reviewed for adequacy and compliance. The auditors should determine if the facility has a written Hazard Communication Standard (HCS) program and if the program is adequate, including training of employees. Oftentimes, the adequacy of RCRA hazardous waste management training is assessed as a health and safety audit topic. Special attention should be given to training documentation records. It should be noted that many states have their own health and safety statutes that might be more stringent than the federal OSHA requirements.

RADIOACTIVE SUBSTANCES. The manufacture, use, and disposal of radioactive substances is regulated by a variety of programs by U.S. EPA, the Nuclear Regulatory Commission, the Department of Energy, and various states. Depending on the facility, it may be that health physicists or nuclear engineers are required to conduct this portion of the audit. These substances include both radioactive materials, such as those used in nuclear power operations, X-ray machines, and various laboratory/testing instruments, and radioactive wastes. Wastes typically are differentiated by potential radiation dose level into a category that includes spent nuclear fuel, high-level and transuranic radioactive wastes, and a low-level category that includes low-dose items such as contaminated clothing and soil.

OTHER CONCERNS. If the audit team is to evaluate compliance with corporate and facility standard operating procedures and policies, such as ISO 14011, it may be necessary to add a full-time auditor to the team for this particular topic. U.S. EPA has the authority to regulate certain types of noise sources under the Noise Control Act. Before the site visit, the audit team leader should determine if this particular concern needs to be addressed during the audit.

Audits of analytical and testing laboratories used by the facility are often conducted as a separate exercise from the overall environmental audit. Experienced analytical chemists should conduct these audits. At certain facilities, specialized topics may be added to the audit agenda, such as marine protection and coastal area regulations.

Finally, it should be realized that there are certain topic areas that go beyond the focus of the traditional multimedia environmental audit of a facility. This would include review of premanufacture notifications under TSCA or pesticide registration under FIFRA. In many cases, the records and personnel that will be of importance to these topics will be at corporate offices rather than at site locations.

Methods To Obtain Necessary Information. The audit team can use a number of effective methods to obtain information and evaluate the facility. First-hand, visual site inspections are extremely important. It is suggested that the auditors initially take a quick walk-through of the entire facility. This is followed by detailed inspection visits of the various units being audited. Site personnel interviews constitute one of the major methods of information gathering; auditors should be careful to document in their notes the date and time of the interview and the name and title of the person being interviewed. Historical photographs and maps are useful in understanding the evolution of a particular facility. Photographs taken during the audit site inspection are useful in refreshing the memory of the audit team as well as in presenting the results to management.

At a minimum, the types of files to be reviewed during the on-site audit include permits and permit applications, correspondence with regulatory agencies, internal and external reports, laboratory results, shipment manifests, and engineering and scientific investigatory reports. There are a variety of published checklists in the literature that the audit team might find quite useful. In some cases, regulatory compliance information may be gathered by reviewing agency files; however, auditors should be discreet to maintain the confidentiality of the audit. Obviously, it is extremely important that any additional regulatory interpretations that are required be obtained. The auditors should not hesitate to seek outside assistance in these matters when necessary.

EVALUATION PHASE. Before writing the audit report, all the information obtained during the site visit and hard copies of the documentation need to be reviewed. The review may be performed by the audit team leader or as a joint effort with the various team members addressing their respective areas of expertise. The primary focus in evaluating the data is the determination of compliance or noncompliance of the facility with local, state, and federal regulations. Next is the determination of the effectiveness of the physical plant, policies, and practices of the facility in those areas relating to the environmental, health, and safety concerns. In addition to determining the effectiveness of the policies and practices, the auditor needs to address whether the employees actually are in compliance with these policies, especially if the effectiveness appears to be less than desired. Along with evaluating the environmental and health effects, the auditor needs to determine if some of the impact in these areas reflects the physical condition of the facility and operating equipment, rather than the policies and practices themselves.

After determining the compliance or noncompliance with regulations, the efficacy level of the policies and practices, and the condition of the facility's equipment, the auditor needs to evaluate the risks and potential liabilities associated with the existing situation. Frequently, the risks and associated liabilities can be reported on a percent probability basis. The potential costs for liability and remediation can then be estimated accordingly, if desired.

AUDIT REPORT AND FOLLOWUP. The audit report should provide a concise summary of the results of the environmental audit. Facility description information is often summarized in appendices to the document or is presented in a very brief form in the document. The various sections of the audit report should be prepared by the auditor(s) responsible for a given topic area. After the draft report is complete, it is advisable that the team leader distribute copies to other members of the audit team to ensure consistency and to allow for a more refined product. After the draft audit report is approved for release, it should first be distributed to facility personnel so that they may correct any misconceptions or errors. Most importantly, they need to understand what is required to correct deficiencies and provide schedules for these corrective actions. In most companies, the corrective action list and its implementation schedule is subject to approval by upper management.

The final report includes the corrected draft and the implementation schedule for required actions. The final audit report should be issued as soon as possible after completion of the field work so that there is minimal loss of interest and momentum.

Finally, the company should provide for followup review to ensure that corrective actions are completed properly and on schedule. This followup step is absolutely essential to making the audit an effective management tool.

PRESERVING THE CONFIDENTIALITY OF ENVIRONMENTAL AUDITS

The environmental audit is a valuable tool for organizations to use in identifying and evaluating potential environmental problems. Keeping the results of environmental audits confidential within the organization helps encourage the free flow of information regarding past and present environmental practices and thereby enhances the accuracy and effectiveness of the audit process.

Facilities must undertake an environmental audit recognizing that no environmental audit possesses an absolute guarantee of confidentiality. Many of the facts discovered during an audit (as distinguished from legal opinions and legal advice) are afforded little or no protection from ultimate disclosure in connection with government investigations and government and third-party litigation. Nevertheless, there are certain procedures that an entity can institute that can heighten the possibility of keeping the environmental audit results and reports confidential and privileged.

Using legal counsel to perform the audit may allow the organization to take advantage of attorney—client privilege or the attorney work-product doctrine. When upper management assumes primary authority for performing an audit, the organization's ability to keep the information confidential and privileged from compelled disclosure, while still possible, is less likely.

Typically, communications made to an attorney by the client are protected by the attorney–client privilege. However, for the privilege to apply, the attorney must be used to provide legal advice. Thus, merely submitting factual information to an attorney will not preserve the confidentiality of the material submitted. The communication must be made with the attorney for the purpose of receiving advice or counsel. In determining whether or not counsel has been retained for the purpose of providing legal advice, courts focus on the authority given to the attorney in connection with the environmental audit and the attorney's responsibilities with respect to analyzing the data contained in the report. Legal counsel should be authorized to undertake a review of the entity's compliance with and potential liabilities for violations of environmental laws by investigating all environmental matters relating to an entity's operations.

To ensure that legal counsel has the proper authority to implement an environmental audit, the board of directors of the operation should authorize the performance of the audit by counsel through a corporate resolution. The resolution should set forth the objectives to be accomplished by having the environmental audit performed. In adopting such resolutions, the corporation cannot only assert its environmentally sensitive goals in performing the audit, but it may also set the stage for a subsequent claim of confidentiality. The corporation can resolve to seek legal advice in connection with all findings developed during the course of the environmental audit and to request the recommendations of counsel. The board's resolutions are also the appropriate time to make known the board's desire that the audit and its results be kept confidential and privileged. The board of directors' resolutions also present a good opportunity to instruct management to comply with the audit program and provide the information requested by counsel. Of course, before undertaking any audit or setting forth such an objective, it is essential that the company or facility decide to correct the deficiencies noted in the audit and comply with notification and reporting requirements. Otherwise, an audit serves only to underscore the willfulness of any violations.

The next requirement for establishing a claim of attorney—client privilege is a communication between a client and an attorney. In the context of a corporation, the definition of client is not restricted solely to the most senior employees or board members. The clients may include any corporate employee whose communications to counsel are necessary for the attorney to render legal advice and who assists in the investigation that management has asked the counsel to undertake. Thus, the facility operation manager's communications to counsel are entitled to the same privilege as those of the director of the environmental office.

The privilege, however, only protects disclosures of communications. It does not protect disclosure of the underlying facts by those who communicated with the attorney. Therefore, the facts of the violations known to the company employee are discoverable from that employee; communication of those violations to the attorney and the attorney's advice may not be discoverable in the first instance.

Many corporations use their in-house environmental staff or environmental consultants to conduct the audit. If the audit is structured such that the in-house staff or consultants are conducting the audit to "assist" in providing legal advice, communications to these nonlawyers may also be treated as privileged. In-house staff and consultants may assist by conducting the fact-gathering portions of the audit, with the company's lawyers supplying the conclusion and advice regarding compliance. A third requirement for the attorney–client privilege to apply is that the communication to counsel be kept confidential by the company. If the audit report itself or the investigation notes are permitted to be circulated among a corporation to anyone other than those with a need to know, a court will be reluctant to consider the reports confidential. Thus, the disclosure to the attorney's in-house "assistant" and the attorney may be privileged, as would the attorney's response back down the chain.

In contrast, the corporation's random or inadvertent disclosure of the attorney's advice with respect to the violations may not be privileged. Clearly, senior corporate management and those who need to implement corrective action are entitled to see the reports, and such disclosure should not put any privilege at risk.

Stamping documents as confidential and privileged and maintaining them as such within the confines of the corporation would greatly assist the corporation in later establishing a claim of confidentiality. Segregating the results of the audit into a nonconfidential report of fact that uses the nonprivileged information and a confidential report that relies on both nonprivileged and privileged information that has been kept confidential will also aid in showing that the requisite care has been taken in protecting the privilege.

After it is determined that the materials are to be kept confidential and privileged, care must be taken to ensure that the privilege is not waived. By publicizing a confidential report for certain purposes, the corporation may lose its ability to keep it confidential for other purposes. For example, an audit report for one facility should not be volunteered to customers if the corporation later may want to refrain from disclosing an audit report in some litigation.

In addition to the protections afforded by the attorney–client privilege, the work-product doctrine may aid in providing protection to the audit. Simply stated, the work-product doctrine provides a qualified immunity from discovery of materials prepared by or for counsel in anticipation of litigation. The protection extends to investigations conducted by nonattorneys to assist counsel; however, the materials that contain the mental impressions, opinions, or legal theories of an attorney warrant the most protection. The work-product doctrine does not extend to facts or even the existence of the audit report. Opposing counsel can overcome the protection by showing "substantial need" for the materials and "undue hardship" in obtaining the equivalent from other sources.

To avail oneself of the work-product doctrine, the threshold requirement is that the audit is undertaken in anticipation of litigation. Litigation does not have to be imminent; a court will look to the specific situation surrounding a document to determine whether it was actually prepared in anticipation of litigation. The most important inquiry for a court is to determine the primary motivating purpose behind the creation of the document. Litigation that is reasonably predictable or that is identifiable from specific claims that have already arisen typically will satisfy the anticipation of litigation test. Thus, if fear of or preparation for litigation does prompt the initiation of the audit, the board of directors' resolution and correspondence to counsel should document the reasons for the corporation's anticipation of litigation.

In summary, a corporation increases the likelihood that an environmental audit will be considered confidential and privileged by authorizing counsel to conduct the environmental audit for the purpose of giving legal advice and by ensuring that a corporation's employees are assisting counsel in providing legal advice in any auditing activities. Moreover, all communications to and from counsel should be made in confidence, stamped "confidential and privileged" and kept confidential.

Finally, U.S. EPA issued its Environmental Auditing Policy Statement on July 9, 1986. In this statement, U.S. EPA asserts in that policy that it

> has broad statutory authority to request relevant information on the environmental compliance status of regulated entities. However, [U.S.] EPA believes routine agency requests for audit reports could inhibit auditing in the long run, decreasing both the quality and quantity of audits conducted. [U.S.] EPA will not routinely request environmental audit reports.
>
> [U.S.] EPA's authority to request an audit report or relevant portions thereof will be exercised on a case-by-case basis where the Agency determines it is needed to accomplish a statutory mission, or where the government deems it to be material to a criminal investigation.

Thus, except in extenuating circumstances such as a pattern of violations, it is unlikely that U.S. EPA will test the confidentiality of a corporate environmental audit. Alternatively, the mere existence of an auditing program will not insulate a corporation from liability if U.S. EPA discovers the existence of violations through monitoring, compliance reports, or other information or records required to be made available to U.S. EPA. An environmental audit is not a substitute for the corporation's legal responsibility to monitor, record, or report information discovered during the audit.

CONCLUSIONS

The information resulting from a well-performed IDD audit can be a very powerful tool for ensuring compliance and reducing liability. Audits can save many times their costs by preventing penalties and fines and by reducing time lost because of accidents or downtime for equipment. Audits used to identify (potential) problems can save money in legal and disposal costs. Audits can identify operations to streamline and improve practices to increase production. Audits can also be used to identify (and prevent) environmentally deleterious discharges and improve or maintain community relations.

SUGGESTED READINGS

American Society of Testing and Materials (1994) *Standard Practice for Environmental Site Assessments: Phase 1, Environmental Site Assessment Process.* ASTM E 1527-94, Philadelphia, Pa.

Bowman, V.A., Jr. (1988) *Checklists for Environmental Compliance.* Pudvan Publishing Company, Inc., Northbrook, Ill.

Government Institutes, Inc. (1987) *Environmental Audits.* Rockville, Md.

International Organization for Standardization (1995) *ISO 14000 Draft International Standards, Environmental Management Systems and Environmental Auditing.*

U.S. Department of Energy (1987) *The Environmental Survey Manual.* Vol. 1, DOE/EH-0053, Washington, D.C.

U.S. Environmental Protection Agency (1984) *Current Practices in Environmental Audits.* EPA-230/09-83-006, Washington, D.C.

U.S. Environmental Protection Agency (1987) *Multimedia Compliance Audits.* EPA-330/9-87-001-R, National Enforcement Investigation Center, Denver, Colo.

U.S. Environmental Protection Agency (1988) *Annotated Bibliography on Environmental Auditing.* 7th Ed., Washington, D.C.

Chapter 3
Prevention and Reduction of Hazardous Waste

Industrial pollution prevention has become an increasingly effective approach toward environmental regulatory concerns and environmental management costs. An environmental audit or the results of reporting requirements, such as those of the Superfund Amendments and Reauthorization Act (SARA) Title III, or planning requirements under the pollution prevention laws of several states may encourage an organization to consider the advantages of operational changes leading to reduced hazardous waste generation. Regulatory burdens can be reduced, process economics can be improved, and long-term liability costs can be decreased. The increased public interest in and concern about the quantity and risk of waste produced by industrial operations is likely to result in the rewarding of well-planned and measurable efforts to reduce waste with significant dividends in community relations and acceptance.

Only a few years ago, the concept of preventing pollution at the source during manufacturing seemed strange in light of decades of attention to treatment of waste at the "end of the pipe." However, a series of reports of successful programs has focused attention of many companies on the possible benefits of this approach. Other new initiatives including environmental management programs such as ISO 14000 and federal procurement regulations that give preference to suppliers who practice pollution prevention have provided additional reasons to consider seriously implementing this approach. Pollution prevention as a concept addresses the reduction of waste and emissions at the source, that is, during manufacturing. More broadly, source reduction opportunities exist throughout the life cycle of the product including materials selection, manufacturing, use, maintenance, repair, and ultimate disposal.

The principles of source reduction apply not only to existing industrial operations but also to projected new facilities. With new facilities, the economic effects can be particularly favorable because initial equipment selection and return-on-investment determinations can be made based on optimized material use, reduction of waste generation, and minimized waste disposal costs. A pollution prevention assessment during the design phase similar to the approach discussed in this chapter can have significant favorable effects.

This chapter discusses examples of proven approaches to hazardous waste prevention and reduction in specific industries and generic approaches to waste minimization. The chapter concludes with a description of the process for identifying opportunities for waste reduction in existing facilities, as well as in proposed facilities and operations. The federal Pollution Prevention Act (PPA) provides a pyramid approach to waste management that gives first emphasis to pollution prevention. The second stage, for waste that cannot be reduced at the source, is to implement strategies to reduce the volume and toxicity. The last steps to be considered include treatment and disposal strategies. The focus of this chapter is on the first two steps.

INTRODUCTION TO HAZARDOUS WASTE POLLUTION PREVENTION AND REDUCTION

Liquid, solid, or gaseous waste materials are generated during the manufacture of products. In addition to environmental problems, these wastes represent losses of valuable materials and energy from the production process and often require a significant investment in pollution control. Traditionally, pollution control focuses on "end of the pipe" and "out the back door" viewpoints. The control of pollution in this way requires manpower, energy, materials, and capital expenditures. Such an approach removes pollution from one source, such as wastewater treatment or air pollution abatement, and places it somewhere else, such as a landfill.

Pollution prevention should be a key component of any cost-effective, comprehensive waste management program. Prevention techniques do not have to be based on high technology or large capital expenditures. Many techniques require simple changes in the way materials are handled in the production process. The techniques used in a management program to reduce waste generation at the source are discussed in detail in the following section. An approach for developing and implementing a multimedia pollution prevention program using these techniques is presented later in this chapter.

HAZARDOUS WASTE PREVENTION AND REDUCTION TECHNIQUES

Hazardous waste prevention and reduction techniques can be applied to any manufacturing process, from something as simple as making a paper clip to as complex as assembling a space shuttle. Available techniques range from easy operational changes to state-of-the-art recovery equipment. The common factor in these techniques is that they reduce operational costs.

Waste prevention and reduction techniques can be broken down into four major categories as shown in Table 3.1. Because the classifications are broad, there may be some overlap. In actual application, these techniques typically are used in combination to achieve the maximum effect at the lowest cost.

INVENTORY MANAGEMENT. The proper control of raw materials, intermediate products, final products, and associated waste streams is an

Table 3.1 Categories of waste reduction techniques.

Inventory management
Inventory control
Material control
Production process modification
Operation and maintenance procedures
Material change
Process equipment modification
Volume reduction
Source segregation
Concentration
Recovery
On-site recovery
Off-site recovery

important pollution prevention technique (Hunter, 1987). In many cases waste streams simply result from out-of-date, off-specification, contaminated, or unnecessary raw materials. Wastes can also be spill residues, damaged or returned final products, product samples retained for future comparisons, or "free" samples provided by vendors. The cost of disposing these materials not only includes the actual disposal costs but the cost of the lost raw materials or product. Methods for controlling inventory range from simple changes in ordering procedures to implementing just-in-time (JIT) manufacturing techniques. Many companies can reduce waste generation by tightening up and expanding their current inventory control programs. This will significantly affect the three major sources of hazardous waste resulting from improper inventory control: excess, out-of-date, and no-longer-used raw materials.

Purchasing only the amount of raw materials needed for a production run or for a set period of time is one of the keys to proper inventory control. Excess inventory often must be disposed because it becomes out of date. More effective application of existing inventory management procedures should reduce this problem. This should be coupled with education programs for purchasing personnel and for those who make purchasing requests on the problems and costs of disposing of excess materials. Additionally, the expiration dates should be evaluated, especially for stable compounds, to determine if they are too short. Occasionally, the expiration dates provided by suppliers represent a conservative estimate of potential performance degradation for the most critical use of the product. When performance requirements are less severe, an extended expiration date should be developed.

Developing review procedures for all materials purchased is another step in establishing an inventory control program. Standard procedures should require that all materials be approved before purchase. In the approval process, all production materials are evaluated to determine if they contain hazardous constituents and, if so, what alternative nonhazardous substitute materials are available. Development of the review procedures can be made either by one person with the necessary environmental background (with an appropriate knowledge of the performance requirements of the chemicals involved) or by a committee made up of people with a variety of backgrounds. It may be possible to obtain needed information from the material safety data sheets (MSDSs) provided by the chemical supplier. Many companies, from electronics to textiles, have established successful material review programs (Governor's Waste Management Board, 1988, and Dadak, 1987). The details of the review may require different levels of input and discussion. Process-critical changes may depend on pilot examination of alternatives before recommendations for change can be made. On the other hand, some types of materials, such as janitorial supplies, may need less evaluation before change. Consideration of change should be based on an evaluation of all aspects of the use of the material, including the procedures for disposal after use.

The ultimate in inventory control procedures is JIT manufacturing because this eliminates the existence of any inventory. This is done by directly moving raw materials from the receiving dock to the manufacturing area for immediate use. The final product is then shipped out without any intermediate storage. Just-in-time manufacturing is a complex program to implement and cannot be used by all facilities; however, it can reduce waste significantly. For example, using JIT techniques, the 3M Company reduced waste generation by 25 to 65% in its individual plants (Hunter, 1987).

PRODUCTION PROCESS MODIFICATION. Improving the efficiency of a production process can significantly reduce waste generation. Many of the most cost-effective reduction techniques are included in this category; some are simple and relatively inexpensive changes to production procedures. Available techniques range from eliminating leaks from process equipment to new state-of-the-art production equipment. The waste prevention and reduction techniques in this category can be divided into improved operation and maintenance, material change, equipment modification, and process redesign.

Operational Procedures. A wide range of methods is available for operating a production process at peak efficiency. These methods are neither new nor unknown and typically are inexpensive to institute because little or no capital cost is necessary. For example, a producer of breaded foods instituted a number of operational changes such as using dry cleanup, installing or modifying drip trays under process equipment, and developing better systems for collecting and handling waste material. These changes decreased water usage by approximately 30%, wastewater generation by approximately 80%, and allowed the company to sell 235 900 kg/a (520 000 lb/yr) of solids to recovery firms (Waynick et al., 1988).

Improved operational procedures are methods that make optimum use of the raw materials used in the production process. Some examples are shown in Table 3.2. The first step in instituting such a program is to review all current operational procedures and examine the production process for ways to improve efficiency. A review should include all segments of the process, from the delivery area through the production process to final product storage. Procedures on how to conduct such a review are presented later in this chapter.

One important area typically overlooked or not given proper attention is material handling procedures. Proper material handling will ensure that the raw material reaches the production process without loss of material through spills, leaks, or contamination. It also will ensure that the material is efficiently handled during the production process.

After proper operating procedures are established, they must be fully documented and made part of the employee training program. A comprehensive training program is a key element of any effective waste prevention and reduc-

Table 3.2 Examples of operational changes to reduce waste generation.

Reduce raw material and product loss as a result of leaks, spills, drag-out, and off-spec process solution.
Schedule production to reduce equipment cleaning. For example, formulate paint from light to dark so that vats do not have to be cleaned out between batches.
Inspect parts before they are processed to reduce number of rejects.
Consolidate types of equipment or chemicals to reduce quantity and variety of waste.
Improving cleaning procedures to reduce generation of dilute mixed waste by using methods such as dry cleanup techniques, mechanical wall wipers or squeegees, pigs or compressed gas to clean pipes, and increased drain time.
Segregate waste to increase recoverability.
Optimize operational parameters (such as temperature, pressure, reaction time, concentration, and chemicals) to reduce byproduct or waste generation.
Develop employee training procedures on waste reduction.
Evaluate the need for each operational step and eliminate those that are unnecessary.
Collect spilled or leaked material for reuse.

tion program. For example, through training, a dairy plant, a semiconductor manufacturer, and a furniture plant reduced waste by 14, 40, and 10% respectively (Kalenowski and Keon, 1987, and Kohl et al., 1986). For a program to be effective, all levels of personnel should be included, from the line operators to the corporate executive officer. The goal of any educational program is to make the employee aware of waste generation, its effect on the company and the environment, and ways it can be prevented and reduced. Written materials should be prepared and used in conjunction with hands-on training. This should be an ongoing process with review updates and interaction between employees and supervisors on a regular basis. Attention should be given to providing similar training for new employees as part of their introduction to the company.

Maintenance Programs. One company found that 25 to 50% of its waste load resulted from poor maintenance (Shober, 1988). A strict maintenance program that stresses corrective and preventive maintenance can reduce waste generation caused by equipment failure. Such a program will help identify potential sources of release and correct the problem before any material is lost. A good maintenance program is important because the benefits of any waste reduction program can be wiped out by just one process leak or equipment malfunction.

A maintenance program can include cost tracking and preventive maintenance scheduling and monitoring. To be effective, a maintenance program should be developed and followed for each operational step in the production process, with special attention given to potential problem points. A strict schedule and accurate records of all maintenance activities should be maintained. Computer-based maintenance scheduling and tracking programs are available from a variety of vendors. A comprehensive program should also include predictive maintenance. This approach provides the means to schedule repairs or replace equipment based on the actual condition of the machinery. Such an approach is valuable in light of alternative maintenance approaches that dictate activity on a regularly scheduled basis regardless of need. In some cases, maintenance operations themselves are contributors to waste streams—paint stripping and recoating, for example. Performing these types of activities only on an as-needed basis can be a positive contributor to pollution prevention. A number of nondestructive testing technologies are available for making the needed evaluations for this approach.

Material Change. Hazardous materials used either in a product formulation or in a production process may be replaced with a less-hazardous or nonhazardous material. Reformulating a product to contain less-hazardous material will reduce the amount of hazardous waste generated during the product's formulation and end use. Using a less hazardous material in a production process typically will reduce the amount of hazardous waste produces. This can, in turn, reduce the cost of capital equipment needed to meet environmental regulatory limits. Some examples of material change to reduce waste generation are given in Table 3.3.

Product reformulation is one of the more difficult pollution prevention techniques, yet it can be very effective. Examples of product reformulation include eliminating pigments containing heavy metals from ink, dye, and paint formulations; replacing phenolic biocides with other less toxic compounds in metal-working fluids; and developing new paint, ink, and adhesive formulations based on water rather than organic solvents. Making such changes will require evaluating any effects on product performance and may require communicating with and educating clients. Because other organizations are examining options for change leading to pollution prevention,

Table 3.3 Examples of waste reduction through material change.

Industry	Technique
Household appliances	Eliminate cleaning step by selecting a lubricant compatible with the next process step
Textile	Reduce phosphorus in wastewater by reducing the use of phosphate-containing chemicals
Aerospace	Replace cyanide cadmium plating bath with a noncyanide bath
Ink manufacture	Remove cadmium from product
Textile	Use UV light instead of biocides in cooling towers

reformulated products with lowered levels of hazardous chemicals can enjoy a commercial advantage.

Hazardous chemicals used in the production process can also be replaced with less- or nonhazardous materials. Changes can range from using purer raw materials to replacing solvents with water-based products. This is a very widely used source reduction technique and is applicable to many industries. Many of these changes involve switching from a solvent to a water-based process solution. For example, a diesel engine remanufacturing facility switched from cleaning solvents and oil-based metal-working fluids to water-based products. This change reduced its coolant and cleaning costs by approximately 40%. Additionally, the company was able to eliminate one cleaning step, and machine filters lasted twice as long, thus reducing material and labor costs (Johnson, 1988).

One important area that is sometimes overlooked in making a material change is its effect on the total waste stream. Switching from a solvent-based to a water-based product can increase wastewater volumes and concentrations of contaminants. This can adversely affect the current wastewater treatment system, cause effluent limits to be exceeded, and possibly increase wastewater treatment sludge production. Therefore, before any change is made, its effect on all discharges must be evaluated.

Process Equipment Modifications. Waste generation may be reduced by installing more efficient process equipment or by modifying existing equipment to take advantage of better production techniques. New or updated equipment can use process materials more efficiently, producing less waste. Additionally, higher-efficiency systems may reduce the number of rejected or off-specification products, thereby reducing the amount of material that has to be disposed of or reworked. Some examples of process modifications are given in Table 3.4.

Modifying existing process equipment can be a cost-effective method to reduce waste generation. In many cases, relatively simple and inexpensive changes in the way materials are handled within the process can ensure that they are not wasted or lost. This can be as easy as redesigning parts racks to reduce drag-out in electroplating operations, installing better seals on process equipment to eliminate leakage, or installing drip pans under equipment to collect leaking process material for reuse. One chemical company reduced the waste from a sump in a production area from 31 750 to 1360 kg/a (70 000 to 3000 lb/yr) by installing a sight glass, using better pump seals, and purchasing a broom (Beck, 1987).

Installing new, more efficient equipment and, in some cases, modifying current equipment will require capital investment in equipment, facility modifications, and employee training. The extent of the investment will vary depending on the type of equipment used. These investments, however, can have a rapid payback. For example, a power tool manufacturer replaced a spray solvent paint system with a water-based electrostatic immersion paint-

Table 3.4 Examples of production modifications for waste reduction.

Process step	Technique
Parts cleaning	Enclose all solvent cleaning units
	Use refrigerated freeboard on vapor degreaser units
	Improve parts draining before and after cleaning
	Use mechanical cleaning devices
	Use plastic bead blasting
Surface finishing	Prolong process bath by removing contaminants
	Redesign part racks to reduce drag-out
	Reuse rinse water
	Install spray or fog nozzle rinse systems
	Properly design and operate all rinse tanks
	Install drag-out recovery tanks
	Install rinse water flow control valves
	Install drip racks and drainboards
Surface coating	Use airless air-assisted spray guns
	Use electrostatic spray-coating systems
	Control coating viscosity with heat units
	Use high-solids coatings
	Use powder coating systems
Equipment cleaning	Use high-pressure rinse systems
	Use mechanical wipers
	Use countercurrent rinse sequence
	Reuse spent rinse water
	Use pigs to clean lines
	Use compressed gas to blow out lines
Spills or leaks	Use bellow-sealed valves
	Install spill basins or dikes
	Use sealless pumps
	Maximize use of welded pipe joints
	Install splash guards and drip boards
	Install overflow control devices

ing unit. This new equipment paid for itself in a little more than 1 year by reducing raw material costs by $600 000/yr, reducing waste disposal costs by 97%, and greatly increasing productivity (Huisingh, 1985).

VOLUME REDUCTION

Volume reduction includes techniques to separate toxic, hazardous, or recoverable wastes from the total waste stream. These techniques typically are used to increase recoverability; reduce the volume and, thus, disposal costs; or increase management options. The available techniques used range from simple segregation of wastes at the source to complex concentration technologies, as shown in Table 3.5. These techniques can be divided into two general

Table 3.5 Examples of waste reduction through volume reduction.

Industry	Technique
X-ray film	Segregate polyester film scrap from other production waste and recycle
Printed circuit boards	Use filter press to dewater sludge to 60% solids and send sludge to metal recovery
Pesticide formulation	Use separate bag houses for each process line and recycle collected dust into product
Paint formulation	Segregate and reuse tank cleaning solvents in paint formulations
Furniture	Segregate and reuse solvents used to flush spray-coating lines and pumps as coating thinner

areas: source segregation and waste concentration. Only those methods that are actual waste reduction techniques will be discussed in this section.

SOURCE SEGREGATION. Segregation of wastes is, in many cases, a simple and economical technique for waste reduction. For example, by segregating wastes at the sources of generation and handling the hazardous and nonhazardous wastes separately, waste volume and, thus, management costs will be reduced. Additionally, the uncontaminated or undiluted wastes may be reusable in the production process or sent off-site for recovery. Specifically, there is significant advantage in keeping nonhazardous waste streams from mixing with hazardous waste streams.

This technique is applicable to a wide variety of waste streams and industries and typically involves simple changes in operational procedures. For example, in metal finishing facilities, wastes containing different types of metals can be treated separately so that the metals in the sludge can be recovered. Keeping spent solvents or waste oils segregated from other solid or liquid waste may allow them to be recycled. Wastewater containing toxic material should be kept separate from uncontaminated process water, reducing the volume of water that must be treated.

A typically used waste segregation technique is to collect and store wash water or solvents used to clean process equipment (such as tanks, pipes, pumps, or printing presses) for reuse in the production process. This technique is used by paint, ink, and chemical manufacturers as well as printers and metal fabricators. For example, a printing firm segregates and collects toluene used for press and roller cleanup operations. By segregating the used toluene by color and type of ink contaminant, it can be reused later for thinning the same type and color of ink. The firm now recovers 100% of the waste toluene, totally eliminating a hazardous waste stream (Huisingh, 1985).

CONCENTRATION. Various techniques are available to reduce the volume of a waste through physical treatment. Such techniques typically remove a portion of a waste, such as water. Available concentration methods include gravity and vacuum filtration, evaporation, ultrafiltration, reverse osmosis, freeze vaporization, filter press, heat drying, and compaction. Many of these are actually recovery techniques and will be discussed further in the next section. Unless the material can be recycled, just concentrating a waste so more can fit into a drum is not waste reduction. In some cases, concentration of a waste stream may also increase the likelihood that the material can be reused or recycled. For example, filter presses or sludge dryers can increase the concentration of metals in electroplating wastewater treatment sludge to such a level that they become a valuable raw material for metal smelters. A printed circuit board manufacturer dewaters its sludge using a filter press to 60% solids. The company receives $7200/yr from the sale of the dewatered sludge for copper reclamation (Beck, 1987).

RECOVERY

Recovering wastes can provide a very cost-effective waste management alternative. This technique can help eliminate waste disposal costs, reduce raw material costs, and possibly provide income from sale of the waste. Recovery of wastes is a widely used practice in many manufacturing processes and can be done on-site or at an off-site facility, as shown in Table 3.6.

The effective use of recovery depends on the segregation of the recoverable waste from other process wastes or extraneous material. This ensures that the waste is uncontaminated and that the concentration of recoverable material is increased. Some companies have assigned the responsibility for the handling, collection, and scheduling of recovery of waste material to one individual (Kohl and Currier, 1987, and Waynick et al., 1988). This helps ensure that the maximum value of the waste can be recovered.

Table 3.6 Examples of waste reduction through recovery and reuse.

Industry	Technique
Mirror manufacturer system	Recover spent xylene using a batch distillation
Printed circuit boards	Recover copper and tin or lead from process wastewater using an electrolytic recovery system
Power tools	Recover alkaline degreasing baths using an ultrafiltration system
Hosiery	Reconstitute and reuse spent dye baths
Pickles	Transfer waste brine pickle solution to a textile plant as a replacement for virgin acetic acid
Chemical manufacture	Use spent electrolyte from one division as raw material for another

ON-SITE RECOVERY. In most cases the best place to recover process wastes is within the production facility. Wastes that are simply contaminated versions of the process raw materials are good candidates for in-plant recycling. Waste can be recovered most efficiently at the point of generation. The possibility of contamination with other waste materials is reduced, as is the risk involved with handling and transporting waste materials. There may be some regulatory advantages for on-site recovery and recycling. If materials can be directly recovered and immediately returned to the manufacturing process, the technique may be considered as in-process recycling and considered pollution prevention. This approach to pollution prevention requires that the waste stream never leave the process.

Some waste streams can be reused directly back into the original production process as raw material. This typically is accomplished when the waste material is lightly contaminated or is excess raw material. Examples include cleaning waste from printers, coaters, and chemical or product manufacturers; electroplating drag-out solutions; process solutions from filter changes; and dust collector residue from pesticide manufacturers. Lightly contaminated waste can sometimes be reused in operations not requiring high-purity materials. For example, spent high-purity solvents generated during the production of microelectronics can be reused in less critical metal degreasing operations, or a caustic waste material can be reused to treat an acid waste stream.

Some waste may have to undergo purification before it can be reused. A number of physical and chemical techniques can be used to reclaim a waste material. These range from simple filtration to state-of-the-art techniques, such as freeze crystallization. The method of choice will depend on the physical and chemical characteristics of the waste stream, recovery economics, and operational requirements. Most on-site recovery systems will generate some type of residue (that is, the contaminant removed from the recovered material). This residue can either be processed for further recovery or be disposed of properly. The economic evaluations of any recovery technique must include the management of these residues.

OFF-SITE RECOVERY. Wastes may be recovered at an off-site facility when the equipment is not available on-site, when not enough waste is generated to make an in-plant system cost effective, or when the recovered material cannot be reused in the production process. Off-site recovery typically entails recovering a valuable portion of the waste through chemical or physical processes. Some materials that are typically reprocessed off site are oils, solvents, electroplating sludges and process baths, lead-acid batteries, scrap metal, food processing waste, plastic scrap, and cardboard. The cost of off-site recycling will depend on the purity of the waste and the market for the waste or recovered material.

In some situations a waste may be transferred to another company for use as raw material in its manufacturing process. This exchange can be economically advantageous to both firms because it will reduce the waste disposal costs of the generator and reduce the raw material costs of the user. For example, an X-ray film manufacturer produces a salable product from waste film stock. The company installed equipment that flakes and bales waste polyester-coated film stock. This is then sold as raw material to another firm. More than 9 million kg (20 million lb) of film stock is exchanged each year, representing a $200 000 annual savings in collection, transport, and processing costs and an annual profit of $150 000 from the sale of the materials (Huisingh, 1985).

The upgrade of a waste into a product requires a strong commitment from the generator to find markets for the waste material both inside and outside the company. In some cases, the production process or the waste may have to undergo some modification to make a more salable product. Regional waste exchanges have been set up in a number of states to help companies find markets by acting as information clearinghouses of wastes available and wanted. The service typically offered is a listing of wastes available from generators and wanted by users provided in a catalog or computer database form.

INDUSTRIAL SECTORS WITH HIGH POLLUTION PREVENTION POTENTIAL

U.S. EPA has evaluated the potential of various manufacturing sectors to achieve significant rates of pollution prevention, based in part on the current rates of waste generation for the sector as a whole (Ashford and Stratikopoulos, 1997). Among the industrial sectors with this degree of potential are electroplating, paints, chemical manufacturing, printing, dry cleaning, and metal working. Included in pollution prevention technologies applicable to these industries are ultrasonic cleaning to replace solvents, raw material recovery systems, introduction of new equipment to reduce the frequency of required cleaning, use of nonsolvent techniques for paint removal, and use of powder coating techniques. The report assumes that opportunities for implementation of pollution prevention are greater in industrial sectors where there is a slow rate of change in the manufacturing technology. Thus, in any particular plant, pollution prevention and waste reduction opportunities may be greater in the oldest and most conservative parts of the process.

IMPLEMENTING A HAZARDOUS WASTE REDUCTION PROGRAM

Incentives for implementing a waste prevention and reduction program include, as previously mentioned, new regulations, increased waste disposal costs, the potential for long-term liability, or simply a desire to operate a business in a more efficient way. The potential approaches to waste reduction already described demonstrate the diversity of techniques used to initiate a program.

The most effective programs are the result of planning developed from knowledge of the operation in question and setting priorities for waste prevention and reduction based on the goals and results expected from the program. For example, a different approach may be taken initially if a primary goal is to adjust to new regulatory concerns rather than if the major goal is to reduce disposal costs. Whatever the goal, the planning requires the gathering of information about the processes and operations underway at the facility.

A factual and accurate understanding of the location and operations where wastes are generated is essential before steps leading to source reduction can be taken. A process for information gathering and option identification must be developed. Although there may be several approaches to carrying out the process, the U.S. EPA has prepared a *Waste Minimization Opportunity Assessment Manual* that outlines a process for gathering information, setting priorities, and selecting the most promising options for waste reduction.

Although experience has shown that this formal approach can produce an effective pollution prevention/waste reduction program, other variations can work as well. In some companies, a smaller feasibility or pilot program in one area of the facility may be required to obtain the high-level corporate commitment required. Recent experiences with volunteer programs such as the U.S. EPA 33/50 program show that they can also be a useful tool in obtaining the necessary commitment and support. The 33/50 program was established by U.S. EPA to encourage volunteer companies to identify and implement strategies to reduce by 33% and then by 50% the quantities of 17 specific toxic chemicals released from their facilities. Frequently, the person who has successfully accomplished the smaller project can become the focal point for the larger program.

Many states now require a formal pollution prevention planning process by companies. This requirement can be a strong motivating force leading to the beginning of an effective source reduction program. However, strong leadership and corporate commitment and continuing support will be necessary to achieve long-term success. Table 3.7 presents a summary of common features in pollution planning programs that have been successful in the past. These activities and components are discussed later in this chapter.

CORPORATE COMMITMENT. When waste reduction has been most successful, an important contributor has been a steadfast corporate commitment to the program. Success is derived from agreement throughout the organization that waste reduction is an important objective and that the necessary time and effort will be taken to identify opportunities and initiate changes. Personal involvement from top management, as well as from the person working on the manufacturing line, is important in making the process work well. Depending on the size and complexity of the organization, a formal cor-

Table 3.7 Common features in successful pollution prevention planning programs.

Obtain corporate commitment to support the program with time and money
Establish challenging, but achievable, goals
Identify all waste-generating sites and processes
Determine quantities of waste generated at each point of the process
Conduct detailed site inspection to determine causes of waste generation
Propose alternative technology or process modifications to prevent waste
Analyze the feasibility of each proposed alternative
Consider financial aspects
Consider availability of new technology
Implement the selected options
Measure effectiveness of new procedures and strive for continued improvement

porate policy committing the company and its employees to waste prevention and reduction may be an effective way to initiate the program.

PROGRAM ORGANIZATION. In a typical organization, waste is created throughout all parts of the operation. Because a waste reduction program will affect personnel and operating groups throughout the company, it is important to have representatives from all the potentially affected groups involved in the information gathering, assessment, and implementation aspects of the waste reduction program. Often a waste prevention and reduction task force is an effective type of organization to initiate a program. Any such task force should include representatives from all portions of the company that may be affected. Some examples of such groups are production, maintenance, engineering, quality assurance, research, marketing, purchasing, and legal. The exact composition of the task force will depend on the size, formality, and complexity of the particular plant or line operation initiating the program. In a very small operation, the task force may include the entire workforce. In establishing the task force, a key to success is ensuring that all groups are represented and that the members of the task force are enthusiastic.

To ensure that the task force develops and maintains momentum and has a point of contact with company management there should be a leader acknowledged by all levels in the company organization. In some cases, the leader may be the person who felt most compelled to investigate the potential for waste reduction at the facility and worked to achieve recognition by employees and management of the incentives and needs. In other cases, the leader may be designated by management. The leader must provide motivation, secure continued management support, overcome inertia and suspicion, and serve as a source of new ideas and suggested approaches.

The waste reduction task force should have clearly understood responsibility and authority for organizing and carrying out the waste reduction program within the organization. A systematic approach for discharging this responsibility may include the steps outlined in the following sections.

CORPORATE COMMITMENT OR POLICY STATEMENT. The task force should strive to obtain a written policy statement or corporate commitment supporting the waste reduction program. This ensures management interest and allocation of resources to accomplish waste reduction. Moreover, it sends a message to all employees that waste reduction efforts are important and cooperation with the task force is "acceptable" behavior. Ideally, a full corporate commitment should be obtained before implementation of the program. In some situations, it may be obtained after some successes are demonstrated in one area. Such a commitment or policy statement, and the support and approval it conveys, is an essential cornerstone of the program. Without it, even the best efforts will not accomplish all that is possible.

ESTABLISHMENT OF PROGRAM GOALS. The task force should establish some overall waste prevention and reduction program goals. Such goals are important because they serve to characterize the company expectations for the waste reduction program. Different organizations may set different types of goals based on varying needs and experiences. Whatever they are, it is important that they be measurable goals. The goals may be defined in terms of disposal costs avoided, other manufacturing costs saved, increased use of raw materials, reduction in the quantities of waste materials leaving the site, or perhaps net reduction in "risk" because of the operation of the plant. Obviously, some of these goals are easier to measure than others. Also, they should be indexed in some way to plant output to ensure that waste reduction is occurring while maintaining production rather than achieving minimal waste generation by slowing production. Some organizations may establish percentage reduction goals; others may establish absolute quantity reduction goals, or zero emission goals. The goals may be changed when more is learned about the waste-generating activities at the facility. It is important to have an initial set of goals, however, because they serve as the context for assessing, evaluating, and setting priorities for waste reduction options.

IDENTIFICATION OF WASTE-GENERATING SITES AND PROCESSES. The task force should develop a plan and procedures for understanding waste generation at the facility. Comprehensive waste reduction will require an understanding of all locations and processes within the manufacturing site that produce waste. All waste should be defined, including such less-obvious categories as fugitive emissions to the air, aqueous waste streams, and nonhazardous solid wastes. The task force should organize a systematic investigation of the operations to develop a database defining the types and quantities of waste streams in the facility and carefully linking the waste streams with the particular processes or operations that produce them. Further subcharacterization according to whether the waste streams are hazardous or nonhazardous and the reasons for this description are also important.

Equally important is an understanding of the types and quantities of the raw materials entering each process. The task force should be particularly interested in balancing the quantities of material entering and exiting the process because large discrepancies in this material balance indicate fugitive or "invisible" losses that should be addressed. The task force should also gather information about the existing process controls, waste segregation, and housekeeping practices that may affect the quantity and composition of the waste streams.

After information about the nature, quantity, and sources of the waste streams has been assembled, the task force should agree on which streams have first priority for further assessment. Ideally, in a small facility, all waste streams should be considered for waste reduction possibilities. In a larger installation, limitations of personnel, time, or budget may require a sequential approach, with those waste streams having the greatest potential effect on waste reduction being considered first. The priorities should be established on the basis of the goals established for the program. Some factors that should be considered in setting priorities include compliance with regulations, waste treatment and disposal costs, quantity of waste currently generated, and hazardous properties of the waste stream.

DETAILED SITE INSPECTION. The task force should designate an assessment team with responsibility for conducting a detailed site inspection. Although the waste reduction task force has a responsibility for the entire facility, the designated assessment team should have intimate knowledge of the prioritized processes or operations. The assessment team may be a subset of the task force augmented with other experienced employees or even outside consulting experts. In smaller organizations, the assessment team may be the same as the task force. In many states, there is a pollution prevention technical assistance program that is equipped to provide low-cost or free assistance with this part of the assessment process. The assignment of the assessment team should be to conduct a "walk-around," "inside-out" survey of the entire specific operation. The survey should follow the process from the entrance of raw materials to the exit of products and wastes. The team should identify potential sources of waste including any waste treatment operations and possible fugitive emissions. The survey may produce some tentative ideas about waste generation causes. These preliminary conclusions may require additional information, comparison with industry standards or benchmarks, or additional site visits to be confirmed.

After the probable causes of waste generation in the process have been identified, the assessment team should turn to identifying potential options for waste prevention and reduction. The options can be developed internally, or outside sources such as trade associations, equipment vendors, environmental agencies, consultants, or technical assistance programs can be contacted for information about possible options. Although specific waste prevention options exist for specific processes, they typically fall within the basic categories described earlier in this chapter. Many options are possible, and the assessment team should consider using a "brainstorming" process that encourages group creativity in listing options.

The final aspect of this portion of the assessment team's activity should be to screen the options identified and select some for further consideration. Because a detailed study of feasibility should be the next step, selection at this stage should be on the basis of elimination of options that are impractical or show obvious marginal or inferior waste reduction potential. The informal evaluation and selection may be performed by the entire assessment team or a smaller group. Ideally, it should be done using some weighted, quantitative method.

FORMAL FEASIBILITY ANALYSIS. The task force should be responsible for conducting a formal feasibility analysis. Any such analysis requires gathering additional specific information about the identified options. Normally two types of evaluative questions arise—technical (Can it be done?) and economic (How much will it cost? What will be the return on investment?). Specifically, the technical questions may involve consideration of safety, product quality, compatibility of changes with present practices, availability of necessary equipment, and new environmental complications. The economic questions may involve consideration of factors such as capital costs, operating costs, payback periods, and predictable profit. Typically, profitability is improved with successful waste reduction options and frequently the payback period can be relatively short. In comparing projected costs with present costs it is important to consider all of the charges involved in current waste management and treatment, including costs that may be incorporated into the overhead charges for the facility as a whole. The U.S. EPA has encouraged the exploration of "green accounting" tools that may aid in the determination of full costs for waste production and management (U.S. EPA, 1995).

The total feasibility study may require expertise that is not found on the assessment team or on the task force. The commitment by management should include provision for this necessary expertise to complete this phase of the program.

STARTING TO USE THE FAVORED OPTIONS. The task force has a major obligation to insist that steps be taken to incorporate the favored options into the routine practices and processes of the facility. By this point in the program, waste reduction options capable of furthering the company goals in a technically and economically feasible manner have been identified. Because the program involves a broad spectrum of company personnel and stresses the importance of waste reduction, implementation of low-cost options should be met with wide and easy acceptance. Options requiring substantial capital or increased operating costs may have to compete with other projects with similar financial needs. At this stage, the task force leader has an important role to play in reminding management of the commitment by the company to waste reduction, the need for funding, and the favorable economic effects to be expected from the investment. When the funding is obtained, actual acceptance of the option should again be relatively easy because of the preparatory work of the task force.

MEASUREMENT OF EFFECTIVENESS AND CONTINUING ASSESSMENT. The task force has a final and continuing responsibility to measure the effectiveness of the implemented waste reduction options. An accurate determination of the effects of waste reduction activities is important to provide a sense of reward to those involved in the program, to monitor the progress of waste reduction, to justify and demonstrate the financial effects of the investments made, and to serve as new assessment data for continued waste reduction activities. Although all of these are important reasons for measuring the success of waste reduction initiatives, the last illustrates a final important point. A waste reduction program should not be a one-time, isolated activity. Ideally, it should become a philosophy, a corporate and personal way of life that continually seeks new options and opportunities to further reduce the generation of waste and to accrue the resulting benefits.

This type of corporate philosophy fits well with newer approaches to corporate environmental strategies based on standards programs such as ISO 14000. Such initiatives help the company work toward continuous environmental improvement. Regular monitoring of environmental performance and a commitment to pollution prevention are important aspects of such improvement.

REFERENCES

Ashford, N.A., and Stratikopoulos, D.M. (1997) Identification of Pollution Prevention (P2) Technologies for Possible Inclusion in Enforcement Agreements Using Supplemental Environmental Projects and Injunctive Relief. EPA-300/R-97-001, U.S. EPA, Washington, D.C.

Beck, W. (1987) Waste Minimization—A Plant Approach to Getting Started. *Proc. CMA Waste Minimiz. Workshop.* Vol. 1, Chem. Manufact. Assoc., Washington, D.C.

Dadak, P. (1987) Waste Minimization: The Hewlett Packard Experience. In *Waste Minimization Manual.* Government Institutes, Inc., Washington, D.C.

Governor's Waste Management Board (1988) *Governor's Award for Excellence in Waste Management.* Raleigh, N.C.

Huisingh, D. (1985) *Profits of Pollution Prevention: A Compendium of North Carolina Case Studies.* N.C. Board Sci. Technol., Raleigh, N.C.

Hunter, J., II (1987) Minimizing Waste by Source Segregation and Inventory Control. *Tech. Strat. Hazard. Waste Prevent. Control Sem.* Government Institutes, Inc., Washington, D.C.

Johnson, A. (1988) Experiences in Getting Rid of Solvent-Based Decreasing in a Diesel Engine Remanufacturing Plant. *Proc. Waste Reduct. Pollut. Prevent. Conf.* N.C. Dep. Natural Resour. Comm., Raleigh, N.C.

Kalenowski, T., and Keon, M.A. (1987) *Waste Generation and Disposition Practices and Currently Applied Waste Minimization Techniques Within the Semiconductor Industry.* Dep. Health Serv., Sacramento, Cailf.

Kohl, J., and Currier, J. (1987) *Managing Waste Oils.* Ind. Exten. Serv., N.C. State Univ., Raleigh, N.C.

Kohl, J., et al. (1986) *Managing and Recycling Solvents in the Furniture Industry.* N.C. Board of Sci. Technol., Raleigh, N.C.

North Carolina Department of Natural Resources (Year Unknown) *Case Study Files.* Pollution Prevention Program, Division of Environ. Manage., Raleigh, N.C.

Shober, R. (1988) Water Conservation and Waste Loan Reduction in Food Processing Facilities. *Proc. 1988 Food Proc. Waste Conf.,* Georgia Inst. Technol., Atlanta, Ga.

U.S. Environmental Protection Agency (1995) Environmental Cost Accounting for Capital Budgeting: A Benchmark Survey of Management Accountants. EPA-742/R-95-005, Washington, D.C.

Waynick, J., et al. (1988) A Breaded Foods Processor Does It Too! *Proc. Waste Reduct. Pollut. Prevent. Conf.,* N.C. Dep. Nat. Resour. Comm. Develop., Raleigh, N.C.

Chapter 4
Hazardous Material and Waste Handling and Storage

INTRODUCTION

There are specific handling standards for raw materials storage and for the treatment, storage, and disposal of hazardous waste. Regulations require that ignitable, reactive, and incompatible materials be handled in a manner that will not generate extreme heat, pressure, fire, or violent reactions or will not produce uncontrolled toxic mists, fumes, dusts, or gases that pose a risk of fire or explosion or threaten human health or the environment. Therefore, raw material storage must be well planned, and those industries that generate waste must manage their materials in a manner that will not violate established regulations. When considering chemical storage, the containment and ancillary equipment must be able to withstand the forces acting on it and must be noncorrosive for the specific application. Some general considerations include the site location, nearby structures and storage areas, existing plant operations, location of existing utilities, total container volume, type of material stored, and type of containers.

COMPATIBILITY

One of the most important considerations is the compatibility of the material to be stored in a tank or container. Table 4.1 lists some incompatible chemical compounds or mixtures. This list may be used as a guideline but is not comprehensive, and before use, material and corrosion properties should be cross-checked with more complete sources. There are a number of sources, including *Perry's Chemical Engineer's Handbook* (Perry et al., 1997), Sax's *Dangerous Properties of Industrial Materials* (Sax and Lewis, 1989), and a number of publications from the National Association of Corrosion Engineers. These guides should be used on a site- and chemical-specific basis. For materials that are outside the range of the published values, test cupons should be run. The test cupon process involves placing the materials in contact with each other under laboratory conditions and observing the degradation of the container

Table 4.1 Incompatible chemical compounds or mixtures.

Tank or container lining material	Incompatibility
Steel	Mineral acids, nitric acid, hydrochloric acid, dilute sulfuric acid, aluminum alkalis, potassium hydroxide, sodium hydroxide, mineral acids
Magnesium	Mineral acids
Lead	Acetic acid, nitric acid
Copper	Nitric acid, ammonia
Zinc	Hydrochloric acid, nitric acid
Tin	Organic acids, alkalis
Titanium	Sulfuric acid, hydrochloric acid
Fiberglass-reinforced plastics	Sulfuric acid 95%, nitric acid 50%, hydrofluoric acid 40%, aromatic solvents, ketone solvents, chlorinated solvents
Alkyds	Strong mineral acids, strong alkalis, alcohol, ketones, esters, aromatic hydrocarbons
Vinyls (PVC)	Ketones, esters, aromatic hydrocarbons
Chlorinated rubbers	Organic solvents
Epoxy (amines, polyamide, or esters)	Oxidizing acids (nitric acid), ketones
Coal tar epoxy	Strong organic solvents
Polyesters	Oxidizing acids, strong alkalis, mineral acids, ketones, aromatic hydrocarbons
Silicones	Strong mineral acids, strong alkalis, alcohols, ketones, aromatic hydrocarbons

materials. Procedures developed by the National Association of Corrosion Engineers and the American Society of Testing and Materials (ASTM) should be followed for this testing. Manufacturer's literature can, with caution, be used to predict the resistance of a tank or container material to certain chemicals. Consult the manufacturer of the tank or container. Unfortunately this probably will not work with steel drums because finding a manufacturer who will warrant the steel drum may be difficult unless the drum is lined.

A word about lined containers is necessary here. Certain products such as steel drums and other containers are manufactured with a factory-installed lining. In most cases, the lining is integral with the drum, and the drum cannot be cut apart without destroying the liner. In other instances, manufacturers will line the drum with a special coating or perhaps even put a plastic bag liner in the drum. These materials are for shipping purposes and quality control purposes only. They are not to be used in lieu of integral container linings and, if used improperly, may expose the user to certain regulatory penalties for improper packaging and shipment of materials. Above all, a garbage bag does not equate with a drum liner.

For example, acids with a pH of 0 to 2 are highly corrosive, and alkalies with a pH of 12.5 to 14 are also highly corrosive. These liquids, under some conditions, such as temperature and high velocity (tank agitation), and with the use of incompatible construction materials can uniformly dissolve a significant amount of the steel tank walls even when the walls are lined. Corrosive substances rapidly attack carbon steel tanks, but most plastics have excellent resistance to acids and bases. When chloride or fluoride solutions are stored in a metal tank, the compounds will form soluble salts with the tank walls, weakening the tank. This form of tank wall deterioration may or may not be uniform over tank walls and may be concentrated in areas of stress (that is, joints, welds, corners, and bends) and create leaks in these places long before the overall strength of a tank is noticeably reduced. Oxidation of some types of nonmetallic tank and liner materials, such as rubber and some chlorinated rubbers and glass linings, can cause softening, dissolution, or decomposition of the liner. In the case of strong bases, caution must be exercised with fiberglass tank materials because the strong base can attack the glass materials in the tank wall causing embrittlement and softening.

Some organic acids and solvent compounds and mixtures can attack both metals and plastics. This is especially true of some mixed wastes. The combination of the products of cleaning compounds such as carbon tetrachloride and methylene chloride and perchloroethylene can have a low pH and be corrosive to plastics. Solvents recovered by distillation can have this characteristic unless they are treated to prevent breakdown during solvent reclamation.

Corrosion protection may be required against galvanic corrosion, stray current corrosion, and soil corrosiveness. Metal corrosion is accelerated when the metal contacts oxygen and moisture in the ground, thereby causing the metal to oxidize. There are a number of corrosion protection methods, some of which are listed below.

- Cathodic protection. These systems reverse the current through sacrificial anodes or impressed dc current, which is good for protecting buried tanks and foundations against the effects of corrosive soils. Cathodic protection should only be used for steel underground tanks that are subject to corrosion. It cannot be used on clad tanks or fiberglass-reinforced plastic tanks, unless the cladding is on the inside of the tank. Cathodic protection may be a preferred method of protection, it is often more reliable, and many tanks have been removed from the ground after 30 years of service without so much as a blemish.
- Fiberglass-reinforced plastic (FRP). This is a noncorrosive construction material that is subject to attack only by some strong solvents and strong caustic wastes.
- Steel clad with noncorrosive material. This involves coating a steel tank with a nonconductive coating, for example, coal tar epoxy (15 mils) or FRP (125 mils). Care must be exercised to touch up any blemishes or scratches in the coating systems before burial. As a standalone coating, it is preferable to no coating, but the likelihood of scratching a tank during burial is too great to permit it to be the principal coating.
- Tank liners. In some instances, it is possible to re-line steel tanks with inert linings that extend the effective tank life and reduce the potential for corrosive attack on the tank. This should be investigated carefully because the materials application will require a tank entry and worker safety precautions to be applied during the work.

Some states require specific corrosion protection for specific types of containers, and state regulations should be checked to see if they are different from federal regulations. The states always have the authority to be more restrictive than the federal government but may not be less restrictive.

Product and waste storage in aboveground tanks or drums should be stored in groups by chemical class. Examples of potentially incompatible wastes are provided in 40 CFR Part 265, Appendix V. Table 4.2 provides a partial list of incompatible wastes. This list is not intended to be exhaustive and serves only as an example of potential incompatibilities. Again, if there are questions, see the appropriate codes, contact the fire marshal, obtain corrosion data from the chemical manufacturer, run tests, or check the literature.

The number of tanks and types of materials to be stored are factors to consider when designing aboveground storage containment areas. Local, state, and federal regulations should be consulted for each site-specific application. In some communities in which flammable materials are involved or there is a potentially large consequence to tank failure, the fire marshal should be consulted.

As a part of the rational design process, an additional word of caution is necessary here. Plan for tank failure. Provide adequate spill control and provisions against the day when the aboveground tank fails. Consider soil cont-

Table 4.2 Potentially incompatible wastes.

Group 1-A	Group 1-B
Acetylene sludge	Acid sludge
Alkaline caustic liquids	Acid and water
Alkaline cleaner	Battery acid
Alkaline corrosive liquids	Chemical cleaners
Alkaline corrosive battery fluid	Electrolyte, acid
Caustic wastewater	Etching acid, liquid or solvent
Lime sludge and other corrosive alkalis	Pickling liquor and other corrosive acids
Lime wastewater	Spent acid
Lime and water	Spent mixed acids
Spent caustic	Spent sulfuric acid

Potential consequences: heat generation or violent reactions.

Group 2-A	Group 2-B
Aluminum	Any waste in Group 1-A or 1-B
Beryllium	
Calcium	
Lithium	
Magnesium	
Potassium	
Sodium	
Zinc powder	

amination and water removals as potential issues that may have to be dealt with when there is a tank failure, and plan accordingly. Traffic and access are always a consideration, and it is surprising to find how often acid tanks are placed by the side of a plant roadway in locations where they can be struck by trucks or fork-lift vehicles.

If the tank is to be loaded by truck or tank car, it is advisable to look at the consequences and potential for leaks (pumps, piping, and couplings) and failures in the loading area. If the tank is improperly designed, a failure of the pump or a valve can create a back syphon and empty the entire contents of the tank through the loading line outside the diked area. Plan containment for the loading area, and determine what could happen if an emergency happened. For example, plan containment strategies for a truck full of hexamethyl chicken fat or dimethyl barbwire that fails while it is inside the plant, or what would happen if and when the driver fails to disconnect the loading hoses and drives off with them still connected. Do not pretend that these things will not happen. The surest course is to plan for failure modes during the design and placement of the tanks.

Any product or waste storage in U.S. Department of Transportation- (U.S. DOT-) approved containers should be managed as carefully as tanks. The inventory should be accessible for tracking, and the area manager and plant safety engineer have to make sure that the only people allowed to handle these tanks are those with the necessary training and experience.

Realistically, any worker who has to handle chemicals in the plant (not wastes) must have, as a minimum, an 8-hour training course on the properties and hazards of chemicals (HAZCOM) and 8-hour refresher training annually. If workers are to handle and respond to spills and clean them up, additional training requirements will be between 24 and 40 hours, with an additional 8-hour refresher training annually.

From the standpoint of management, the materials entering and leaving a storage area must be tracked. If hazardous wastes are involved, the log information must be available for at least 5 years after the wastes have been moved.

STORAGE AREAS

One of the greatest areas of regulatory interface with an operating facility is in the management of hazardous wastes. The regulations are detailed and specific. Perhaps the greatest number of fines and citations for the mismanagement of hazardous wastes are generated in the handling, storage, and labeling operations. Most of the citations issued are for improper labeling of hazardous waste drums or for failure to adequately inspect and keep records.

Raw materials may be required to be stored at several locations throughout a manufacturing facility. These materials typically are received in large quantities in a centralized receiving area, warehouse, or tank farm and distributed to work areas on the production floor in smaller quantities. Specific details for hazardous waste storage are discussed later in this chapter.

There are two distinctive types of waste storage areas—central storage areas and satellite accumulation areas. A central storage area is an area of a generator's facility where more than one drum of hazardous waste is stored at a time and which is not in the area where the hazardous waste is generated. For example, a company might accumulate three or four drums of waste paint thinner in a flammable storage room before they are picked up by the waste hauler at the end of each quarter. Another generator might have a paved, fenced, roofed, and bermed area outside where 40 drums of various types of hazardous waste accumulate before they are picked up at the end of each month. Both of these areas would be called central storage areas.

Satellite accumulation areas are locations where one container, typically a drum, is kept to collect the waste as it is being generated. An example of this would be in a research laboratory of a manufacturing facility where a drum is used to collect waste chromic acid that was used to clean glassware. Another example would be a drum kept in the painting area to collect solvent from cleaning paint guns.

When the satellite accumulation drum is full, it must be marked with the date it became full and moved to a central storage area within 3 days. The drum must then be shipped off-site within 90 days of the date it became full (unless the generator is a small-quantity generator) and the waste must be stored in compliance with all applicable regulations for a central storage area.

The reason for the distinction between these two types of storage areas is in 40 CFR Part 261.34(c)(1), which allows generators of hazardous waste to accumulate up to 55 gallons of hazardous waste at or near the point of generation (in satellite accumulation areas) without complying with the applicable time limit (90 or 180 days), as long as the specific requirements for satellite accumulation are being met. This differs from the old system (before December 1984), which specified that the 90-day time limit for on-site storage by generators started when the first drop of waste was put in the drum.

All drums containing hazardous waste are to be inspected as required by state and U.S. EPA policy. Drums are inspected for correct labeling and integrity of the container (that is, leaks, signs of corrosion, deterioration, pitting, bulging, and secure closure of all bungs and top rings). Adequate space should be available between each drum to allow for a thorough inspection of each drum. During these inspections each drum is visually observed and inspected.

Before shipping (within 90 days of the accumulation start date) each drum is inspected to make sure that all hazardous wastes are packaged, labeled, and marked in compliance with U.S. DOT regulations (49 CFR Parts 172, 173, 178, and 179). The results of each inspection are entered in the required inspection log.

Container Storage

Tanks containing hazardous waste must be managed in accordance with U.S. EPA regulations. These regulations require that tanks be inspected daily to ensure that discharge control equipment and monitoring equipment are operating properly and to check and record the level of waste in the tank. The regulations also require weekly inspections of tank construction materials for corrosion and deterioration and weekly inspections of the area surrounding the tank for signs of leakage and erosion. The results of the daily and weekly inspections are recorded in a separate inspection log. Many of the requirements for hazardous waste storage and handling are also the "best management practices" for raw materials and chemical products.

On July 14, 1986, U.S. EPA promulgated new regulations for the management of hazardous waste in underground tanks. These regulations went into effect January 12, 1987. These regulations can be found in 40 CFR Parts 260.10 and 264.190. A major provision of these new regulations is that secondary containment will eventually be required for all tanks, including underground tanks. Secondary containment systems are designed to temporarily contain leaks or spills, preventing them from contaminating the surrounding environment. A secondary leak detection system is necessary and should be located in the bottom of the containment below the tank. Some of the more typical forms of secondary containment are double-walled tanks, vaults, and underground lining systems. Existing tanks that are not currently equipped with secondary containment must be retrofitted in accordance with the following schedule:

- Tanks used to store F020, F021, F022, FO26, or FO27 wastes by January 12, 1989.
- Existing tanks of known and documented age by January 12, 1989, or when the tank system reaches 15 years of age, whichever is later; and
- Existing tanks of unknown or undocumented age by January 12, 1995, unless the facility where the tank is located is older than 7 years, in which case the time limits of known-age tanks apply.

The regulations also require that the integrity of existing tanks that do not have secondary containment be assessed every year by a qualified registered professional engineer and be certified to be nonleaking. Detailed design standards for new tanks are included in the new regulations.

Tanks and containers used by generators to store or accumulate hazardous waste—no matter where they are located—must be clearly marked and labeled throughout the period of accumulation with the words "Hazardous Waste." The drums must be properly labeled using the prescribed label format, and the label must also contain the U.S. DOT shipping name for the waste and the date on which each period of accumulation began in the tank or container. Marks and labels should be placed on the sides of each tank or container and must be clearly visible for inspection. The drums must also be logged in and their movement and placement accounted for.

In addition, if the generator is taking advantage of the satellite accumulation regulation, full drums should also be marked with the date the drum became full along with the accumulation start date. Before drums are shipped off-site, they must be marked and labeled in accordance with all applicable U.S. DOT regulations. Many generators label containers with preprinted labels that are designed to meet U.S. DOT's marking requirements. These preprinted labels should be reviewed carefully to determine if they meet U.S. EPA and U.S. DOT labeling requirements.

U.S. EPA prescribes special regulatory standards for managing hazardous waste in containers and tanks. Each kind of vessel has separate standards. Containers that are portable have standards that address the risks arising from their portability. Tanks, which are higher capacity stationary installations, have standards that apply to those characteristics. The design of a tank system requires substantial documentation and engineering certification before the tank's construction.

All containers used to store hazardous waste, raw materials, or product chemicals must be in good condition, nonleaking, and compatible with the material stored in them. The content of leaking containers must be transferred immediately to a nonleaking container. Those containers used to ship waste off-site as well as to store it on-site must also meet U.S. DOT packing regulations. U.S. DOT packaging regulations are also a good guideline to use when determining what type of drum to use to store waste on-site. Another guideline is that waste material can typically (although not always) be stored in the drum that the product from which the waste was generated came in. This may not hold true if the waste material is very different from the product.

There are specific requirements for the management of containers of hazardous waste. Containers of hazardous waste must be kept closed at all times except when waste is actually being added to or removed from the container. Containers of hazardous waste must not be handled or stored in a manner that would cause them to rupture or leak. (For example, containers that are stacked must be separated with pallets.) Containers of hazardous waste must be labeled at all times, and the labels must be legible and visible.

Containers must be stored on an impervious surface that is bermed to prevent leakage in case of a spill. Storage areas should meet the requirements of 40 CFR Part 264.175, unless the area is inside a building and it can be demonstrated that a spill cannot escape to the environment and can be easily cleaned up, 40 CFR Part 264.175 requirements outline what is needed in a container storage area. The goals of the regulations are to provide a method of preventing spills and, when and if they occur, to clean them up promptly and thoroughly. With some planning and proper forethought, establishment of cleanup protocols should not be difficult or expensive.

All generators who store hazardous waste in containers must inspect the containers and the area surrounding the containers at least once per week. The inspector should look at the condition of the containers and look for signs of leaks, spills, and deterioration. It is also important to ensure that labels are in place and correct and that all hazardous waste management requirements are being met. The inspections must include containers in central storage areas and all satellite accumulation containers.

All inspections must be documented by recording the results in an inspection log. The inspection log should include spaces for the name or initials of the inspector, the date of the inspection, the inspector's observations, and a notation to indicate that any observed problems have been corrected.

REGULATIONS

CONTINGENCY PLANS. Any industry that handles hazardous materials (raw materials or stored waste) must have a contingency plan (hazardous wastes contingency plan or HWCP) and a hazard communication program. These are required for the protection of human health and the environment.

The HWCP or any contingency plan should be a written plan detailing the measures to be taken to reduce harm to public health, safety, and welfare and to the environment. Specific equipment may be required under these regulations. Copies of the contingency plan must be sent to the local police and fire department, emergency response contractors (if needed), the local hospital, and local, state, and federal regulatory agencies for their review and approval of the emergency response arrangements. One current trend is to have an emergency response action plan (ERAP) as part of the facility spill control plan. U.S. EPA will permit a combination of hazardous waste action plans with the spill control plan as routine practice. If properly written, the ERAP will contain contacts and basic cleanup instructions common to both plans. To satisfy the submission requirements, the ERAP can be submitted as a fundamental part of the HWCP.

The primary purpose of a written contingency plan is to clearly describe the actions to be taken by facility personnel and the equipment to be used in response to potential or actual fires, explosions, or any other unplanned releases of hazardous waste or hazardous waste constituents to the air, soil, surface water, or groundwater. All employees are subject to testing on proper behavior during an emergency. The specifics of the plan for hazardous wastes are detailed in 49 CFR Part 265.52.

Generators are required to send copies of their contingency plans to the following agencies or authorities:

- Local fire departments,
- Local police departments,
- Local hospitals,
- State and local emergency response teams, and
- State and federal regulatory authorities.

Part of the rationale for submission of the HWCP to the various regulatory agencies is to enable them to plan not only for what they may have to face in the cleanup or fire from a hazardous waste incident, but also to be able to anticipate what precautions may be necessary to perform field decontamination of injured personnel or decontamination of equipment. For this reason, HWCPs should include any information about specific chemical constituents and a summary of the analytical properties and appropriate material safety data sheets (MSDSs).

It is very important that the contingency plan be clear and detailed so it can be understood by someone who is not familiar with the plant. A general description of the facility (type of business or manufacturing operations) and a site plan typically are included for this purpose. It is particularly important for firefighters to know what kinds of wastes are stored at the plant, the hazards that may result from a fire (such as toxic gases released in a fire and explosion dangers), and where the wastes are stored. All of this information should be included in the plan. The plan should be kept up to date, and it should be amended whenever:

- The facility license is revised,
- The plan fails in an emergency,
- The list of emergency coordinators changes,
- The list of emergency equipment changes, or
- There is any change in maintenance or operation at the facility.

Updated copies should be sent to the agencies mentioned earlier. The issue of the privacy of the plant personnel may come into play when listing contact phone numbers. Often the regulatory agencies want personal contact information, including home telephone numbers and addresses for various plant management personnel. One possible solution is for plant managers and spill response supervisors to carry pagers.

It may be necessary to make specific arrangements for emergency response with the police department, fire department, hospital, and state agencies. A cover letter should be sent with the plan describing what agreements are being requested. A self-addressed response letter should be included in the package so the emergency response agency can sign it and mail it back. In addition, it is suggested that all contingency plans be sent by certified mail with return receipt requested for proof that the plans were received by the agencies, even if a response is not received. Many emergency response agencies will respond to all requests.

Before any individual handles waste, employers are required to provide certain information and training to that individual. The methods of providing this information and training must be described in the required written hazard communication program. The information has to be provided to employees when they are initially assigned to a position or when a new hazard is introduced to their work area.

HAZARD COMMUNICATION STANDARD. The hazard communication standard (29 CFR Part 1910.1200) was promulgated by the U.S. Department of Labor's Occupational Safety and Health Administration (OSHA). This standard became effective November 25, 1985, for manufacturers and distributors of hazardous chemicals and May 25, 1986, for employers who use hazardous chemicals in the workplace and who are included in Groups 20 to 39 of the Standard Industrial Classification (SIC). On August 24, 1987, the hazard communication standard was amended to include all employers with employees who may be exposed to hazardous chemicals in the workplace. This change became effective September 23, 1987, for all chemical manufacturers, importers, and distributors because they must include MSDSs with their next shipment of hazardous chemicals.

The hazard communication standard requires a comprehensive program of dissemination of information about hazardous chemicals by manufactures, distributors, and employers. Information is transmitted by labeling of containers, MSDSs, and formal training of employees.

Employers must develop a written hazard communication program to describe how they will comply with labeling, MSDSs, and employee training requirements. The most important part of any training session is to teach personnel the correct emergency response procedures, as described in the contingency plan. This training must include an explanation of the hazards of the wastes at the plant, the use of emergency equipment, and all other aspects of the contingency plan.

Training can be conducted either by qualified personnel from within the plant or by outside consulting firms. The regulations do not (at this time) specify what is required of a person to be "qualified" to teach personnel about hazardous waste management. Typically, the qualifications of trainers are documented by attaching certifications of training from seminars or hazardous-waste-related courses that the trainer has attended. Obviously, any person or firm selected to conduct training should be extremely knowledgeable in state and U.S. EPA hazardous waste regulations and their application. Many companies prefer to have a consulting firm conduct the initial training in a classroom setting to ensure that all important areas are discussed.

SUPERFUND AMENDMENTS AND REAUTHORIZATION ACT. In addition to the new OSHA standards, on October 17, 1986, the U.S. Congress enacted the Superfund Amendments and Reauthorization Act (SARA) of 1986. This legislation governed many areas including Title III, the Emergency Planning and Community Right-to-Know Act of 1986, which affects many companies in several ways such as

- Requiring companies to report to state and local officials whether or not certain chemicals are used on-site;
- Requiring companies to notify state and local officials of emergency situations that arise;
- Allowing community access to information about chemicals used at plants; and
- Requiring companies to submit emissions inventories to state, federal, and local officials.

All of these amendments are aimed at making the community more aware and industries more responsible for the handling of their waste.

PACKAGING, LABELING, MARKING, AND PLACARDING. U.S. EPA and each state have hazardous waste regulations that refer directly to U.S. DOT hazardous materials regulations for proper packaging, labeling, marking, and placarding requirements. U.S. EPA and U.S. DOT have worked together in promulgating certain segments of the hazardous waste regulations so that both agencies' regulations are uniform and consistent. Table 4.3 gives examples of U.S. EPA hazardous waste regulations that directly reference

Table 4.3 U.S. EPA regulations that reference U.S. DOT shipping requirements.

	U.S. EPA regulations	U.S. DOT regulatory reference
Packaging	40 CFR Part 262.30	49 CFR Parts 173, 178, and 179
Labeling	40 CFR Part 262.31	49 CFR Part 172
Marking	40 CFR Part 262.32	49 CFR Part 172
Placarding	40 CFR Part 262.33	49 CFR Part 172, Subpart F

U.S. DOT shipping requirements. There are three categories of items regulated by U.S. DOT: hazardous wastes, hazardous materials, and hazardous substances, each with a specific and different definition.

Hazardous wastes are items that are listed or classified as such by U.S. EPA. These may be found in 40 CFR Part 261, Subparts C and D. For the purpose of U.S. DOT regulations, the term hazardous waste does not include state-regulated hazardous waste, which would not be considered hazardous under the U.S. EPA system.

A hazardous material is any material listed in the U.S. DOT hazardous materials table in 49 CFR Part 172.101. These materials have been determined by U.S. DOT to be capable of posing an unreasonable risk to health, safety, and property when being transported. They include used and unused materials.

Finally, hazardous substances are those designated by the Clean Water Act that, when in sufficient quantity, pose a significant danger if spilled to a waterway. These materials were identified by an "E" in Column 1 of the hazardous materials table, but this was amended in the November 21, 1986, *Federal Register.* Currently, hazardous substances are items listed in an appendix to the hazardous materials table. Wastes generated at a plant may meet any individual or all of these hazardous definitions. After the appropriate designation has been located, the facility must comply with the specified requirements.

It is important to distinguish between U.S. EPA and U.S. DOT regulations. The Resource Conservation and Recovery Act (RCRA), which is administered by U.S. EPA, regulates hazardous waste. The Hazardous Materials Transportation Act (HMTA), administered by U.S. DOT, regulates hazardous materials. It is important to note that the universe of hazardous materials includes the subset of hazardous wastes. Although the U.S. EPA regulations governing packaging, labeling, marking, and placarding directly refer to the U.S. DOT regulations, all the definitions are not the same. Some examples are shown in Table 4.4.

The hazardous materials table found in 49 CFR Part 172.101 is the heart of U.S. DOT regulations. It is an alphabetical listing of proper shipping names and related information that is extremely important. It is, therefore,

Table 4.4 U.S. DOT and U.S. EPA definitions.

U.S. DOT		U.S. EPA
Shipper	is equivalent to	Generator
Carrier	is equivalent to	Transporter
Destination	is equivalent to	Designated facility
Shipping paper	is equivalent to	Manifest

very important to learn how to use this table and to become familiar with the information it contains.

Perhaps the most important step to take in complying with 49 CFR is to select the correct "proper shipping name." If this name is not properly selected, much of the remaining information required will be incorrect because it is based on the shipping name. Part 172.101(c) describes the necessary procedures in selecting a shipping name.

Certain guidelines are recommended when determining the proper U.S. DOT shipping name. First, the shipping name must be obtained directly from the hazardous materials table. The table has an alphabetical listing of all proper shipping names, each of which has an associated hazard class and UN/NA number. The shipping name selected is used on all manifests and markings. Second, the U.S. EPA waste code or unlisted characteristic of the waste must be looked up in the appendix to 49 CFR Part 172.101. Finally, the reportable quantity (RQ) in the third column of the appendix must be compared with the quantity being shipped in each container. If the quantity being shipped in one container exceeds the RQ, the letters "RQ" and the U.S. EPA waste code or characteristic must be added to the shipping name that is used on the manifests and markings to indicate that the material is a hazardous substance.

The U.S. DOT maintains its own set of labeling requirements. U.S. DOT labels are color-coded, diamond-shaped stickers that attach to the shipping container. The purpose of the labels is to identify the principal hazard or hazards of the material to alert anyone who handles the waste of the potential hazards associated with the material. Labeling requirements are delineated in 49 CFR Part 172, Subpart E. The 49 CFR Part 172.400 requirements set forth general labeling requirements and indicate specific instances when labels are not required.

Another responsibility of the shipper is to provide the required placards to the transporter (49 CFR Part 172.500). A placard is a color-coded, diamond-shaped sign placed on the outside of a transport vehicle; one on each end and one on each side. Placards warn of the potential hazards of the material being transported.

As when selecting a proper shipping name, there is a stepwise procedure that must be followed in determining the correct vehicle placards. These requirements are found in 49 CFR Part 172, Subpart F.

*H*AZARDOUS WASTE FACILITY AUDITS

Many industries have various people involved in the use and management of hazardous materials. It is important to maintain a single uniform plan for the entire facility. Conducting self-audits ensures best management and provides a concise overview of what is going on at the facility. Internal due diligence facility audits are described in Chapter 2.

The first step in conducting a self-audit inspection of a plant is to determine which areas and operations in the plant generate hazardous wastes. To do this, each operation in the plant and all the waste generated must be reviewed and listed, respectively. The wastes that are obviously nonhazardous, such as paper, wood, scrap metal, and other ordinary trash, must be eliminated. However, be careful not to eliminate potentially hazardous wastes such as paint booth filters, spill clean-up residue, and other industrial wastes.

If the environmental manager at the facility is performing the job satisfactorily, the items previously mentioned will already have been taken care of, and a plan and a routine will be available from within the plant. If this type of detailed determination is necessary by the audit personnel because it is not available from the facility, it is safe to assume that the wastes are being mismanaged. The information should already be available but should be thoroughly reviewed.

Next, the list of potentially hazardous wastes is reviewed to determine whether they are hazardous or not. Material safety data sheets and other information from the product suppliers may be helpful if they name the chemical components and physical characteristics (flash point, pH, and so on) of the product.

This type of information may indicate that the product contains a listed waste or exhibits one of the characteristics of a hazardous waste. If this is the case, it is often assumed that the waste will also be hazardous. This is not always true. A hazardous product does not always yield a hazardous waste, just as a nonhazardous product may result in a hazardous waste. At this point, it may be necessary to collect samples for laboratory analysis to complete the hazardous waste determination. Any analyses indicating that the wastes may contain specific hazardous substances should be summarized and the results included in the preparation of the HWCP, which is sent to the emergency response facilities.

After completing the hazardous waste determination, an inspection that covers the major requirements for large- and small-quantity generators should be conducted. Those large-quantity generators who store waste for longer than 90 days are subject to a more stringent and extensive list of requirements than those large-quantity generators who store waste for less than 90 days. The checklist used in an audit should be specific to the site being audited.

Again, any checklists prepared should be compared with the regulations before the audit is conducted, and if in doubt the corporate legal services should be consulted. The audit checklist may devolve into a visual inspection checklist with initials or may not generate a formal report—especially if corporate legal services are involved. Again, the difference between an informal inspection and a formal audit may not be substantially different in nature except that the formal audit will generate a paperwork trail that may be handled at a higher level within the corporation.

Self-audits are one tool that may be useful in the management of hazardous wastes, but they should be conducted with caution to the points previously discussed because a formal audit can be a "smoking gun" in the hands of a hostile regulatory inspection. The point is that some type of internal or external review should be conducted periodically to ensure that the plant is maintaining compliance with the regulations. The potential for violations is dramatically reduced and the costs of operation are also reduced when better management and handling practices of hazardous wastes are followed.

SUGGESTED READINGS

Applied Environmental Technologies Corporation (1988) *Hazardous Waste Rules and Regulations Reference Guide.*

Brady, J.G. (Ed.) (1988) *The Supervisor's Complete Guide to Hazardous Waste and Materials Management.* Revised 2nd Ed., Business and Legal Reports, Madison, Conn.

Couture, T. (1986) Practical Methods for Minimizing Waste Generation and Handling Storage Under New Regulation. Paper presented at Litton Ind. Hazard. Waste Conf., St. Louis, Mo.

Newton, J. (1988) *A RCRA Generator's Compliance Program.* Pudvan Publishing, Northbrook, Ill.

Perry, R.H.; Green, D.W.; and Maloney, J.O. (Eds.) (1997) *Perry's Chemical Engineers' Handbook.* McGraw-Hill, New York.

Sax, N.I., and Lewis, R.J. (1989) *Dangerous Properties of Industrial Materials.* Van Nostrand Reinhold, New York.

Shields, E.J. (1985) *Pollution Control Engineer's Handbook.* Pudvan Publishing, Northbrook, Ill.

U.S. Code of Federal Regulations (1988) Title 29 CFR Parts 1–799, Occupational Safety and Health Administration. U.S. Dep. Labor, Washington, D.C.

U.S. Code of Federal Regulations (1988) Title 40 CFR Parts 1–700, Protection of Environment. U.S. EPA, Washington, D.C.

U.S. Code of Federal Regulations (Year Not Given) Title 49 CFR Parts 100–199, Transportation. Research and Special Programs Administration. U.S. Dep. Transp., Washington, D.C.

Chapter 5 Biological Treatment of Hazardous Wastes

INTRODUCTION

GENERAL. Aitken and Irvine (1987), in their introductory remarks to their chapter on biological treatment of hazardous wastes, stated "…most, if not all, 'toxic' organic chemicals can be degraded biologically. There is, however, a tremendous difference between what a microbe can do under controlled laboratory conditions, and what it does do in the environment or in the waste treatment system."

Biological treatment technologies have been successfully used for the treatment of industrial and municipal wastes containing hazardous constituents. There are many examples of the successful use of conventional and emerging technologies on the bench- and pilot-scale level and on the number of full-scale plants using these technology increases each year. The advantages of biological treatment over other methods are that when complete, or nearly complete, biodegradation occurs, and the waste stream can become nonhazardous. Successful biological treatment can completely remove contaminants by the action of the microbes breaking down the wastes into nonhazardous byproducts. The cost is often less expensive than many other alternatives. When complete biodegradation is not possible, removal of the hazardous components can be accomplished by other technologies and methods, such as activated carbon, ion exchange, and air stripping.

This chapter is divided into seven major sections. These sections will look at the biological metabolism and how contaminants are biodegraded, potential applications in all types of matrices, how to characterize waste streams and determine the feasibility of biodegradation, general and specific treatment technologies and methods, the conditions and limits of the biological processes, how to scale up from pilot tests, and the current and emerging technologies that hold promise for biological treatment.

Few generalizations can be made when discussing biodegradation of toxic and hazardous materials by conventional biological treatment. The removal capabilities of any given system are strongly influenced by its physical configuration, regardless of the inherent biodegradability of the compounds of interest. However, the fate of any compound in a biological treatment system will depend on its inherent biodegradability (Grady, 1988). Recalcitrant compounds will not be removed by biological mechanisms regardless of the engineering effort applied. Biodegradable compounds may or may not be removed, depending on the reaction environment. The primary goal of engineering biological treatment systems for the removal of toxic and hazardous chemicals is to allow maximum expression of inherent biodegradation potential. The biodegradability depends on the environmental conditions the compound encounters. Other than the type of compounds targeted for biodegradation, factors influencing their removal include the concentration of the compounds, synergistic chemical reactions, nutrient availability, oxygen concentration, pH, temperature, retention time, and other operational factors.

This chapter will discuss the treatment of hazardous wastes in liquid, solid, sludge and slurry, and air. Successful treatment of hazardous wastes in all of these media has been accomplished. The discussions that follow are an attempt to consolidate the research and designs conducted on biological treatment of hazardous wastes. Review of these results can help determine the mechanisms that remove the contaminants, predictions for removal efficiencies, whether the constituents in their waste stream are amenable to biological treatment, and some successful design and operational strategies.

BIOLOGICAL METABOLISM. Knowledge of the mechanisms by which microorganisms use food sources for growth and cell maintenance is fundamental to the understanding of biodegradation. These mechanisms are collectively referred to as a "metabolism."

All cells must have sources of carbon and energy to live and grow. In waste treatment, engineers take advantage of this fact by providing an environment in which the waste serves as the carbon or energy source. In this way, the growth of a biomass culture is dependent on the utilization of the same constituents in the medium targeted for removal. In domestic aerobic wastewater treatment this process results in the conversion of organic compounds to carbon dioxide, water, and new cell mass. The use of biological treatment for the destruction of hazardous wastes is based on the same principles. However, a more complicated set of parameters needs to be analyzed. This section will offer a brief overview of the fundamentals of metabolism before discussing the specialized application to hazardous waste treatment.

Metabolic Processes. Energy conservation in biological systems requires the transfer of electrons between donor and acceptor molecules. The primary energy source is the electron donor. The electron acceptor provides the means of microbial classification. It also provides for chemical transformations.

If electrons are accepted by a molecule present in the extracellular medium, the metabolism is termed a respiration. If the electron acceptor is molecular oxygen, the respiration is termed aerobic. If it is some other inorganic compound such as nitrate or sulfate, the respiration is termed anaerobic or anoxic. Microorganisms that respire oxidize the energy source by transferring electrons to the acceptor. If the energy source is an organic compound, oxidation goes to carbon dioxide and water and all available electrons are removed. The transformation is mineralization. Often, in the degradation of complex organic compounds (such as some hazardous wastes), the oxidation is only partial, resulting in organic compounds remaining in the medium. This concept is important because the products of incomplete biodegradation can be just as toxic as, or even more so than, the original compound. Therefore, any analysis concerning the biodegradation of a toxic organic compound must consider byproduct formation.

In some cases, no external electron acceptor is present in the growth medium. This environment would not allow growth of organisms by respiration. In these cases, only those organisms capable of transferring the electrons to an acceptor present inside the cell would grow. This type of metabolism is termed fermentation. Fermentation does not result in a net oxidation of organic carbon present in the growth medium because all electrons are transferred to other organic molecules. It can, however, transform a specific organic compound by this electron transfer. Indeed, some research has found organic priority pollutants that are transformed only by fermentative pathways.

Metabolic Classification. An understanding of the meanings of microbial classification is required for the determination of the types of transformations that a specific organism or culture can catalyze. Because much hazardous waste engineering involves the use of treatability studies or novel processes, a background in this area is needed.

Microorganisms are classified by the type of metabolism they use for growth. These classifications are based on the carbon source and energy source. All cells need an external source of carbon for biosynthesis because carbon typically composes approximately 50% of the total dry mass of cells. Only two such sources exist in nature, organic compounds and carbon dioxide (CO_2). Microorganisms that can use carbon dioxide as their primary source of carbon are known as autotrophs. Many microbes, however, do not possess the specialized apparatus necessary for the utilization of carbon dioxide as a principal carbon source. They must rely on organic compounds for this purpose. These organisms are known as heterotrophs. Because many hazardous wastes are organic in nature, the heterotrophs are the most important group for biodegradation of hazardous wastes.

Cells may obtain energy either from organic or inorganic compounds assimilated from the extracellular environment or from harnessing energy present in visible light through a specialized cellular apparatus. Few microorganisms can use both energy sources. Those that use chemical compounds (organic or inorganic) for their primary energy source are known as chemotrophs, while those that extract energy from visible light are called phototrophs. A widely used classification scheme uses these terms to divide microorganisms into four metabolic groups.

- Chemoheterotrophs are those organisms that use organic compounds as an energy source and must rely on organic compounds as the primary carbon source. This group includes all fungi (yeasts and molds), protozoa, and many bacteria. They are the primary organisms responsible for organic compound degradation.
- Chemoautotrophs are those organisms that use inorganic compounds as an energy source and can use carbon dioxide as their primary carbon source. Only certain highly specialized bacteria, including the nitrifying bacteria, are chemoautotrophic.
- Photoautotrophs are those organisms that use light as an energy source and can use carbon dioxide as their primary carbon source. Most of the algae and the cyanobacteria (blue-green bacteria) belong to this group.
- Photoheterotrophs are those organisms that use light as a principal energy source and must rely on organic compounds as their primary carbon source. Only a few algae and cyanobacteria are classified in this small group, which is dominated by anaerobic, photosynthetic bacteria.

Co-metabolism. As previously stated, all organisms require sources of carbon and energy for growth. If a particular carbon source also happens to be a compound that is undesirable, exploitation of microbial metabolism will serve a useful purpose. In some cases, however, an organic compound cannot by itself serve as a carbon and energy source. This may be because of a lack of adequate available chemical energy (electrons), a lack of requisite systems to allow uptake by biochemical pathways, or a lack of sufficient quantities of the compound to allow proliferation of an active culture. It has been observed, however, that microorganisms can transform substances present in the growth medium while actively growing on a different substance. This process is referred to as cometabolism, defined as "the concomitant oxidation of a nongrowth substrate during the growth on a utilizable carbon and energy source" (Horvath, 1972).

The concept of cometabolism is especially important in the evaluation of treatment technologies because the absence of biodegradation of a specific organic compound may not preclude it from treatment with a cosubstrate. The evaluation of biodegradation in the presence and absence of cosubstrates should be evaluated before biological treatment is abandoned as a feasible alternative.

Biochemical Principles of the Anaerobic Processes. The molecules resulting from hydrolysis are used as carbon and energy sources by the bacteria that carry out fermentations. The oxidation end products of these fermentation processes are primarily short-chain volatile acids such as acetic, propionic, butyric, valeric, and caproic. Their production is referred to as acidogenesis, and the responsible organisms are called acid-producing bacteria. The reduced end products of the fermentation depend on the nature of the culture and the environmental conditions in the reactor. Some of the acid-producing bacteria possess a specialized enzyme system that allows them to oxidize reduced coenzymes without passing the electrons to an organic acceptor, thereby releasing hydrogen gas (H_2) to the medium. The collective activity of these hydrogen-producing bacteria is called hydrogenesis.

The distinction between acid- and hydrogen-producing bacteria is not clear. Because hydrogen-producing bacteria typically produce acids but acid-producing bacteria do not all produce hydrogen, it is probably best to think of the hydrogen-producing bacteria as a subset of the acid-producing group. The combined groups of acid- and hydrogen-producing bacteria are typically referred to as nonmethanogenic bacteria and their integrated metabolism results primarily in formic acid, acetic acid, CO_2, and H_2. If no hydrogen is formed, the nonmethanogenic phase results in insignificant reductions in chemical oxygen demand (COD) because all electrons released in the oxidation of organic compounds are passed to organic acceptors that remain in the medium. Consequently, the energy level of the entire system is lowered only by losses because of microbial inefficiency. When hydrogen is formed, however, it represents a gaseous product that escapes from the medium, causing a reduction in the energy content, and thus the COD, of the liquid (McCarty, 1964).

The products of the nonmethanogenic phase (that is, formic acid, CO_2, and H_2) are used by methanogenic bacteria to produce methane gas. Almost all of the energy removed from the liquid is recovered in the methane.

The complications involved in the process kinetics during the catabolic activities of the consortium of anaerobic bacteria amplify when considering hazardous waste treatment. However, researchers consider that methanogenesis is probably the rate-limiting step because of the following reasons.

- Production of methane, essentially insoluble in water, drives the overall reaction to completion.
- During hydrolysis and acidogenesis, only a molecular rearrangement occurs. During methanogenesis, stabilization occurs.
- Methane bacteria are slower and more sensitive to the environment than their counterparts. Therefore, an upset to methanogens will cause upset of the entire treatment system.

Figure 5.1 shows the multistep anaerobic reactions.

POTENTIAL APPLICATIONS

GENERAL. Biological treatment of hazardous wastes is many times mistakenly assumed to not be possible because it is assumed that the hazardous constituents are toxic to living microorganisms. Therefore, it is assumed that biological treatment is not a viable alternative. To overcome this perception, a more thought out and aggressive approach has to be considered to apply biological treatment and exploit its full potential (Freeman, 1988).

Biological treatment can be used to treat various types of media including industrial wastewater, surface waters, groundwater, sludges, sediments, soil, gases, and air. It is possible to treat any individual or a combination of these media with biological treatment. Soils, sediments, and sludge biological processes include aboveground piles, land farming, composting, washing, and in situ processes. Liquids, wastewater, and groundwater biological processes include aboveground bioreactors and in situ processes. Gases and contaminated air streams include bioreactors and other in situ processes.

WATER. Wastewater treatment is the most common biological technique because it is the easiest to design and control. This is mainly from using the same technologies that have been used for decades in municipal wastewater treatment. The treatment is performed in reactors using either aerobic or anaerobic bacteria. There are several reactor designs available that have been shown to be applicable to industrial wastewater and in situ groundwater applications, including activated sludge, trickling filter, submerged fixed film, fluidized bed, two stage, sequencing batch, and other hybrid designs. These various designs have been adapted to industrial wastewater, surface water, and groundwater applications with various hazardous waste constituents and therefore are capable of treating a wide range of these contaminants using aerobic or anaerobic bacteria. In some cases, biological treatment

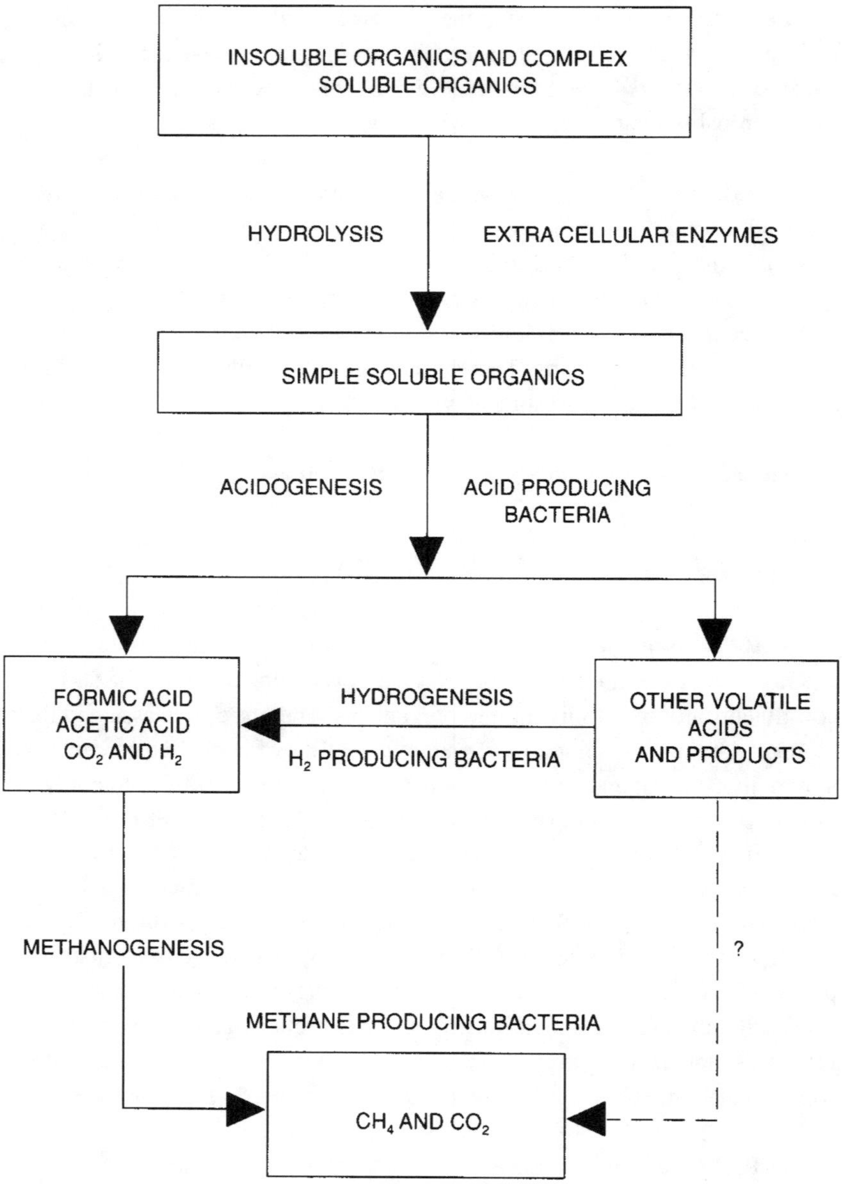

Figure 5.1 The multistep anaerobic reactions.

does not completely degrade the waste, and it is necessary to combine other treatment technologies for complete treatment. These technologies may include air stripping, granular activated carbon (GAC), precipitation, or thermal processes.

The use of aerobic and anaerobic biological reactors for industrial wastewater, surface water, and groundwater will be discussed in detail later in this chapter. Specifically, the different designs, capabilities, and limitations will be addressed.

AIR. Bioreactors have been developed specifically to treat gas and air streams, whether for offgas emissions from other treatment systems such as bioventing, vapor extraction, air stripping, soil reactors, or manufacturing process emissions. When examining an air-emissions control need, a number of technologies typically are evaluated, including activated carbon, incineration, catalytic oxidation, and water and chemical scrubbing. Biofiltration is the newcomer among these conventional treatment technologies (Roy, 1992).

Economic analysis of biofiltration versus alternative technologies often leads to the conclusion that from an operational, maintenance, and capital basis, it is the least expensive technology. However, for that measure to have any value, the technology must first be evaluated using the following parameters: biodegradability, space, destruction efficiency, and air stream continuity.

General biodegradability information can be obtained from vendors and from the literature. It is generally accepted that biofilters can treat H_2S, CS_2, ammonia, small organic sulfides, BTEX (benzene, toluene, ethyl benzene, and xylene), light aliphatic hydrocarbons, ethanol, and styrene. However, process streams vary from facility to facility in both concentration and composition. A mixed substrate stream can slow or inhibit the degradation of a specific target compound, making the process less efficient than expected or requiring a larger biofilter. In addition, some materials are toxic, and high concentrations can have a severe effect on microbial activity. These aspects should not eliminate the use of biofiltration but should suggest that at a minimum a laboratory demonstration of biodegradation be conducted if the process stream can be suitably modeled. A pilot study is a prudent option when a company is investigating the use of biofiltration on complex process streams.

Water solubility plays an important role in biodegradability for biofilter applications. For example, compounds such as aliphatic hydrocarbons are reported as readily biodegradable in the literature but have a limited solubility in water. Most studies looking at these materials are done in solution in which the organisms can respond to a free hydrophobic phase (a layer of oil, for instance) by making biosurfactants. However, in a biofilter, these compounds are mass transfer limited, and biodegradation can be slow. Thus, a biofilter design would have to have extremely low volumetric flow rates or a large bed volume to accommodate slow degradation. It is worth mentioning that many biofiltration vendors have active research programs and that ongoing efforts to make use of organisms specifically adapted to new substrates will likely extend the breadth of application of biofiltration.

The footprint of a biofilter can be large, and available space at a given facility may be a limiting factor. As a rule, the minimum and maximum area required for installing a biofilter can be calculated using a range of values for the rate of contaminant elimination and the maximum air loading rate. Most elimination rates are between 30 to 100 g contaminant/$m^3 \cdot h$. Assuming that the filter is 1 m deep, it is relatively easy to get an idea of the bed area by dividing the anticipated contaminant flow rate by the estimated elimination rate. A second factor is the maximum amount of air that can be applied to a given area of biofiltration medium. Typically, no more than 300 m^3/h may be applied to 1 m^2 of biofilter surface area.

The emissions reduction required for each project can vary and depends on the compound(s), the reasons for treatment (odor or air quality), and local rules and regulations. However, biofiltration can provide 99% removal of many odor-causing compounds, such as H_2S, and at least 90% removal of volatile organic contaminants (VOCs). If greater control levels are required, biofiltration may still prove desirable as part of a hybrid control system.

Biofilters are ideal for applications with continuous air stream flows, which provide a steady source of nourishment for contaminant-hungry microbes. Intermittent air streams can cause problems because of a decrease in biological activity during flow interruptions. Exceptions abound, however. A biofilter treating a xylene-laden stream can recover immediately after a 48-hour, air-contaminant feed cutoff, and similar results occur with various biofilter configurations and target compounds. Longer period feed shutdowns may be managed by providing a sacrificial substrate to keep the biofilter healthy and operational until the feed can resume.

SOILS AND SLUDGES. Soils and sludges can be biologically treated whether in a solid phase or slurry phase. Composting is one form of solid phase biodegradation but has conventionally been used for nonhazardous solid wastes. Organic contaminants in sludges and soils can be biologically treated in situ or excavated and treated by solid- and slurry-phase bioremediation processes. Solid-phase processes are continually being developed and, in many cases, have been used to treat a wide range of contaminants such as pesticides, diesel, gasoline, fuel oil, creosote, pentachlorophenol, and halogenated volatile organics. Enhanced in situ biodegradation is being used for sites having soil and groundwater contaminated with readily biodegradable organics such as gasoline and diesel. This same technology is being developed and used to treat contaminants that are more difficult to degrade (U.S. EPA, 1988a).

Soil bioremediation is a process that treats soils in situ or ex situ in above-grade systems using conventional soil management practices to enhance the microbial degradation of contaminants. In situ treatment for soils typically consists of bioventing, which is the movement of air through the contaminated soil. Another biodegradion technology involves the treatment of contaminated soil or sludge in a large mobile bioreactor. This system maintains mixing and contact of microorganisms with the hazardous compounds and creates the appropriate environmental conditions for optimizing microbial biodegradation of target contaminants (U.S. EPA, 1988a).

SPECIFIC CHEMICAL COMPOUNDS. The majority of biological treatment to date has focused on treating petroleum hydrocarbons, which typically are natural organic compounds. There are many synthetic organic compounds that can be degraded by microorganisms. Current technology has shown these capabilities, with research and development of new technologies still ongoing. Specific compounds of interest that can appear in one or more of the environmental media discussed above are pesticides, including organohalogen insecticides, chloroaromatic compounds like chlorobenzenes, (polychlorinated biphenyls) PCBs, chlorophenols and chlorobenzoates like pentachlorophenol, chloroaliphatic compounds like trichloroethylene, nitroaromatic compounds like nitrobenzenes and nitrotoluenes, dibenzodioxins and dibenzofurans, and metals. The treatment of these compounds by biological treatment has been developed and implemented in lab-, pilot-, and full-scale applications. The use of aerobic and anaerobic bacteria, bacteria-derived enzymes, as well as other microorganisms, such as fungi, are being used to treat these specific compounds in water, soil, sludge, and air.

WASTE CHARACTERIZATION, TREATABILITY STUDIES, AND MODELING

GENERAL. The definition of a hazardous waste, under RCRA, is any waste that is corrosive, toxic, ignitable, or reactive or is a listed waste. The first step in assessing if a waste can be treated by biological treatment (or other methods) is to characterize the constituents of the waste.

WASTE CHARACTERIZATION. An evaluation of processes that contribute to the waste stream can assist in identifying the constituents. When the source is a leachate or a waste stream of unknown composition, laboratory analyses should be conducted to determine the type of contaminants and their concentrations. After assessing the character of the waste stream, the feasibility of biological treatment can be initially assessed on paper. Typically, heavy metal concentrations should be less than 5 to 10 mg/L. This varies with the metal and, as discussed subsequently, acclimated biomasses have tolerated much higher levels.

TREATABILITY STUDIES. After an evaluation, initial laboratory screening using dilutions in a 5-day biochemical oxygen demand (BOD_5) test, specific dissolved oxygen (DO) uptake rates, Microtox, adenosine triphosphate, or respirometry can give a gross indication of the feasibility of continuing with biological treatment.

If the results of the laboratory screening are favorable, the next phase is to set up bench-scale (0.5 to 5 L) reactors and test various loadings and operational configurations. The length of the tests should ensure stabilization of the system. These bench-scale tests are followed by larger scale pilot studies. The larger reactors are run to more closely define operational requirements and refine design parameters. If all of the above prove successful, full-scale design and implementation can proceed with a much higher degree of confidence than if these steps are abbreviated or deleted. A later section of this chapter presents considerations for scale-up to full-scale treatment.

Regulatory Considerations. Before treating hazardous wastes and wastewater, permitting by the regulatory authority (state or federal) may be required. Analysis of samples may be conducted by a laboratory to characterize the wastes without a permit if the sole purpose of testing is to determine the wastes' characteristics (U.S. CFR, 1989). Some states may require certification of the laboratory for acceptance of the results.

According to federal regulations (not adopted uniformly by all states), a permit is required before undertaking pilot- or bench-scale investigations to evaluate treatment alternatives (treatability studies). These federal regulations [40 CFR Part 261.4(e)] make it possible for the people who generate or collect samples for treatability studies to be exempt from 40 CFR Parts 261 to 263 and the notification requirements of Part 3010 of RCRA. Certain conditions that must be met for treatability studies include limitations on the quantity of waste treated and transported; packaging, labeling, and shipping requirements; and reporting and recordkeeping requirements. The U.S. EPA regional administrator or the state director (if in an authorized state) may grant requests for variations.

Samples undergoing treatability studies and the laboratory or testing facility conducting such treatability studies (to the extent such facilities are not otherwise subject to RCRA requirements) are not subject to any requirement of 40 CFR Parts 124, 262 to 266, 268, and 270 or to the notification requirements of Section 3010 or RCRA, provided specific conditions are met. A mobile treatment unit may also qualify as a testing facility. Conditions to be met include notification of the appropriate state or U.S. EPA officials; procurement of a U.S. EPA identification number; limitation on the mass of hazardous waste treated, received, or stored per day; and requirements that specific time limits for completion are not exceeded, no open burning or direct land application occurs, records are kept and properly maintained, a report is submitted to the state or U.S. EPA, and residues are properly disposed.

It should be noted that if the wastewater containing hazardous waste constituents is treated by a wastewater treatment unit that is an integral part of the process generating the waste stream and is permitted under the Clean Water Act, with an NPDES or state permit, then it may be exempt from the RCRA (hazardous waste) requirements.

Seeding, Acclimation, and Bioaugmentation of Bioreactors. *SEEDING.* This section discusses the alternative for development of a viable culture for start up of an activated-sludge plant (Reitano, 1981). The following sources were considered for seed cultures:

- Mixed liquor from another activated-sludge plant,
- Naturally occurring microbes in the wastewater to be treated (self-seeding),

- Soil with naturally occurring microbes, and
- Enriched freeze-dried cultures.

The use of mixed liquor from another activated-sludge plant is a popular method for developing an acclimated seed. Major problems typically are associated with the different species of microorganisms developed for particular wastewater. A specific type of substrate will result in the development of a specific microbial population. To minimize acclimation problems, mixed liquor should be obtained from an activated-sludge plant that is treating wastewater with similar characteristics to the wastewater that will be treated. A municipal sludge required 4 to 8 weeks to develop a stable mixed liquor suspended solids (MLSS) concentration of 3000 to 4000 mg/L for a petrochemical wastewater (Reitano, 1981).

Another typical method for an activated-sludge startup is to develop an acclimated culture from naturally occurring microorganism in the wastewater. The self-seeding approach was used to develop mixed liquor cultures during startup of activated-sludge plants at several refineries worldwide. At one refinery, 1 month was required to reach a MLSS concentration of 2 000 mg/L, while at two others, approximately 2 months were required to develop a high concentration of microorganisms (Reitano, 1981).

Soil seeding reportedly is an economical method for activated-sludge startup that can be substantially faster than the previous methods discussed. In addition to supplying a source of bacteria, the soil particles can result in nucleation for development of microbial flocs (Reitano, 1981).

Freeze-dried cultures of bacteria, acclimated to specific waste components, are being used. The advantages claimed are that the addition of this active microbial mass results in a rapid growth rate and a short period necessary to develop a viable mixed liquor population. This method would be especially attractive when a mixed liquor treating a similar wastewater is not available in the geographic area and a short startup period is required (Reitano, 1981).

During a recent seminar on treatment of hazardous wastes. Aitken and Irvine (1987) presented the following information on the sources of seed cultures:

> The most appropriate source of inoculum for a biological treatability study is any local environment to which the waste has been introduced, if microorganisms are also likely to be present. Such sources can include contaminated soil or materials from waste storage impoundments, biological sludges from systems treating similar wastes, or sludges from municipal treatment plants receiving a substantial input of industrial wastes. Primary treatment sludges can be a source of aerobic, anaerobic, or facultative organisms. Secondary treatment sludges may be good sources of aerobic inocula. Primary and secondary anaerobic digestion sludges often are used to start up anaerobic systems. Other sources of anaerobic inocula include manure, or fluids obtained directly from the rumen of animals such as sheep or cattle. It is well

worth the effort to collect inocula from a variety of sources. Sources can be tested individually in shake flasks or serum bottles, or can be combined to increase the probability of success.

ACCLIMATION. The need for patience during treatability testing stems from the long acclimation periods often required before a mixed culture "takes off" and responds aggressively to a particular waste. A major factor affecting the length of the acclimation period is the selection of an appropriate seed culture. If the required genetic information is not present in the seed culture (or in the waste), it is not likely to develop from spontaneous mutation under normal laboratory conditions. As a result, compounds that are initially toxic or nonbiodegradable will remain so, independent of the acclimation period length. When the desired genetic information is present in the startup culture or in the waste, an acclimation period will be productive if the organisms with the appropriate genes can compete. Alternatively, the necessary genes can be transferred to other more competitive organisms, provided the proper operating strategies are implemented (Aitken and Irvine, 1987).

Standard microbiological enrichment techniques involve the introduction of a relatively small inoculum to a relatively large volume of solution containing the material to be treated. Alternately, engineers typically begin with the addition of a relatively small volume of waste to a massive population of organisms in the reactor to be evaluated. Because the toxicity or inhibitory properties of the waste constituents are often a function of their concentration, the enrichment techniques must take such effects into account. Enrichment techniques are more likely to develop competitive cultures in a short period of time if concentration effects are overcome. In the engineering approach, the diversity of the genetic pool is likely to be greater, which can increase the probability that some members of the "audience" will be satisfied. However, it may take several weeks after the desired treatment activity begins before nonessential or unwanted members of the culture are reduced to insignificant numbers (Aitken and Irvine, 1987).

For wastes with a significant amount of potential growth substrates, the initial success of the enrichment techniques is judged by noticeable increases in the microbial population, either observed visually or measured as optical absorbency or turbidity. The activity of larger populations in a reactor can be judged by measuring changes in parameters such as COD, total organic carbon (TOC), or gas production. Activity of aerobic cultures can also be evaluated using respirometric techniques. Respirometry is particularly useful in determining the activity of the culture toward specific waste constituents. For acclimation periods aimed at the conversion of low concentrations of xenobiotics, direct measurement of the xenobiotic or readily analyzed byproduct is necessary (Aitken and Irvine, 1987).

Using two or more reactors operated in parallel allows evaluation of a number of variables, both during acclimation and after continuous operation is established. One of the important variables to evaluate during the acclimation period is the waste feed pattern or loading rate. Many biological treatment evaluations begin with a period of batch treatment, even if the treatment process to be evaluated is continuous. It should be noted here that sustained batch feeding techniques in laboratory systems intended to evaluate continuous processes may be subject to unreliable scale-up. Intuition might suggest that slow initial feeding is most appropriate because removal kinetics are of secondary importance to population establishment during acclimation. However, a more rigorous feeding strategy may be appropriate in some instances. Initial feeding patterns often allow for accumulation in reactor volume. This can help maintain all potentially vital components of the culture, provided they do not allow for the accumulation of an unwanted inhibitor.

Abundant data collection is very important during the acclimation period. Feeding should be reduced substantially or terminated if indicators such as gas production, oxygen consumption rates, or TOC removal drop off suddenly. The pH of the culture (or effluent) should be monitored carefully because large pH changes are quite possible as a result of biological activity and are a frequent cause of upsets. Process control strategies such as sludge wasting should be varied slowly, until the desired operating conditions are reached. Among the advantages of operating multiple reactors is that more creative or risky operating techniques can be evaluated. Failure of a reactor can provide as much information as success if the reason for failure is understood (Aitken and Irvine, 1987).

BIOAUGMENTATION. In an operating bioreactor used for waste treatment, an indigenous bacterial population arises that is unique from the standpoint of species diversity. This population is the result of physiological and genetic adaptation to the waste characteristics and process operation parameters. In bioreactors treating a variable mixture of toxic or inhibitory wastes, bacterial diversity is reduced by the increased selective pressure on the population because of the presence of selected waste components. Reducing the types of bacteria present can diminish the genetic pool of the reactor population and decrease the ability of the population to respond to changes in the environment or waste composition. Because of the diminished capacity for adaption, biological hazardous waste treatment processes are often plagued by upsets and are unable to degrade new compounds entering the waste stream. Bioaugmentation is the process of adding nonindigenous bacterial supplements to a bioreactor for the purpose of artificially increasing the bacterial diversity and activity of the population by adding bacteria that have higher metabolic rates. Many bacterial supplements are available commercially to augment existing populations of biological hazardous waste treatment processes. To date, no clear consensus has been reached on the relative merits of bioaugmentation (Lange et al., 1987).

A compilation of findings from published papers on bioaugmentation are presented in Table 5.1. The location of the testing, the waste type, and

Table 5.1 Compilation of current supplementation reports.

Test setting	Waste type	Bioaugmentation effective	
		Yes	No
Field	Pharmaceutical	X	
Field	Petrochemical	X	
Field	Refinery	X	
Field	Refinery	X	
Field	Municipal	X	
Field	Dairy	X	
Laboratory	Dairy		X
Laboratory	Hazardous		X
Laboratory	Synthetic		X
Laboratory	Chlorinated organics		X

whether or not treatment performance was enhanced by bioaugmentation are given. Based on the information presented in Table 5.1, enhanced biological treatment performance has been attributed to bioaugmentation for those studies conducted under field conditions, while those studies conducted in the laboratory have been generally negative.

It is also important to note that even when bacterial supplements were acclimated to test conditions they did not perform better than the indigenous population. This result raises further questions about the utility of commercial supplements when compared to an indigenous population.

In summary, published results from recent research dealing with the effectiveness of bacterial supplementation have been inconclusive. However, on the basis of a literature review, a number of factors that may affect the potential success of bioaugmentation were identified as follows: (1) the viability of the bacterial supplements should remain high when added to the indigenous reactor population, (2) once supplements are added they should initiate biodegradation of the target compound rapidly or degrade compounds not degraded by the indigenous population, and (3) added supplements should have the necessary characteristics to maintain their population in the reactor (Lange et al., 1987).

Inhibition and Shock Loading. Important factors determined during laboratory-, bench-, pilot-, and full-scale testing are the concentrations of hazardous components that can be tolerated by the biological treatment system and the responses that can be expected. The specific dissolved oxygen uptake rate (SDOUR) and adenosine triphosphate (ATP) can be used as indicators of microbial activity (Powell and Reitano, 1976).

Changes in a system such as substrate leaking, reactor appearance, or biomass characteristics are occurrences that operators need to watch for to identify when an inhibitory situation is occurring. For example, the work of Rozich and Gaudy (1985) showed that for both inhibitory and noninhibitory substrate quantitative shock loads, the effluent COD increases following a shock, although little or none of the original substrate may appear in the effluent. The primary effect of the shock is, in both cases, an apparent increase in production and release of metabolic products. The products released by the cells in the phenol system are almost certainly less toxic then phenol (Rozich and Gaudy, 1985).

Rozich and Gaudy (1985) noted that the similarity in the reactions to severe shock loads in systems treating inhibitory and noninhibitory wastes was the difference in the duration of the predicted and observed transients. In both types of systems, the transient, as judged by fluctuations in COD and solids, continues for much longer than predicted by kinetic equations. This prolonged instability is attributed to ecological shifts in the biomass that result from the change in operational conditions. Research has not yet led to a prediction of the occurrence of the changes in predominance characteristic of the delayed or ecological response that ultimately correlated the complete system failure occurring in the phenol system reported. For a nontoxic substrate, this secondary response, while leading to severe COD leakage, did not lead to wash out of the system. The biomass was able to recover, but when phenol was the carbon source, no recovery ensued. The analysis of the model, using experimentally determined biokinetic constants, gave some insight as to why recovery ensued and why recovery may not be possible, even if changes in predominance do not occur and cause wash out. If the specific growth rate exceeds peak or critical growth rate, with the substrate (S) diluting into the system, the biomass can no longer respond to the increased Si, and the resultant system failure cannot be avoided. In the phenol system described, when COD in the reactor increased drastically several days after the original shock, essentially all of the COD was due to phenol; the cells were no longer taking up phenol and using it for either replication or conversion to metabolic products. Likely, once phenol began to accumulate in the reactor because * and S* had been exceeded, the toxic effects of phenol caused cell lysis, exacerbating the decrease in solids level because of decreased growth rate.

From a practical engineering point of view, Rozich and Gaudy (1985) conclude that successful treatment calls for designing low, rather than high, specific growth rates and for providing the operator with the means to keep the specific growth rate low, regardless of the changing influent conditions.

Considerations for Treatability Studies for Anaerobic Processes. Waste characterization requirements for anaerobic treatment studies are similar to aerobic treatment. However, special attention should be given to the pH conditions of the waste and whether the waste is contaminated with toxic organics or heavy metals. See the section on Treatment Methods, Anaerobic Processes, for more details.

The equipment required for determining the design parameters for anaerobic degradation of hazardous wastewaters consists of various types of reactors that may be batch or continuously fed systems. The reactors can be completely mixed or plug flow. The experimental systems must have the capability to contain the off gases from the system. A typical completely mixed, continuously fed reactor is presented in Figure 5.2. An operational startup procedure follows:

(1) Set up the apparatus, as shown in Figure 5.2.
(2) Obtain an actively digesting seed from a municipal plant and immediately place it in the reactor. If possible, maintain the contents at 35 °C during startup and acclimation.
(3) If a significant quantity of air is trapped in the digester, purge with an inert gas, preferably helium.
(4) Adjust the suspended solids concentration to a level of approximately 6000 to 10 000 mg/L.
(5) Begin feeding a starter substrate to the reactor contents at a preestablished rate. This starter substrate should be readily degradable by the anaerobic culture present.
(6) Adjust the flow of the gas recirculation pump to completely mix the contents of the reactor. Alternatively, use a mechanical mixer or magnetic stirrer.

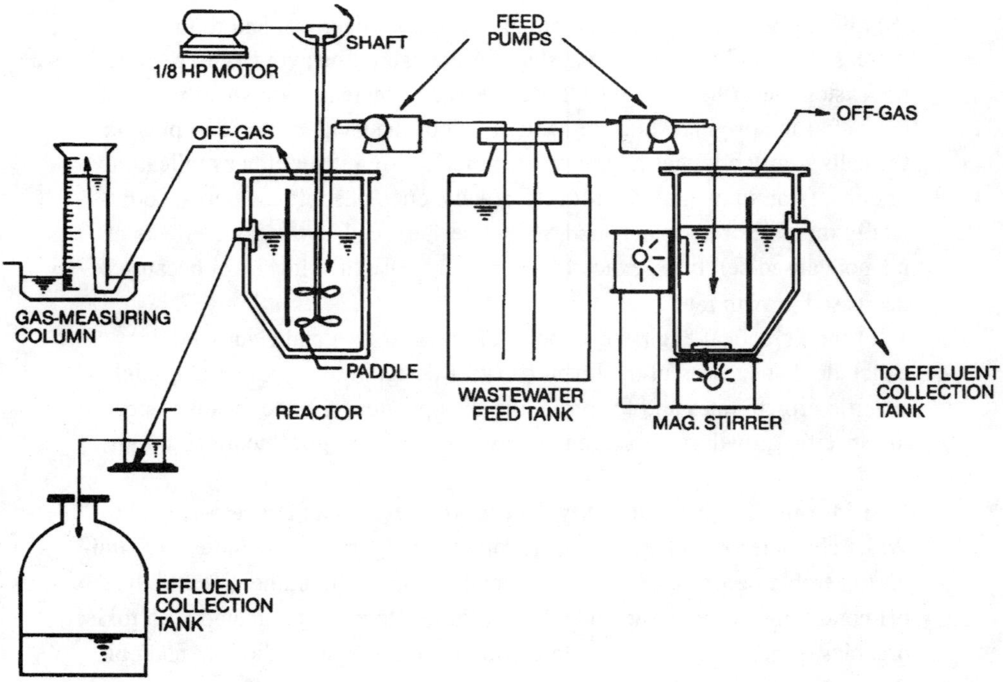

Figure 5.2 Schematic of the complete mix anaerobic systems ($hp \times 0.7457 = kW$).

(7) Check the entire system for gas leaks by putting small quantities of soapy water or mineral oil around the joints and watching for bubbles.

(8) After gas production is noticed and appears to be consistent [approximately 5 to 6 cu ft of gas/lb of COD removed (0.35 ml of gas/mg COD removed) at standard temperature and pressure conditions], begin to introduce the hazardous liquid wastewater diluted by the starter substrate. It is recommended that the initial feed contain only a 10% dilution (by COD, not volume) of the wastewater. Cautiously, increase the dilution factor to 100% hazardous wastewater during a period of 5 to 20 days, depending on such factors as the relative degradability of the waste and the presence of inhibitors. During this acclimation period, it is desirable to maintain the organic loading to the system at less than 1.0 mg biochemical oxygen demand (BOD)/mg VSS d. Another control parameter during acclimation is to feed the waste at a rate of less than 3.2 $kg/m^3 \cdot d$ (200 lb of BOD/1000 cu ft/day) of reactor volume.

(9) Withdraw samples periodically for analysis of BOD or COD removals, pH, VSS level, and volatile acids:alkalinity ratio. During startup and acclimation, the volatile acids:alkalinity ratio should be carefully observed and controlled to ensure that it stays above 0.4 (Grady and Kim, 1980). This ratio is often the first indicator of an anaerobic reactor being upset because of the overproduction of organic acids. Alkalinity should be added to maintain the pH within the range of 6.6 to 7.6.

(10) After stabilized conditions are obtained, the systems can be operated at the desired test conditions and samples should be withdrawn daily for analysis of BOD, COD, VSS, and pH. In addition, the gas production should be measured and recorded daily.

MODELING. Steady-State Assumptions. Marshall et al. (1982) observed the limitations of assuming steady-state conditions and applied the assumption in the use of empirical models. This skepticism is an appropriate means of introducing this section on modeling.

According to Marshall et al. (1982), a major assumption in the development of design relationships is that the biological system is in steady state. The inability to reach steady state has also been observed by others, in spite of using constant wastewater flows and compositions. The authors also state that full-scale biological treatment systems rarely operate in a steady-state fashion and that most laboratory activated-sludge reactors never reach a true steady state. If steady-state conditions exist, then the average effluent TOC and MLVSS concentrations will be representative of the operating conditions over the entire data collection period. On days when the reactors produced an effluent TOX equal to the average effluent TOC in these studies, the corresponding MLVSS in the reactor were not the same as the average MLVSS during the data period. Without steady state, the average values lose their significance because they may not be representative of actual operating conditions, and difficulties in applying design models may be expected.

Modeling Equations—Monod and Haldane. Philbrook and Grady (1985) evaluated biodegradation kinetics for priority pollutants. They presented modeling equations describing kinetics for priority pollutants. They presented modeling equations describing the biokinetics and techniques for measuring kinetic parameters after a "competent" biomass was established. Their work is representative of the many discussions of modeling and should be referenced for a more detailed discussion.

The most typically used rate equation for the noninhibitory case is by Monod:

$$\mu = \frac{\mu_m S}{K_s + S}$$

Where

μ = specific growth rate, time^{-1};
μ_m = maximum specific growth rate, time^{-1};
S = substrate concentration, ML^{-3}; and
K_s = saturation constant, ML^{-3}.

Engineers typically use the specific substrate removal rate (Q):

$$Q = \frac{Q_m S}{K_s + S}$$

or

$$Q = \mu / Y_g$$

Where Y_g = true growth yield, million cells formed^{-1}. The saturation constant represents the substrate concentration at which Q is one-half of its maximal value.

A widely accepted expression capable of fitting the data for an inhibitory substrate is the Haldane equation:

$$\mu = \frac{\mu_m S}{K_s + S + S^2/K_i}$$

or

$$Q = \frac{Q_m S}{K_s + S + S^2/K_i}$$

Where K_i = inhibition constant, ML^{-3}.

When K_i is very large, the Haldane equation reduces to the Monod equation.

Determining Kinetic Parameters. Although many methods have been developed for measuring the kinetic parameters μ_m, Q_m, K_s, and K_i, Philbrook and Grady (1985) grouped them into five categories, three using batch procedures and two relying on continuous culture.

A typically used method has its origin in the work of Monod and has been used on a number of compounds, both noninhibitory and inhibitory. In this method, batch reactors containing the pollutant of interest at different concentrations are inoculated with small quantities of biomass capable of performing the desired biodegradation. The increase in biomass concentration in each reactor is measured as a function of time and analyzed. The values are then correlated with the initial substrate concentration to obtain m and K_s of the Monod equation or m, K_s, and K_i of the Haldane equation.

Another batch technique focuses on substrate removal rather than growth and thus provides data for determining Q_m, K_s, and K_i. As typically used in engineering studies, the batch reactors are inoculated with large quantities of biomass and substrate removal is measured as a function of time. If a specific test is available for the target pollutant, it need not be provided as the sole substrate, and thus some data may be obtained on the effects of other waste constituents, although care must be used when doing this or the data may be misinterpreted. The data on substrate concentration as a function of time must be differentiated for conversion into data on Q as a function of substrate concentration. It is then analyzed to evaluate the parameters.

Both of the above techniques have been widely applied to wastewater treatment problems using general measures of waste strength such as BOD_5 and COD, but they lead to several problems when they are applied to compounds such as the priority pollutants. The first, and most serious, is associated with the fact that the K_s values for many individual organic compounds are quite low, often less then 100 g/L. To use a simple measure of cell concentration in the growth technique, such as absorptance measurement, it is impossible to detect any growth at the substrate concentration range over which it will be affected. In that case, the only way to measure is with viable cell counts, which greatly increases both the effort and the error associated with the procedure. In the substrate removal experiment, the combination of a high cell concentration with a low K_s value results in a linear substrate removal curve, which can lead to the erroneous conclusion that removal of the pollutant is zero-order. It must be recognized that completely mixed bioreactors will be operating at very low pollutant concentrations at which zero-order kinetics are not valid and that accurate prediction of the effluent concentration requires accurate assessment of K_s.

Infinite Dilution Technique and Modifications. To reduce the time required to evaluate kinetic parameters and to alleviate the problems associated with shifts in the composition of the microbial community, Williamson and McCarty (1975) developed a pseudo-steady-state procedure that they called the infinite dilution technique. The test was predicated on the self-regulating nature of microbial metabolism. If a microbial culture was placed into a batch reactor and an organic substrate was added continually at a rate less than the maximum rate at which the organisms can use it, they will match their rate of utilization with the rate of addition by regulating the substrate concentration surrounding them to a value consistent with that rate. If the volume of feed added was negligibly small with respect to the reactor volume and if the mass of cells formed was small with respect to the mass carrying out the removal, then a pseudo-steady state would be achieved during which the substrate concentration in the reactor was constant and consistent with the rate of substrate utilization. Such a reactor has been termed a "fed batch" reactor. Measurement of the pseudo-steady-state concentrations in several reactors receiving feed at different rates provides data on Q versus S that can be analyzed to estimate Q_m and K_s.

The procedure has several distinct advantages over the traditional procedures. First, it is a fact that if a culture is available, its kinetics can be evaluated in as little as 1 day. This means that during pilot-scale testing, a culture can be sampled and analyzed several times giving an accurate measure of any time-dependent changes in its kinetics. Second, the parameter values obtained are specific for a given microbial community because the amount of growth that occurs during the test is small enough to minimize changes in species predominance. Third, the procedure can be applied to any culture. This means that it can be used to obtain a "snapshot" of the kinetics associated with the sludge in a full-scale plant at any time and a series of such "snap shots" provides the engineer and operator with a moving picture of how the plant responds over time (Philbrook and Grady, 1985).

Philbrook and Grady (1985) observed that even under carefully controlled conditions the kinetics of biodegradation of priority pollutants are subject to considerable variation. Additional research must be conducted to determine how best to minimize this variability. Application of the modified infinite dilution technique (MIDT), outlined by Philbrook and Grady (1985), provides a rapid and effective way of evaluating variability. If the MIDT is used on a routing basis during treatability studies, its use provides the engineer with statistical knowledge of the variability likely to be encountered in the full-scale system. Variability can then be considered quantitatively during design. When used on biomass from a full-scale plant, the MIDT can provide data both on the variability of the kinetic parameters and the effect of other waste constituents. Such data will lead to a better understanding of the factors affecting priority pollutant removal and will allow the operator to take corrective action to maintain the desired degree of removal.

TREATMENT TECHNOLOGY

APPLICABILITY AND EFFICIENCY OF TREATMENT FOR METALS. General. Physical-chemical means are most often used to remove metals from a hazardous waste stream, but aerobic suspended growth technology has been assessing bench-, pilot-, and full-scale activated-sludge processes for the removal of heavy metals. Activated-sludge processes are followed by secondary clarification, and the physical and chemical aspects of the removal of metals must be considered.

Removal of metals from a waste stream by activated sludge depends on the individual metal, the wastewater matrix, and the operational characteristics of the treatment system. Metals may cause operational problems at an activated-sludge facility by producing toxic effects, accumulating on sludge solids to an extent that the sludge becomes hazardous, thus complicating and restricting disposal. They can also pass through the plant, limiting the discharge or causing increased penalties.

Characteristics of the treatment process that may influence metal removal efficiency include the MLSS concentration; wastewater pH, temperature, and hardness; solids retention time (SRT); metal solubility; partitioning of the metal between soluble and insoluble fractions; and the presence of chelants or other complexing agents.

Removal Mechanisms. Stephenson et al. (1987) identified five basic types of metal removal mechanisms: direct settling of precipitated insoluble metals, interaction of insoluble metals with flocs of settling mixed liquor solids, uptake of soluble metals by the bacterial cells of activated-sludge biomass, adsorption of metals into the biological solids, and volatilization of the metals. The latter is considered to occur to an insignificant extent.

The results of Elenbogen et al. (1985) on the effects of solubility, removals, and SRT showed cadmium and copper removals were always greater then 75%, while nickel was less than 40%. From Figure 5.3, it can be seen that metals with higher solubility were not removed as well as those with lower solubility. The SRT had little effect on overall metal removal. Solubilities of nickel were consistently close to 100%. The percentage solubilities of the three metals increased dramatically in the final effluent compared to settled wastewater (primary effluent). It appears that insoluble metal was removed to a greater extent than soluble metal.

For cadmium and copper, and to a lesser extent nickel, the removal of insoluble metal was responsible for most of the overall percentage removed. The metals were removed by their interaction with the settleable biological solids, a physical process that could occur in the primary and secondary sedimentation tanks. Direct settling of particulate associated metal was of minor importance for cadmium and copper but was slightly more significant for nickel, which was always poorly removed.

Stephenson et al. (1987) concluded that the partitioning of metals between soluble and insoluble phases was important in determining the extent of removal during the activated-sludge process. Metals that are the most soluble in the influent settled wastewater are the least well removed.

Lowe and Gaudy (1985) evaluated removal of cadmium in a pilot extended aeration process. Changes in system pH of a few tenths of a pH unit in the range of 6 to 7 brought about substantial changes in cadmium solubility. The pH changes were the result of microbial activity. Kodukula and Patterson (1983) conducted batch activated-sludge adsorption experiments

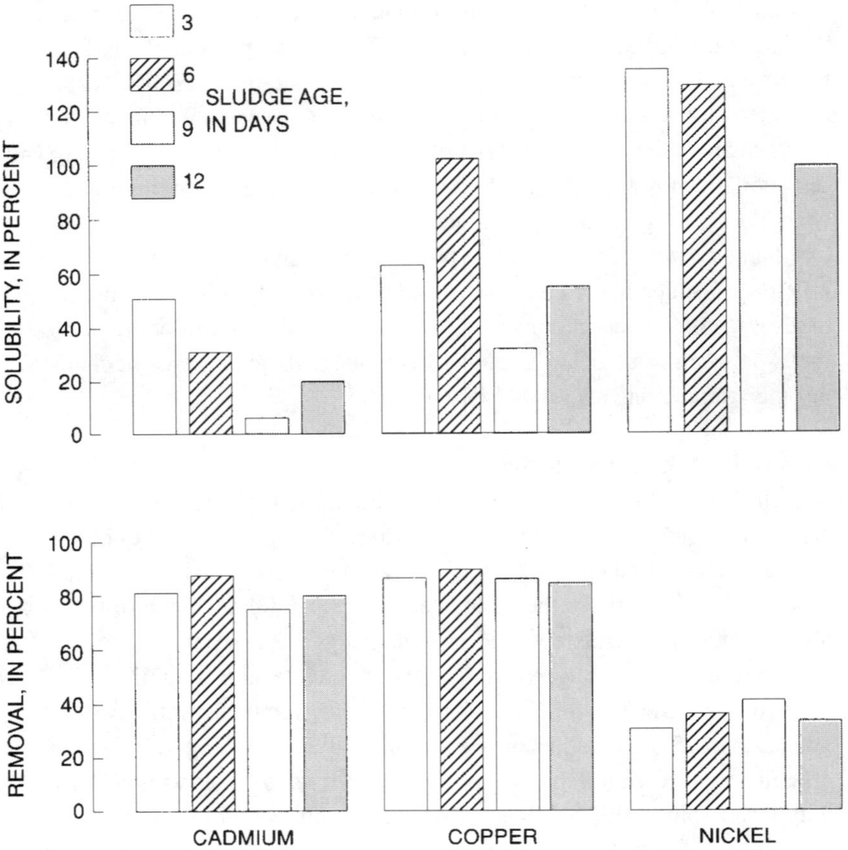

Figure 5.3 Percent removals and solubilities in final effluent in an activated-sludge pilot plant for cadmium, copper, and nickel at 3-, 6-, 9-, and 12-day sludge ages.

and evaluated pH, adsorption, and precipitation effects. For affecting adsorption of metals to activated-sludge solids, pH seems to be a controlling variable. The percentage of metal adsorption increased by more than 70% over 1 to 2 units of pH change. The effect of suspended solids on distribution of metals at a high pH level (8.0) seems to be minimal but is quite significant in low pH conditions. At pH levels below 7.0, the influence of suspended solids on metal distribution is pronounced. Thus, a combination of low pH levels and low suspended solids concentrations yields adverse results when high removals of metals, particularly nickel, are desired.

Removals of metals by activated sludge have been studied by numerous workers. Cormack et al. (1983) spiked a powdered activated carbon (PAC) plug-flow, pilot-plant influent with organic and inorganic pollutants. The removals of metals were compared with DuPont and U.S. EPA data. The U.S. EPA data were for biological treatment without PAC addition. These data are presented in Table 5.2.

Table 5.2 Metal data for pilot-plant operation.

Priority pollutant					PACT removal, %	
Metal	Range of influent concentrations, mg/L	Detection limit, μg/L	Range of effluent concentrations, μg/L	Cormack and Hsu's study	DuPont	U.S. EPA secondary treatment
Arsenic		10	ND–10	9	0	—[b]
Cadmium	ND[a]–11	5	ND–20	>78.4	4	55
Chromium (T)	ND–60	10	ND–140	>49.7	42	69
Copper	30–500	10	ND–200	>81.4	0	75
Lead	11–2100	50	ND–480	>59.1	52	63
Mercury	ND–11.2	0.5	ND–2.8	82.9	50	—[b]
Nickel	30–710	30	ND–280	54.9	23	24
Silver	8–40	10	ND–40	>40.6	—[b]	76
Zinc	140–3400	20	30–700	79.1	40	70

[a]ND = not detected.

[b]Not analyzed or data not available.

The removal of inorganic pollutants ranged from 9% for arsenic to 82.9% for mercury. These were higher than those obtained in DuPont's PACT plant. This was reportedly because of the difference in the influent concentration. Influent concentrations of the inorganic pollutants in DuPont's PACT plant were all much lower than those observed by Cormack et al. (1983). The percentage of removal obtained in the study was also slightly higher than the average national removal rate for secondary treatment processes published by U.S. EPA. Cormack et al. believed this could also have been attributable to the use of carbon in the wastewater treatment process because activated carbon can adsorb inorganics, although the removal efficiency may not be as high as for organic compounds.

Shock Loads and Acclimation. Inhibitory metal effects were evaluated by Lowe and Gaudy (1985). They found that extended aeration biomass could retain extremely high concentrations of cadmium (in this study in excess of 2000 mg/L) without ill effects on the biochemical efficiency of purification of a synthetic waste. There was evidence that acclimation played a protective role in the system.

These results are of value because they demonstrate that relatively large concentrations of cadmium can build up in the sludge without hampering removal of soluble substrate, and they offer some indication that even higher levels of cadmium might be retainable in the sludge without ill effects. At levels up to 320 mg/L, the system responded extremely well, and results indicated that the process flow scheme for the biological removal and recovery of heavy metals (depicted in Figure 5.4) offers a viable alternative for

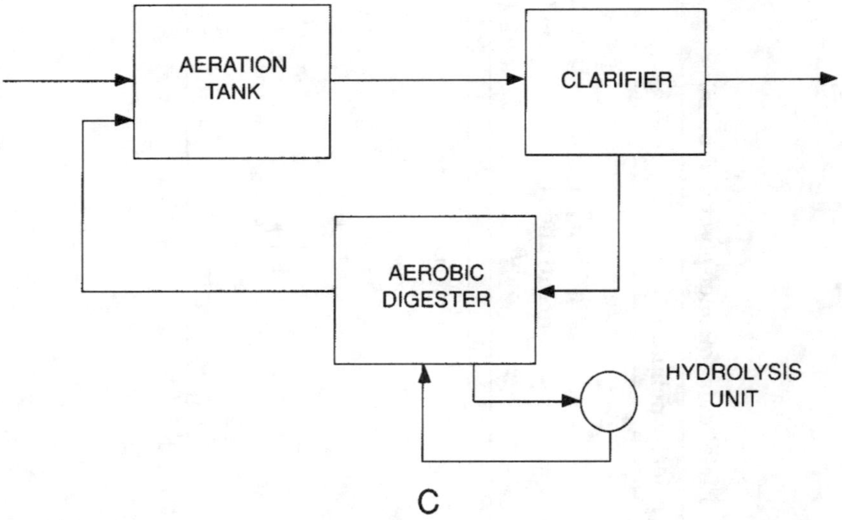

Figure 5.4 Schematic comparison of conventional extended aeration flow schemes with proposed modification incorporating external recycle and the hydrolytic assist.

treatment of wastes containing very high concentrations of heavy metals. The similarity between the process in Figure 5.4 and the contact stabilization process should be noted. Lowe and Gaudy (1985) concluded that long-term exposure to cadmium does select for a biomass that exhibits tolerance to rather high concentration of the metal.

When the cadmium dosage was rapidly increased, eventually to 320 mg/L, there were no ill effects on the biochemical performance of the system. Previous long-term contact (1 year) at 1mg/L may have imparted acclimation, permitting tolerance of the high dosages.

Further evidence of the importance of acclimation was seen for the qualitative shock with copper of 100 mg/L on Day 385. Copper was chosen for the qualitative shock because it has been reported to have a greater affinity for, and toxicity to, sludge solids than cadmium, despite the fact that it typically is considered to be even less soluble than cadmium at neutral pH values. Regardless of solubility considerations, exposure of the nonacclimated biomass to copper resulted in prompt and severe deterioration of system operation. Therefore, it would appear acclimation played a role in the response of this system to heavy metals.

Variability. Unger and Claff (1985) used statistical methods to evaluate percentage of removal from U.S. EPA's 40 POTW study (U.S. EPA, 1982). Data for all POTWs indicated that nickel, arsenic, and selenium were removed most poorly, while copper, chromium, and zinc were removed to the greatest extent. They concluded percentage of removal of priority pollutants vary significantly among the 40 POTWs tested for both metals and organics. Metal removals in activated-sludge plants ranged from 43 to 81%. Percentage of removal for a given priority pollutant were highly variable from plant to plant and even from sampling to sampling within a given treatment plant. Metal variability among plants ranged from 21 to 50%. The overall conclusion was that percentage of removal data vary among POTWs and require evaluation on a plant-by-plant basis rather than on the mean percentage of removal derived from all treatment plants. This conclusion may be extrapolated from POTWs to treatment of hazardous wastes in that generalization about removal of a specific metal should not be made.

Comparison of Different Processes. Hannah et al. (1986) conducted a pilot study evaluating removals of toxic pollutants by six wastewater treatment processes. The aerated suspended growth units included were activated sludge, an aerated lagoon, and a facultative lagoon. Metals removal by these pilot processes were calculated from the average concentrations of Table 5.3 and are shown in Table 5.4. The activated-sludge system and facultative lagoon showed the best removals for metals. The high-rate trickling filter system achieved percent removals from 28% for cadmium to 60% for copper. The metals were concentrated in the sludge. Copper, which exhibited the greatest percentage of removal, had the highest concentration in the sludge.

Table 5.3 Concentrations of metals in wastewater feed and effluents.

Metal	Wastewater feed		Activated-sludge effluent		Aerated lagoon effluent		Facultative lagoon effluent	
	μg/L	σ	μg/L	σ	μg/L	σ	μg/L	σ
Chrome	221	88	40	18	65	106	46	34
Copper	345	119	61	40	89	61	71	46
Nickel	141	93	81	45	91	50	81	59
Lead	165	168	58	75	70	76	82	110
Cadmium	25	23	19	17	—	—	17	9

Table 5.4 Percent removal of metals by different processes.

Metal	Activated sludge	Aerated lagoon	Facultative lagoon
Chrome	82	71	79
Copper	82	74	79
Nickel	43	35	43
Lead	65	58	50
Cadmium	24	—	32

Cadmium, which had the lowest percentage of removal, was the least concentrated in the sludge. For all the metals used in this study, the percentage of removal in the trickling filter system increased as the degree of metal concentration in the sludge increased.

Chang et al. (1986) studied the uptake of cadmium and copper in a three-stage rotating biological contactor (RBC). Cadmium concentrations of 5 and 20 mg/L and copper concentrations of 1, 5, 10, 25, and 50 mg/L were used. The reactor consisted of three discs per stage, with a total surface area of 9120 cm^2. The flow rate was 50 L/d resulting in a 54.8 $L/m^2 \cdot d$ surface loading rate. The total detention time was 4 hours. The rotation speed was 25 rpm and the liquid volume:surface area ratio was 9.38 L/m^2.

The system was allowed to reach steady state with a feed consisting of sucrose and mineral supplements. A cadmium dose of 5 mg was added for 24 hours, then discontinued. The organic removal efficiency decreased from 92 to 84% after the cadmium was added. The removal efficiency increased to 90% within 5 days after the cadmium addition was stopped. Cadmium was again added at a rate of 20 mg/L for 3 days. The substrate removal efficiency decreased from 90% to approximately 81 to 85%. The authors concluded that the initial cadmium shock with the addition of 5 mg/L for 1 day acclimated the system for the higher cadmium shock of 20 mg/L for 3 days.

The RBC system did not lose any efficiency when 1, 5, and 10 mg/L of copper were added. The organic substrate removal efficiency decreased by an average of approximately 10% when 25 mg/L of copper were added. The system efficiency returned to approximately 90% within 10 days after the copper feed was stopped. Copper was then added at a concentration of 50 mg/L. The efficiency declined an average of approximately 7%. Again, the conclusion was that the initial copper exposure acclimated the system for the subsequent higher copper loading.

During the copper addition, the biofilm color turned dark blue or black. The amount of sloughing also increased, and the dry density of the biofilm increased. The authors attributed this to the adsorption of copper into the biofilm. The composition of the biomass was examined with a microscope to assess the population changes from the copper addition. In general, the number of protozoa and rotifers decreased when 5 mg/L of copper were added.

APPLICABILITY AND EFFICIENCY OF TREATMENT OF ORGANICS. General. The removal of complex organic compounds by aerobic biological treatment of listed hazardous wastes has been evaluated in lab-, pilot-, and full-scale studies. These organics are measured and classified as the acid extractable, base-neutral extractable, volatile (purge and trap), and pesticide and PCB compounds.

Removal of these compounds by aerated suspended growth can be by biodegradation, air stripping, adsorption by biomass or polymers and incorporation into sludge, or other mechanisms or combinations of mechanisms not easily described or measured in a complex waste stream (for example, photo-oxidation).

Complex organics are degraded by microorganisms with low growth rates (Melcer, 1986 and 1987). Processes that promote the growth of these types of organisms, therefore, would favor biological removal of complex contaminants. The microbial growth rate is inversely proportional to SRT in a biological reactor. Consequently, as the growth rate decreases, the SRT must increase. Control of the SRT and maintenance of an optimum SRT will significantly affect slow-growing microorganisms that are important in degrading complex contaminants. An SRT that is too low will result in slow-growth organisms being washed out of the system or being inhibited by other faster-growing populations (Melcer, 1986 and 1987). Biomass concentrations are also high in fixed film processes promoting the removal of highly adsorbable compounds (Melcer, 1986).

The removal of toxic trace contaminants from wastewater by fixed film systems can be the result of stripping from solution, adsorption onto biological solids, precipitation from solution, and biological degradation.

Grady (1988) defined characteristics of wastes as biodegradable, recalcitrant, or persistent. Biodegradability and recalcitrance are mutually exclusive intrinsic properties of organic compounds determined by molecular structure. Consequently, each compound is either one or the other. A recalcitrant compound can never undergo biodegradation in a wastewater treatment system, no matter how that system is configured or operated. A biodegradable compound may or may not undergo biodegradation under any given condition, and that is when the concept of persistence rises. Persistence is a conditional property of a biodegradable organic compound in the sense that, if the environment is not suitable for biodegradation to occur, the compound will persist in spite of its inherent biodegradability.

According to a U.S. EPA report by Rogers (1979), different organic compounds at the same concentration will have varying effects on microbes with resultant levels of biodegradability depending on environmental and chemical conditions. Some, but not all, organisms have a preference for peptones, carbohydrates, and other forms of readily usable carbon sources over hydrocarbons or synthetic organic compounds. The sequential utilization of preferred substrates from a mixture has also been demonstrated for various classes of hydrocarbon and synthetic organic molecules. The following generalizations may be drawn concerning diauxic (two-phase) growth in mixed industrial wastes.

- Nonaromatic or cyclic aromatic compounds are preferred over aromatics.
- Materials with unsaturated bonds in their molecules, for example, alkenes, silynes, tertiary amines, and so on, are preferred over materials exhibiting saturated bonding.
- The comparative stereochemistry of the molecules of certain compounds makes them more or less susceptible to attack by microbial enzymes. The *n*-isomers of the lighter weight molecules are preferred over branched isomers and complex, polymeric substances.
- Soluble organic compounds typically are more readily degraded than insoluble materials. Biological waste treatment is most efficient in removing dissolved or colloidal materials, which are more readily attached by enzymes and transported through cell membranes. Readily dispersed compounds typically are degraded more rapidly because of the increased surface area that is presented to the individual microorganisms.
- The presence of key functional groups at certain locations in the molecules can make a compound more or less amenable to biodegradation. Alcohols, for example, are often more readily degraded than their alkane or alkene homologues. On the other hand, halogenation of certain hydrocarbons may make them resistant to degradation.

Rogers (1979) identified the following classes of wastes as generally resistant to degradation:

- Oil,
- Isoprene,
- Methyl vinyl ketone,
- Morpholine,
- Ethers,
- Ethylene chlorohydrin,
- Polymeric compounds,
- Polypropylene benzene sulfonates,
- Tertiary and volatile aliphatics,
- Aromatics,
- Alkyl-aryl groups,
- Tertiary aliphatic alcohols,
- Tertiary benzene sulfonates, and
- Trichlorophenols.

Even those compounds that typically have been thought of, and shown to be, resistant to typical aerobic biological degradation are now being researched and processes developed for their applicability to special strain aerobic and anaerobic bacteria, as well as fungal technologies.

Biodegradability in complex waste mixtures is not identical to laboratory experiments. Some organic materials can polymerize or react in a synergistic manner on contact with other wastes. Acclimation of microorganisms can also affect biodegradability.

Ford et al. (1971) determined that the ratio of BOD_5 and COD of wastewater compounds can be used to indicate the fraction of dichromate-oxidizable materials amenable to biological degradation. A high ratio indicates that many of the dissolved organic materials can be degraded biologically, while a low value indicates the presence of a significant fraction of bioresistant organic constituents. Table 5.5 shows the measured BOD and COD for several classes of chemicals. The relatively high degradability of most unsubstituted aliphatic compounds should be noted.

Tabak et al. (1981) conducted biodegradability and acclimation studies on 96 organic priority pollutants using a static culture flask screening procedure. These included phenols, phthalate esters, naphthalenes, monocyclic aromatics, PCBs, halogenated ethers, nitrogenous organics, halogenated aliphatics,

Table 5.5 Comparisons of chemical oxygen demand (COD) and biochemical oxygen demand (BOD) of selected organic chemicals.

Chemical group	Measured COD, mg/mg	Measured BOD_5, mg/mg	BOD, COD, %
Aliphatics			
Methnol	1.05	1.12	107
Ethanol	2.11	1.58	75
Ethylene glycol	1.21	0.36	30
Isopropanol	2.12	0.16	8
Maleic acid	0.80	0.64	80
Acetone	2.07	0.81	39
Methylethyl ketone	2.20	1.81	82
Ethyl acetate	1.54	1.24	81
Oxalic acid	0.18	0.16	89
Aromatics			
Toluene	1.41	0.86	61
Benzaldehyde	1.98	1.62	82
Benzoic acid	1.95	1.45	74
Hydroquinone	1.83	1.00	55
o-Cresol	2.38	1.76	74
Nitrogenous organics			
Monoethanolamine	1.27	0.83	65
Acrylonitrile	1.39	0	0
Aniline	2.34	1.42	61
Refractory			
Tertiary Butanol	2.18	0	0
Diethylene glycol	1.06	0.15	14
Pyridine	0.05	0.06	120

and organochlorine insecticides. Yeast extract was used as a synthetic medium, and the test compounds were incubated for 7 days at 25 °C. This was followed by three weekly subcultures using domestic wastewater as the microbial inoculum. Although limited to screening because degradation and acclimation would differ greatly under process operating conditions, this method can be helpful in predicting compound treatability.

Observed Removal Efficiencies of Selected Organic Compounds. The following organic compounds cover a wide range of hazardous contaminants seen in the environment and its various media. These compounds are typical of those that would require treatment and remediation and generally would be receptive to biological treatment.

PHENOLS. Hickman and Novak (1984) evaluated removal of pentachlorophenol (PCP) in batch tests. They noted that removals of PCP in laboratory-scale tests have not been equaled in full-scale treatment plants. Among the acid extractables, PCP is the most heavily chlorinated compound and has the highest octanol-water partition coefficient. Hickman and Novak (1984) observed that the PCP concentration decreased from 12 to 13 mg/L to virtually 0 after 30 days of feeding PCP and then increased slowly. Apparently, there was a shift in species composition to active PCP-degrading organisms at that time. After about 1 week of nearly complete PCP removal, PCP again began to accumulate in the reactor. However, the activated sludge continued to remove PCP for the rest of the study. Hickman and Novak stated that is was not clear why the PCP concentration began to increase; it was assumed that, during the period of nearly complete degradation, the PCP concentration was so low that there was no selective advantage for PCP degraders and their population declined. Anaerobic processes and fungal technologies are being developed that have shown positive results in the degradation of PCP.

PHTHALATE ESTERS AND NAPHTHALENE. Tabak (1981) observed significant biodegradation, with rapid acclimation observed with the phthalate compound. The phthalate esters, EHPE and DOPE, were the most persistent of the phthalates under the conditions of the static flask screening test.

MONOCYCLIC AROMATICS. Tabak (1981) found significant biodegradation with rapid acclimation with benzene, toluene, and nitrobenzene.

POLYCYCLIC AROMATIC HYDROCARBONS. In the work by Tabak (1981), the polycyclic aromatic hydrocarbons (PAHs) demonstrated varied rates of biodegradation with different acclimation periods, depending on the test compound and the dose of substrate in culture media. The PAHs, in general, have been reported to be potentially biodegradable compounds, particularly by soil microorganisms and in soil systems. Research performed by IGT (Middleton et al., 1991) found that the optimum biological conditions for

PAHs in soil are found in slurry reactors. The biodegradability data have shown that the tricyclic aromatic hydrocarbons are more susceptible to biodegradative action than the tetracyclic and higher polycyclic hydrocarbons.

POLYCHLORINATED BIPHENYLS. The individual PCBs vary widely in their susceptibility to biodegradation. The mono-, di-, and trichlorinated species may be significantly biodegraded or biotransformed, as well as volatilized. The PCBs with five or more chlorine atoms per molecule have a tendency to absorb suspended materials and sediments, bioaccumulate because of very low solubility, photodissociate, and resist biodegradation.

In work by McDermott et al. (1989), PCB biodegradation was demonstrated in the laboratory using selected bacteria and naturally occurring bacteria. They found that the ability to degrade PCB ranges from very poor to outstanding, depending on the bacteria and the type of PCB.

Fungal technologies have shown success in the degradation of PCB species and are being applied to lab-, bench-, pilot-, and full-scale applications.

HALOGENATED ETHERS. In Tabak's (1981) work, the compounds *bis*-(2-chloroethyl) ether, 2-chloroethyl vinyl ether, and *bis*-(2-chloroisopropyl) ether are significantly biodegradable with slightly varied acclimation periods, whereas 4-chlorodiphenyl ether, 4-bromodiphenyl ether, and *bis*-(2-chloroethyl) methane demonstrated insignificant biodegradability.

NITROGENOUS ORGANIC COMPOUNDS. Tabak (1981) showed that the diphenylnitrosamine, N-nitroso-N-phenylbenzamine (N-nitrosodi-phenylamine), was easily biodegradable by microorganisms with rapid acclimation. Diphenylnitrosamine was reportedly more easily degraded by microorganisms than are dialkylnitrosamines.

HERBICIDES. Tabak (1981) found that unsaturated ketone (cyclohexeneone) isophorone exhibited rapid degradation by microorganisms. The herbicide compound, acrylonitrile (vinyl cyanide), was easily degraded. Acrolein (an unsaturated aldehyde) was also shown to be easily dissimilated with rapid acclimation of microbiota to the substrate.

The work of Hill et al. (1986) using an activated-sludge pilot plant showed that activated-sludge treatment proved ineffective in removing chlorophenoxy herbicides (CPHs) from settled wastewater. Statistical treatment of the data suggested neither adsorption nor catabolic biodegradation were responsible for those significant CPH removals. There was no evidence of acclimatization of the mixed liquor to the presence of CPH contaminants. Furthermore, there was no obvious effect of sludge age on the efficiency of CPH removal from settled wastewater.

ORGANOCHLORIDE PESTICIDES. Tabak (1981) observed poor removal efficiency for organochlorine pesticides. Aldrin, dieldrin, chlordane, p,p′-DDT, p,p′-DDE, p,p′-DDD, alpha-endosulfan, beta-endosulfan, endosulfan sulfate, heptachlor, heptachlor epoxide, endrin, hexachlorocyclohexane isomers (alpha, beta, and delta), and lindane compounds were shown to be recalcitrant to biooxidative activity by wastewater microbiota. However, significant recent studies using fungal technologies have indicated success in the reduction and degradation of organochlorine pesticides.

HALOGENATED ALIPHATICS. Tabak (1981) reported on the biodegradation of halogenated aliphatics. Volatilization was accounted for by using controls when media were tested without biological inoculum. The chloroethane aliphataics, 1,1-1,2-dichloroethane and 1,1,1-trichloroethane, are considered to be potentially biodegradable and probably will be destroyed during aerobic treatment after a considerable period of acclimation, whereas tetrachloroethanes will exhibit recalcitrance to microbial activity and might be biooxidized only after very extensive acclimation periods. Because of their lipophylicity, the dichloroethanes should not be bioaccumulated to a large extent, whereas the tri- and tetrachloroethanes, being lipophylic in nature, might have a tendency for accumulation.

In Tabak's (1981) work, the halomethane aliphatics demonstrated different biodegradation rates and acclimation periods, depending on the test compound and substrate concentration in the culture media. Methylene chloride, bromochloromethane, and carbon tetrachloride were shown to exhibit rapid degradation, whereas chloroform showed significant dissimilation with gradual adaptation. The chloroethylene aliphatics, trichloroethylene, and tetrachloroethylene were significantly biodegradable with gradual adaptation observed.

Variability of Removal. Unger and Claff (1985) used statistical methods to evaluate data from the U.S. EPA study (U.S. EPA, 1982) on the occurrence and fate of priority pollutants in 40 POTWs. Table 5.6 shows volatile, acid extractable, and base-neutral removals and the coefficient of variation (CV). The CV is an indication of the variability of removal efficiencies of priority pollutants. The observed percentages of removal varied significantly among the 40 POTWs tested. Organic removals ranged from 52 to 87%. Variability occurred from plant to plant and from sampling to sampling within a plant, ranging from 10 to 55%. It was concluded that because of large variations among POTWs, evaluations should be conducted on a plant-by-plant basis rather than on mean percentages of removal derived from all treatment plants.

Marshall et al. (1982) analyzed reactor performances and models and attributed difficulties in fitting experimental data to empirical models to problems in estimating viable cell concentrations and achieving steady-state conditions. In their study, a considerable amount of day-to-day fluctuation was apparent in effluent TOC. Even greater fluctuations were present in measurements of MLVSS. Achievement of steady state was not demonstrated for any of the reactors, even after operating for periods equal to sev-

Table 5.6 Mean percent removals and coefficients of variation for activated-sludge plants.

Chemical group	Removal, %	Coefficient of variation
Volatiles		
Benzene	74.7	28.2
1,1,1-Trichloroethane	81.9	17.3
Chloroform	65.2	30.7
1,2-*trans*-Dichloroethylene	68.7	31.4
Ethylbenzene	82.4	20.0
Methylene chloride	57.1	41.7
Tetrachloroethylene	72.2	33.4
Toluene	86.3	19.0
Trichloroethylene	84.8	19.0
Acid extractible		
Phenol	83.9	22.7
Base neutrals		
Naphthalene	68.9	33.3
bis(2-Ethylhexyl)phthalate	67.4	32.3
Butylbenzyl phthalate	67.9	33.1
Di-*n*-butyl phthalate	61.9	38.7
Diethyl phthalate	62.2	41.5
Anthracene	69.5	31.4
Phenanthrene	65.9	33.5
Pyrene	86.6	10.0

eral residence times. Variability was attributed to a combination of inherent process variability, hydraulic variations (for example, HART and SRT), environmental fluctuation (for example, pH and temperature), sampling error, and analytical error.

Stover and Kincannon (1981) presented figures showing the typical scatter of data used for determining biokinetic constants. All individual data points from the 2-month operating period were shown to indicate the scatter typically observed during this type of analysis. The actual scatter presented could have been masked by plotting the average values at each sludge age. This is the approach typically used when presenting biological system data and conducting the data analysis because this type of variability is inherent in biological systems.

The variability observed during these type of studies could have been caused by several factors. For example, wastewater that is very biodegradable yields very low effluent BOD_5, TOC, and COD values. Low effluent values create two problems. First, the precision, accuracy, and detection capabilities of the analytical test methods themselves become limited at the effluent levels observed. Second, these very low values allow no distinction to be made with respect to substrate removal. The initial substrate concentration is so large compared to the effluent concentration that the difference becomes negligible at different sludge ages. Variability could also be enhanced based on the amount of specific compound that was stripped on any given day. In the previous analysis, the percentage of stripping was assumed to be constant; however, the amount of compound stripped could have varied on a given day because of fluctuations in the air flow rates to a given system (Stover and Kincannon, 1981).

Mechanisms of Removal. The mechanisms for removal of organic compounds in an aerobic treatment system are air stripping, biodegradation, adsorption onto biological solids, and removal by subsequent settling or photodecomposition. Removal can also occur by addition of polymers or coagulants into the aeration basin with the aim of removal during secondary clarification. Because this represents intentional action by physical-chemical means, it will not be considered here.

SORPTION. Mass balances performed on the influent, effluent, and sludge streams of treatment plants will help determine the extent of removal of contaminants. Detection in the sludge typically indicates sorption of the contaminant has occurred.

A determination of the organic materials adsorbed onto the sludge indicates the degree of oxidation or stripping capability of the system. The type and quantities of organic compounds identified in the sludge are an indication of the sludge as a sink for recalcitrant organics. Compounds identified in the sludge and not present in the influent are the breakdown products from the oxidation of other compounds, or the organics were adsorbed before sampling and were not degraded in the biological system, or the compounds were present at concentrations below the analytical detection limit and have accumulated to above that limit in the biological system.

Dobbs and Cheng (1988) developed a correlation between sorption of toxic organics on biological solids and the octanol-water partition coefficient. The relationship was used to estimate removal of toxic organic compounds by sorption and predicting concentrations of compounds in various sludges. The octanol-water partition coefficient proved useful as a means for prediction of soil adsorption, biological uptake, lipophilic storage, and biomagnification. The isotherms obtained for the toxic organic compounds studied showed a wide range of sorption capacities. Relevant Freundlich parameters for sorption of mixed liquor solids were also determined. The data developed by Dobbs and Cheng (1988) are a useful starting point for determining the degree of sorption.

STRIPPING. Stover and Kincannon (1981) measured and sampled off gases from complete mix bench-scale continuous flow activated-sludge reactors. The reactors were fed a synthetic wastewater containing compounds present in chemical, plastics, petrochemical, and petroleum industry wastewater. All are listed hazardous compounds.

Table 5.7 shows the percentage of each individual compound that was stripped from the biological systems. The amount of compound stripped varied from 0 to 99%, with five compounds showing no stripping, three compounds showing high levels of stripping, and four compounds showing moderate levels of stripping. An important point to note is that several of the compounds that showed little or no stripping from the biological reactors were completely stripped under identical conditions without the biological population.

When the components of a waste stream are known, the degree of stripping that can be expected can be preliminarily screened using the ratio of the compound's vapor pressure to its solubility, or Henry's law constant.

BIODEGRADATION. If stripping and adsorption phenomena can be accurately measured or predicted and if photochemical or chemical reactions are accounted for or neglected, the remaining removals may be assumed because of biodegradation. Biodegradation may not be complete, however, and intermediate compounds may be formed. Thus, reduction of a target hazardous compound may occur by biodegradation, but the breakdown products may also be hazardous and may or may not be amenable to further biodegradation. Determining the route of biodegradation or proving that biodegradation has occurred is a complex problem often conducted by analyses of metabolic intermediates by gas chromatography/mass spectrophotometry (GC/MS). Commercially available chemicals having the identical molecular formula compared with those of the samples in this GC/MS investigation (Masunaga and Yonezawa, 1986).

In the work of Blackburn et al. (1987), the concentration of cells in an activated-sludge system containing a gene known to participate in degradation of naphthalene was experimentally related to the biotransformation and mineralization of naphthalene. The ability to enumerate a critical genotype

Table 5.7 Specific organic compound stripping characteristics.

Compound	Compound stripped from biological systems, %
Tetrachloroethane	93
Nitrobenzene	0
2,4-Dichlorophenol	0
Acrolein	0
Acrylonitrile	0
1,2-Dichloropropane	99
Methylene chloride	5
Ethyl acetate	17
Benzene	15
1,2-Dichloroethane	98
Phenol	0
1,2-Dichlorobenzene	24

and relate it to enzymatic activity in a mixed culture suggests an improved capability for system understanding at the ecological level and the potential for process control at the genotype level. Thus, if the cells that carry out the degradation (or their genotype) are identified, isolated, and made available for use, biodegradation of specific compounds may be hastened. Blackburn et al. (1987) concluded that their method permits dynamic in situ investigations of the microbial ecology of mixed culture systems. They felt the ability to discern the relative effects of ecological factors, population dynamics, genotype concentrations, and enzyme synthesis on the system's performance may emerge with further work.

Hill et al. (1986) discussed the role of cometabolism in biodegradation of CPHs. They stated cometabolism may have been important in CPH removal, particularly if the concentration of each compound is relatively low. Cometabolism supposedly involves the breakdown of chemical compounds by organisms, usually by fairly nonspecific enzyme pathways, but does not yield useful energy to the organism. As a result, the chemical compound in question does not exert a selection pressure on the microbial community, and therefore, although degradation may commence immediately after the substrate is added to the system, the extent of degradation will tend to be small and the degradation rate will remain relatively constant.

TREATMENT METHODS

AEROBIC PROCESSES. Biological treatment using aerobic bacteria is still the most common method of treatment. These bacteria require the presence of an oxygen source, a carbon source, and nitrogen and phosphorus as nutrients for the bacteria. This process breaks down the organic contaminant into CO_2, water, and biomass (additional bacteria). This technology has been used for decades in domestic and industrial wastewater treatment plants. The aerobic system is limited by certain parameters including temperature, pH, oxygen, nitrogen and phosphorus balance, hazardous waste concentrations, and contact time. The optimum temperature range for aerobic digestion is 50 to 90 °F (10 to 32 °C), a neutral pH, a minimum of 1 ppm dissolved oxygen, and nitrogen at 5% and phosphorus at 1% of the concentration of organic contaminants (Calmbacher, 1991).

Aerobic bacteria has been found to be effective in digesting organics such as petroleum hydrocarbons, nonchlorinated solvents, some chlorinated solvents, and, more specifically, compounds such as benzene, toluene, acetone, alcohols, methyl ethyl ketone, tetrahydrofuran, creosote, and pentachlorophenol. Aerobic treatment has been used to treat hazardous wastes in wastewater, groundwater, lagoons, soils, and surface waters.

Biological treatment of industrial wastewater and groundwater contaminated with hazardous wastes uses the same approach and processes as uses for domestic wastewater treatment. These processes involve the use of suspended growth, fixed film, and miscellaneous reactors for digestion of hazardous wastes. Suspended growth reactors have the bacteria growing in the water and mixed with the wastes in the water. Fixed film reactors have the bacteria growing on an inert media with the contaminated water flowing over the bacteria film that is formed. The miscellaneous reactors are special designs that use a combination of the technologies by sequencing, using powdered activated carbon and using a fluidized bed configuration (Nyar, 1992).

Suspended Growth Reactor. This system uses a lagoon, tank, or basin that is aerated by diffusers or surface aerators to supply oxygen. The bacteria are suspended in the water and degrade the hazardous waste as it enters and creates new bacteria in the process. Therefore, the residence time must be sufficient for this process to occur. Oxygen at 2 ppm, along with proper agitation, must be maintained to keep the tank thoroughly mixed. Mixing is critical for long residence time. Typically, 2 days is the shortest residence time that should be maintained to ensure bacteria regeneration and adequate degradation of the waste.

The main problems with aerated lagoons are the degree of treatment of biodegradable wastes, which is limited to 50 to 70% because of limited residence time of the bacteria, and the bacteria that are created in the breakdown reaction flows from the reactor with the water because it is well mixed. Adding a clarifier to remove bacteria solids and returning them to the reactor increases the residence time and increases efficiencies. This process of bacteria solids return is known as activated sludge, which is the most common biological treatment method in wastewater treatment. Activated-sludge processes have been shown to remove 85 to 95% of biodegradable wastes and up to 99+% of specific organic compounds from the influent stream. Effluent concentrations of 10 to 30 ppm of BOD is indicative of an efficient activated-sludge system.

Two other types of suspended growth reactors that are used are extended aeration and contact stabilization. Extended aeration is very similar to activated sludge, but it contains a larger reaction chamber for increased aeration and produces less sludge. Most packaged equipment is of this design. Contact stabilization typically is used for high concentrations of easily biodegradable wastes and contains a small aeration tank, a clarifier, and a large aeration tank, with recycle of the waste stream.

Fixed Film Reactor. This system uses an inert media with a large surface area that is placed in a reactor vessel. Bacteria is grown on the media. When contaminated water flows through the reactor, the wastes are transferred to the attached bacteria, and the biodegradation waste products are transferred back to the water stream. Oxygen is obtained from the atmosphere; therefore, no added oxygen or aeration source is required. This reactor is less maintenance intensive because any surplus of bacteria formed will fall off the media and exit the reactor with the water stream.

Two main types of fixed film reactors are the trickling filters and rotating biological contactors. The trickling filter is designed to have the wastewater pumped to the top of the reactor, distributed over the media, and allowed to trickle down the media in thin films. The transfer of contaminants and waste products takes place through this thin film of water on the bacteria. The media used for small trickling filters is typically plastic media, as used for air stripper packing. The larger filters use large plastic bricks for the media.

The rotating biological contactor is designed to have the wastewater enter one end of the tank and come in contact with rotating cylinders that act as the media for the bacteria. As the cylinder rotates out of the water, a thin film of water forms on the media, and oxygen transfers through the water film from the atmosphere and then, along with the contaminants, is transferred to the bacteria. After the contaminants are degraded, the waste byproducts are transferred back to the water.

Because of the shorter retention and reaction time available in the fixed film reactors, it is sometimes necessary to recycle to lessen the effect of high organic concentrations and increase efficiency. Degradation efficiencies are lower than an activated-sludge system. These systems are capable of removing 75 to 90% of BOD and 85 to 95% of specific organic compounds.

Submerged Fixed Film Reactor. This process is a combination of suspended growth and fixed film designs. The media on which the bacteria grows is fully submerged in the reactor tank. This allows the wastewater and contaminants to be constantly in contact with the bacteria film, not having to rely on the contaminants and oxygen having to pass through the thin water film. Aeration is performed by pushing air up from the bottom of the reactor and creating the oxygen source and the necessary mixing.

The submerged fixed film reactors are designed to handle from 50 to 5000 ppm of influent waste concentration. The removal efficiencies are equal to that of the activated-sludge systems. The major limitations are the system size and cost. These reactors are good for low flow rates and, therefore, good for groundwater and small to large industrial wastewater treatment systems.

Submerged fixed film reactors have been used in full-scale applications to treat acetone, MEK, benzene, chlorobenzene, tetrahydrofuran, and butanol (Nyar, 1992).

Miscellaneous Treatment. Powdered activated carbon treatment (PACT), which uses powdered activated carbon in the activated sludge, increases the removal efficiency of the treatment system. The activated-sludge process is capable of removing the bulk of hazardous wastes by biological treatment, with the PAC removing the slow degrading, nondegrading, and residual contaminants. This system has been proven extremely effective with hazardous organic wastes. It is capable of treating a very wide variety of hazardous compounds by biodegradation and carbon absorption so that wastes that are not amenable to biological treatment are absorbed by the PAC.

Sequencing batch reactors have been used to treat industrial waste during the past decade. This system typically has a two-tank design to allow somewhat continuous flow. While one tank is processing, the other tank is filling. The process consists of a tank with continuous aeration, into which wastewater is pumped. After the tank is full, digestion of the wastes is completed by the bacteria. Then the air agitation is stopped; the bacteria are allowed to settle, and the treated water is decanted. This process is very useful for a wide variety of wastes and concentrations.

A fluidized bed reactor is a submerged fixed film design with very small packing. Water and air flow into and up through the media, which is typically sand, activated carbon, glass beads, or other small particles. The size of the media allows for very high concentrations of bacteria to be formed and maintained on the media. This design allows for treatment of high concentrations of wastes.

ANAEROBIC PROCESSES. General. Anaerobic treatment is a biological process using facultative and strictly anaerobic bacteria to decompose complex organic matter in the absence of free oxygen. Proper application of anaerobic treatment results in virtually complete destruction of putrescible organics, significant reduction in pathogens, conversion of hydrophilic solids to water, minerals, a valuable gas, and a humus-like residue and favorable alteration of the dewaterability of the sludge (Kugelman et al., 1982). Because of its successful application to municipal sludge treatment and its several important advantages, anaerobic treatment is being used for treatment of industrial and hazardous wastes. Anaerobic treatment is just starting to be more widely used in hazardous waste treatment. It has the ability to degrade halogenated compounds by reductive dehalogenation. Reactions typically are performed in a sealed reactor to avoid the influx of air.

Although anaerobic digestion is not a novel concept in waste treatment, recent and important innovations in the technology promote it as an important one for the effective destruction of organic chemicals. Design improvements of modern reactors are related to promoting effective methods for maintaining the microbiological consortium within the reactor and efficiently exposing the organisms to the organic substrate. Several reactor types with these features are commercially available and suitable for hazardous waste treatment.

Anaerobic Filter. The bacteria of this type of reactor attach as a thin biofilm to a surface within the reactor.

Fixed bed reactors contain biological solids that grow on inert, high-surface-area material, and the waste stream is circulated along the vertical length of the reactor. It can be operated either in an upflow or downflow direction, the choice being dependent, in part, on the possible effects of insoluble solids entering or forming within the reactor. High organic loading rates (OLR) of 10 to 20 kg/(m^3/d) [0.6 to 1.2 lb/(ft^3/d)] can be achieved with

COD-removal efficiencies of 80 to 95%. Startup can be exceptionally long, but the reactor is stable after the organisms are established. The reactor is relatively stable to toxic and shock loading.

Fluidized bed reactors have inert materials such as sand that have extremely high surface areas covered with biological growth. These reactors are operated with hydraulic velocities sufficient to maintain an expanded bed but low enough to avoid solids washout. Recycling of some of the effluent stream typically is required to maintain the design velocities, and the reactor can be operated with high efficiency at OLRs greater than 15 kg/(m^3/d) [0.9 lb/(ft^3/d)]. Particulate solids and solids buildup on the media surfaces are not a concern, nor are problems of hydraulic channeling and plugging. The reactor is relatively stable to toxic and shock loading. The cost of pumping power for this type of reactor can be a significant part of the operating costs.

Upflow Anaerobic Sludge Blanket. This type of reactor contains highly settleable sludge particles. The waste stream is fed from the bottom of the reactor and passes through a blanket of sludge granules as it moves toward the top of the reactor. The granules consist of the active anaerobic biomass and tend to entrain biogas and become buoyant. An overhead baffle system is located at the top of the reactor to capture and promote degassing of the rising granules. The granules subsequently settle to the sludge blanket, and in this way the biomass of the reactor is efficiently controlled. Maintaining the organisms while treating high COD concentrations sometimes as high as 40 kg/(m^3/d) [2.5 lb/(ft^3/d)] are not uncommon. Startup can be accomplished by importing the granules from other upflow anaerobic sludge blanket (UASB) reactors. Although the effects of introducing insoluble solids to the reactor must be monitored, problems of toxicity and maintaining stable operating conditions during intermittent and large shifts in OLRs can be managed.

Other Reactors. Various designs and reactor configurations may be appropriate for particular wastes and are commercially available. The special attributes of a two-stage reactor may become exceptionally useful in hazardous waste treatment as anaerobic digestion becomes more commonly used. In this type of reactor, organisms that hydrolyze insoluble organics, and ferment other organic constituents are maintained in the first stage, and their catabolic products become the nourishment for the organisms associated with efficient methanogenesis in a second reactor. The particular usefulness of this type of reactor lies in at least two important areas of interest. When wastes contain high concentrations of organic materials that are either insoluble or slowly degraded during acetogenesis, the first reactor would be operated to degrade these types of compounds to other, more simple forms. This type of reactor may be appropriate when particular compounds, toxic to methanogens, are degraded in the first stage and rendered amenable to methanogenesis in the second stage. No vendor of this technology is known, although ample operating and design information is found in the literature.

Lagoon reactors with full-length biomass contact are constructed with a sealed liner and cover and operate by contacting the waste stream with the biomass along the length of the lagoon. Typically, the reactor is considered slow rate and requires relatively long hydraulic retention times. This type of reactor is especially appropriate in applications in which relatively small volumes of high-COD waste are generated and land costs are minor factors in the choice of treatment. Relatively high solids content can be accommodated, and maintenance costs are minimal.

Recent engineering and design incorporate the features of at least two of the reactors to form a hybrid reactor. For example, a reactor may consist of a UASB on the bottom and a fixed film reactor on top to promote higher treatment efficiencies. Other combinations and configurations may be found to be useful as the number of applications increases and as various types of wastes are degraded in anaerobic digesters.

BIODEGRADATION OF HAZARDOUS WASTE IN SOILS AND SLUDGES. Introduction. Soils and sludges can be biologically treated whether in a solid or slurry phase. Composting is one form of solid phase biodegradation but has conventionally been used for nonhazardous solid wastes.

Organic contaminants in sludges and soils can be biologically treated in situ or excavated and treated by solid- and slurry-phase bioremediation processes. Solid-phase processes are being developed continually and, in many cases, have been used to treat a wide range of contaminants such as pesticides, diesel, gasoline, fuel oil, creosote, pentachlorophenol, and halogenated volatile organics. Enhanced in situ biodegradation is being used for sites with soil and groundwater contaminated with readily biodegradable organics such as gasoline and diesel. This same technology is being developed and used to treat contaminants that are more difficult to degrade (U.S. EPA, 1988a).

Solid-Phase Treatment. Solid-phase soil bioremediation is a process that treats soils in situ or ex situ in above-grade systems using conventional soil management practices to enhance the microbial degradation of contaminants.

In situ treatment for soils typically consists of bioventing, with the possible assistance of air sparging. This process involves the movement of air through the undisturbed subsurface soil, with or without the addition of nutrients. This process typically uses indigenous bacteria, which are stimulated by the increase of oxygen and nutrients, to degrade the contaminants in the soil. These contaminants typically are petroleum hydrocarbons or other readily biodegradable and volatile organic compounds. The use of air injection into the soil assists with the movement of air through the entire area of contamination. Air sparging assists in removing, by volatilization, contaminants in the groundwater, which then migrates into the vadose zone above the water table and is removed or treated by the bioventing system. Air sparging provides moisture laden air, which reduces the potential desiccation of soil as seen in typical soil vapor extraction processes.

These systems typically consist of slotted PVC well screen installed vertically into the subsurface to just above the water table. These screens are then connected to a vacuum blower system to evacuate a designed flow and volume of air from the ground. This air, if treated or acceptable as is, can be reinjected into the soil via screened pipe around the perimeter of the contaminated area. If air sparge wells are used for air injection into the groundwater to remove volatile organics, this is done by installing small diameter pipe with either well screen or a diffuser at the bottom. Air is pumped into the groundwater by use of a regenerative blower or air compressor.

Above-grade treatment of soil and solids typically is designed to contain and treat soil leachate and VOCs. This system can be designed and configured as a soil pile or land farming. The systems are similar in set up, with the major difference being in the overall thickness/height of the soil to be treated. Land farming typically consists of a very thin layer, 6 to 12 in. of soil, while a soil pile can be upwards of 10 ft high. The systems consist of a treatment bed lined with a high-density liner with sealed seams. Clean sand is placed on top of the liner to provide protection for the liner and provide drainage for any contaminated water as it leaches from the contaminated soils placed on the treatment bed. Lateral perforated drainage pipe is placed on top of the synthetic liner in the sand bed to collect the soil leachate. If volatile contaminants must be contained, the lined soil treatment bed is completely covered by a plastic liner or a modified plastic-film greenhouse. A spray irrigation system typically is used to provide moisture control and a means of distributing nutrients and microbial inocula to the soil treatment bed or soil pile (U.S. EPA, 1988a).

Volatile organic compounds released from the soil during processing can be collected, as necessary, from under the liner or cover and processed through an air management system. Biodegradable VOCs can be treated in a vapor phase bioreactor. Nonbiodegradable VOCs can be removed from the effluent gas stream by adsorption on activated carbon or by incineration.

Contaminated leachate that drains from the soil is transported by drain pipes and collected in a gravity flow lined sump and then pumped to an on-site bioreactor for treatment. Treated leachate can then be used as a source of microbial inocula and reapplied to the soil treatment and other environmental parameters.

U.S. EPA evaluated the solid treatment potential of soil for 56 chemicals identified as hazardous (U.S. EPA, 1988b). These chemicals were organized into four categories of substances: polynuclear aromatic hydrocarbons (PAHs), pesticides, chlorinated hydrocarbons, and miscellaneous chemicals. Treatability screening studies were conducted to determine degradation rates; partition coefficients among air, water, soil, and oil phases; and transformation characteristics. The quantitative information developed for a subset of the tested chemicals was put into two mathematical models (RITZ and VIP) specifically designed to describe the soil treatment process. Results of the fate and transport predictions for the models were compared with laboratory and literature results to evaluate the ability of the models to predict the behavior of the selected chemicals in a soil system.

The experimental approach used was designed to characterize degradation, immobilization, and transformation potentials for the hazardous substances evaluated. Results indicated that the significance of volatilization and abiotic loss processes in influencing "apparent loss rates" of substances from soil depends on the class of substance (U.S. EPA, 1988b). These processes were insignificant for the majority of PAH compounds, and biodegradation appears to be the major process for the loss of PAH compounds from soil systems. However, abiotic loss may also be an important process for certain pesticide substances. Volatilization was found to play the primary role in influencing loss rates of volatile chemicals from soil. Very little leaching or air emissions were predicted by either mathematical model for the subset of pesticides selected under the simulated test conditions.

Slurry-Phase Treatment. A second biodegradation technology involves the treatment of contaminated soil or sludge in a large mobile bioreactor. This system maintains mixing and contact of microorganisms with the hazardous compounds and creates the appropriate environmental conditions for optimizing microbial biodegradation of target contaminants (U.S. EPA, 1988a).

The first step in the treatment process is to create the aqueous slurry. During this step stones and rubble are physically separated from the waste, and the waste is mixed with water, if necessary, to obtain the appropriate slurry density. The water may be contaminated groundwater, surface water, or another source of water. A typical soil slurry contains approximately 50% solids by weight; a slurried sludge may contain fewer solids. The actual percentage of solids is determined in the laboratory based on the concentration of contaminants, the rate of biodegradation, and the physical nature of the waste. The slurry is mechanically agitated in a reactor vessel to keep the solids suspended and maintain the appropriate environmental conditions. Inorganic and organic nutrients, oxygen, and acid or alkali for pH control may be added to maintain optimum conditions.

Microorganisms may be added initially to seed the bioreactor or added continuously to maintain the correct concentration of biomass. The residence time in the bioreactor varies with the soil or sludge matrix, the physical and chemical nature of the contaminant (including concentration), and the biodegradability of the contaminants. After biodegradation of the contaminants is completed, the treated slurry is dewatered. The residual water may require further treatment before disposal.

Depending on the nature and concentration of the contaminants and the location of the site, any emissions may be released to the atmosphere or treated to prevent emission. Fugitive emissions of VOCs, for instance, can be controlled by modifying the slurry-phase bioreactor so that it is completely enclosed.

Aside from the biodegradability of a particular compound, other limiting factors include the presence of inhibiting compounds and operating temperature. Heavy metals and chlorides may inhibit microbial metabolism because of their toxicity. The operating temperature range is approximately 15 to 70 °C. Dissolved oxygen is also critical and must be monitored along with pH, nutrients, and waste solubility.

One advantage of treatment in a contained process is that a remediation system can be designed to pretreat waste contaminated with heavy metals as well as biodegradable semivolatile compounds. Soil washing and extraction of metals using weak acids and chelating agents can be combined with biological treatment by coupling two separate slurry-phase reactors in series.

Composting. Composting involves the storage of highly biodegradable structurally firm material (for example, chopped hay or wood chips) with a small percentage (5%) of biodegradable waste. Composting is enhanced by waste size uniformity. Adequate aeration, optimum temperature, moisture, nutrient contents, and the presence of an appropriate microbial population are necessary to enhance decomposition of organic compounds.

There are three basic types of composting: open windrow systems, static windrow systems, and in-vessel (reactor) systems. The open windrow system consists of stacking the compost into elongated piles. Aeration is accomplished by tearing down and rebuilding the piles. The static windrow system also involves long piles of compost. However, the piles are aerated by a forced air system; that is, the piles are built on top of a grid of perforated pipes. Finally, the in-vessel system involves placing the compost into an enclosed reactor. Aeration is accomplished by tumbling, stirring, and forced aeration. In general, compared to in situ biodegradion, composting is relatively insensitive to toxicants. The optimum temperature range for composting is between 10 and 45 °C. When treating hazardous wastes, it is necessary to collect leachate and runoff from the composting beds. Composting has not been widely used but is potentially applicable to hazardous sludges and soils.

Co-composting with Solids Wastes. Co-composting of sludges and municipal solid wastes typically does not require that sludge be dewatered. Feed sludges contaminated with hazardous wastes may have a solids content ranging from 5 to 12%. A 2:1 mixture of solid wastes to sludge is recommended; in fact, any amount of sludge can be mixed with solid wastes for composting, provided that the sludge is dewatered adequately. The solid wastes should be presorted and pulverized in a hammermill before mixing with sludge (Metcalf and Eddy, 1979). Practical application of this method was not, however, found in the literature.

PROCESS LIMITATIONS

GENERAL. The applicability of biological treatment to hazardous waste is highly dependent on many potential limiting factors. These factors include the extent of biodegradability of the waste, the ability to apply conventional technologies, the effects of toxic and nontoxic inorganics and organics, variations in concentrations and flows, requirements for nutrients, pH, oxygen concentrations, and temperatures. All these factors must be considered and addressed when evaluating the applicability of biological treatment.

PROCESSES. Advantages and Disadvantages of the Aerobic Treatment Process. The advantages of the aerobic treatment process include

- Log history of use in municipal and industrial treatment,
- Carbon dioxide and water as the byproducts of reaction,
- Applicability to a wide range of organic wastes,
- Cost competitive with all other technologies, and
- Various technologies and designs for varying conditions.

The disadvantages of the aerobic treatment are that

- It is not applicable to all organic wastes;
- Chlorinated compounds are difficult to, or unable to, biodegrade;
- It requires adequate oxygen, nitrogen, and phosphorus;
- Many potential limiting factors exist; and
- A large sludge volume is possible.

Advantages and Disadvantages of the Anaerobic Treatment Process. The advantages of the anaerobic treatment process include the following:

- Oxygen transfer is not a limiting factor, and therefore anaerobic treatment is amenable to high-strength organic waste. For high-strength industrial wastewaters with BOD_5 exceeding 5000 mg/L, oxygen transfer is impractical and prohibitively expensive.
- There can be a very low cell yield and less sludge production.
- Methane gas is produced.
- There is a reduced sludge volume with good dewaterability.

The disadvantages of the anaerobic treatment process include the following:

- It requires a longer mean cell residence time (MCRT) to maintain stability. This means a larger reactor volume, which accounts for the very high initial cost.
- Heat input is required.

- Oxidizing agents are toxic.
- The reaction chemistry is still mysterious.
- It is highly sensitive to upsets and environmental changes.
- It may require high maintenance.

Aerobic Performance Limitations (Levin and Gealt, 1993). *EXTENT OF BIODEGRADABILITY.* Although many hazardous organics are quite biodegradable, some are not. For example, 2,3,7,8-tetrachlorodibenzo-p-dioxin (2,3,7,8-TCDD) has maintained a somewhat elitist status by remaining at or near the top of the nonbiodegradability list for a rather extended period of time. In addition, the question of biodegradability is very system specific. Unless the proper microbial consortium can be both developed and maintained in a given system, a compound "known" to be biodegradable will not be degraded in that system. Even if good biological removals are achieved, the aerobic biological process may have to be used in conjunction with other physical and chemical destruction technologies to achieve the levels of treatment that are required for a particular cleanup operation. In any event, biological systems can reduce the cost of downstream processes by reducing organic load.

APPLICABILITY OF CONVENTIONAL TECHNOLOGY. Leachates and highly contaminated wastewater containing hazardous materials can be treated in systems that use conventional reactor technology. In such systems, adequate mixing, oxygen supply, nutrient addition, and temperature and pH control can be provided. Although this level of control has not yet been demonstrated for contaminated soils, especially in situ, some rather interesting and innovative designs have been developed that allow reasonably high levels of control. Whether dealing with contaminated soils or water, there are many more full-scale aerobic biological systems in operation today that were specifically designed to remove hazardous organics. As a result, the full extent of the problems that can develop with such systems is becoming better understood. Although the lack of control and understanding contributed to the reluctance of many potential users to adopt biological processes for the treatment of hazardous materials, in the past the current level of technology, control, and understanding of the process has lent to the marked increase in the use of biological processes.

TOXIC AND NONTOXIC INORGANICS. Although a microbial consortium can be developed that is resistant to moderately high levels of heavy metals, the deposition and accumulation of precipitates of these metals on the biomass may severely inhibit the activity of the community developed. As a result, a pretreatment system designed to remove these contaminants will reduce the likelihood of system failure. This is easily accomplished for systems treating leachates and wastewater but may be more difficult to achieve for contaminated soils. As far as nontoxic inorganics are concerned, organism selection and overall treatment efficiency can be adversely affected, especially for those cases in which the salt content can range from 2 to 5% or more. It has been the authors' experience that the salt matrix within which the organism community must work can have a strong effect on system performance. Short-term variations in the salt concentration are of special concern. Proper selection and control procedures can be developed, however, that will allow for the development of a reasonably efficient microbial consortium.

TOXIC AND NONTOXIC ORGANICS. Selective removal of a particular organic toxin by chemical or physico-chemical means is difficult to achieve. Pretreatment with activated carbon, with some "reasonably" selective resin, or with an immobilized enzyme may be useful but is likely to be inefficient and costly. If the organic toxin is biodegradable and if the organisms possessing the specific capabilities needed can be enriched and maintained in the microbial community, then the system can be designed so that the toxin is maintained below toxic levels.

The question of toxicity can be somewhat confusing. It must be remembered that we are dealing with several microbes within the total microbial community. As a result, a toxin may impair the activity of one or more members of the community that are essential for good overall removals while serving as an excellent carbon and energy source for other members of the consortium.

If the toxin is nonbiodegradable, more serious, and perhaps insurmountable, problems can develop. In this case, microbial strains that are "resistant" to the toxin must be enriched; otherwise only limited removals may be possible biologically.

LOAD VARIATIONS. The concept of load variations has limited applicability to soil decontamination systems. It is quite relevant, however, to systems treating leachates and highly contaminated wastewater. Because of rainfall or because of a number of other reasons, the types of wastes received at a hazardous waste disposal site and the mass flow rate of the total load of organics and of the individual hazardous compounds in the waste stream can vary notably with time. This means that the concentration of certain organics may fluctuate above the toxic limit during selected periods, causing the performance of the biological system to deteriorate. Installation of an equalization and storage facility before the biological system would reduce the possibility of upset. Because biological processes are known to be sensitive to load variations even when treating nonhazardous wastewater, care must be taken to incorporate all possible information on load variation into the design of a biological system.

NUTRIENT REQUIREMENTS. Nitrogen and phosphorus are the two most common nutrients that may have to be added to a biological system. If either or both are not present in sufficient quantities in a usable form, biological treatment will be severely retarded. Both nitrogen and phosphorus may be covalently bound to some of the organics in the hazardous materials. Unless hydrolyzed to the inorganic form (for example, ammonia nitrogen and phosphate), a supplemental supply will be required. In the case of phosphorus, some problems may result because the phosphate may precipitate with the cations present in the hazardous materials. This problem can be mitigated somewhat by adding phosphate with a chelating agent.

pH. Most of the organisms capable of treating hazardous substances grow best in the 6 to 8 pH range. Adjustment of the pH of contaminated soils and pH control of biological reactors treating wastewaters may be required if either the biological activity or the hazardous material supply causes a marked change in the pH. Conventional technology is readily available to deal with this problem.

OXYGEN SUPPLY. This is one of the more difficult challenges facing in situ soil and groundwater detoxification. Many hazardous organics are readily degraded aerobically, but, because the vertical and lateral extent of oxygen penetration in soils may be limited, the rate and extent of biological detoxification is also limited. Some of the innovative approaches that have addressed this problem include air injection with sparging wells and vapor extraction points, pure oxygen injection with sparging wells, and the use of peroxides or solid oxygen compounds. As far as biological treatment of leachates and other heavily contaminated wastewater is concerned, conventional aeration equipment and design approaches are directly applicable. It has been noted, however, that the stripping of volatile organics is greater in mechanical aeration systems than in diffused-air systems.

TEMPERATURE. The classical view of the effect of temperature on treatment performance is that the rate of removal of organics increases with increasing temperature (until some maximum temperature is reached) and decreases with decreasing temperature. This view is certainly appropriate for short-term fluctuations in temperature. The effect of long-term or seasonal changes in temperature on removals is less easy to define because the microbial consortium that is established at one temperature may be different than the consortium established at another temperature. As a result, the rate and extent of removal of any individual hazardous organic can vary with temperature much more than would be expected from short-term considerations only. A third, and perhaps more important, effect of temperature on the removal of xenobiotics (manufactured rather than biologically produced compounds that may or may not be on some list of hazardous materials) has been observed by the authors. Some microbial communities experience a dramatic drop in activity at approximately 10 to 15 °C. In one case, the activity was restored by slowly bringing the temperature down below 15 °C (at approximately 1 °C decrease per week) and remaining at each new temperature until "reasonable" activity was observed (Lange et al., 1987). The initial drop in activity was too abrupt to be explained by the need for a shift in the organism community. One explanation for the success of gradual low-temperature acclimation may lie in the observation that the fluidity (and, therefore, the transport properties) of the cell membranes is strongly influenced by temperature. Many, if not all, microorganisms can restructure the lipid content of the membranes if they are to function properly at the reduced temperatures.

The question of temperature reduction is important in colder climates. Although insulated reactors being supplied oxygen with compressed air are able to maintain a temperature some 5 to 8 °C above the temperature of influent wastewater, a heat-exchanger system may be necessary to keep the temperature above critical levels.

Temperature control for systems used to detoxify contaminated soils, especially in situ, is obviously a much more difficult problem. Shifts in microbial populations at elevated temperatures, either in insulated reactors in colder climates or in noninsulated reactors in warmer climates, can reduce performance and should be considered in the final system design.

Anaerobic Performance Limitations. *COEXISTENCE OF TWO GROUPS OF BACTERIA.* Anaerobic processes are basically unstable. This results from the necessity for the two main groups of bacteria to coexist in one reactor. Under the environmental conditions typically imposed, the non-methanogenic populations can respond more rapidly to environmental changes than can the methanogenic populations.

pH. Acetate and fatty acids produced during anaerobic treatment tend to lower the pH of the liquid medium. However, the ion bicarbonate equilibria of the carbon dioxide in the reactor exert substantial resistance to pH change. Thus, the presence of bicarbonate can help prevent adverse effects on methanogens resulting from low pH from excessive production of fatty acids during biodegradation.

Most anaerobic microorganisms grow best under neutral pH conditions because other pH values may adversely affect metabolism by altering the chemical equilibrium of enzymatic reactions or by actually destroying the enzymes. The methanogenic group of organisms is the most pH sensitive.

TEMPERATURE. The metabolic and growth rates of chemical and biochemical reactions tend to increase with temperature, within the temperature tolerances of the microorganisms. Microorganisms exhibit optimum growth and metabolic rates within a well-defined range of temperatures that are specific to each species.

Methanogenic bacteria are more sensitive to changes in temperature than other organisms. All bacterial populations in the reactor are fairly resilient to short-term temperature upsets of up to about 2 hours. They will rapidly return to normal gas production rates when the temperature is restored. However, numerous or prolonged temperature variations can result in unbalanced populations resulting in damage to the biological activities.

Temperature drops as small as 2 °C can have adverse effects on mesophilic (−35 °C) digestion or 0.5 °C with thermophilic (−55 °C) digestion. The usual causes of temperature decrease in an anaerobic treatment system are failure of the heating system or rapid introduction of a large quantity of cold feed.

TOXICITY. Table 5.8 gives data on the concentration ranges in which various substances exert toxicity in anaerobic waste treatment systems. As can be seen in Table 5.8, the concentration of any heavy metal that will produce toxicity is 1 mg/L or less; however, much higher levels of heavy metals in anaerobic waste treatment systems have been reported with no ill effects because sulfides remaining from the breakdown of proteins and sulfates form insoluble metal sulfides and precipitate the heavy metals. If sufficient sulfide is not present to precipitate the heavy metals (0.5 mg/L S_2 per mg/L metal) toxicity will result. Sodium sulfide and ammonium sulfate have been successfully added to avoid toxicity (Lawrence and McCarty, 1969).

The method of toxicity control is removal of the toxic material from solution. Other examples of the use of precipitation to remove a toxicant is the use of cationic detergents to precipitate anionic detergents and calcium to precipitate long-chain fatty acids (U.K. Dep. of the Environ., 1971). Another version of this technique is the use of gas stripping to remove volatile substances such as chloroform and sulfide (U.S. CFR, 1989).

SCALE UP AND CASE STUDIES. Aerobic Suspended Growth. Biological treatment of hazardous wastes in aerobic suspended growth reactors has been done successfully at low and high organic loading rates, high compound loading rates, and variable microbe concentrations, SRTs and hydraulic retention times (HRTs). Table 5.9 presents selected data from plants treating hazardous wastes. From Table 5.9, it is apparent that generalizations about operating conditions are not possible and that determination of operational characteristics must be done on an individual basis depending on the waste components and treatment capabilities.

Aitken and Irvine (1987) stated that the primary objective of a treatability study was to provide information to be used in determining the technical and economic feasibility of scale up, or information to be used directly in design. It also may be useful to continue the operation of bench or pilot systems beyond the evaluation stage to provide more accurate information during design or to evaluate operating strategies before they are implemented in the field.

A summary of the information needed to complete a feasibility evaluation is given below:

- The need for pretreatment. The presence of inhibitory inorganics may not be known until the biological treatability study is underway. Steps

Table 5.8 Toxic concentrations in anaerobic digestion.

Substance	Nonacclimated	Antagonists present	Acclimated	Acclimated and antagonists present
Iron	200	—	200	—
Sodium	4 500	8 000–10 000	8 000	10 000
Potassium	4 000	6 000–8 000	5 500	15 000
Ammonium	1 700	4 000	—	—
Calcium	2 800	4 500–5 500	5 500	8 000
Magnesium	1 200	3 000	1 500	3 500
Ammonia	30–60	—	50–100	—
Sulfide	100	—	200	—
Long-chain fatty acids	—	—	500	—
Heavy metals	1	—	1	—
Anionic detergents[a]	15 000	—	20 000	—
Chlorinated hydrocarbons[a]	10	—	15–30	—

[a]mg substance/kg dry sludge solids.

Table 5.9 Selected operational conditions for treatment plants receiving hazardous wastes.[a]

Waste treated	Parameter **Food:microorganism ratio, g COD/g MLVSS·d**	Value	Reference
2500 mg/L COD		0.4–0.5	Bell and Hardcastle, 1984
88 000 mg/L COD, cellusolves		0.1–0.6	Brown and Weintraub, 1982
Full-scale, organic solvent; 1900–10 900 mg/L COD		0.2	Marston and Woodward, 1986
DMF	g BOD/g MLSS·d	0.04	Carter and Young, 1984
Phenols, 5600 mg/L COD_3		0.04–0.38	Tsai et al., 1984
DMF	Organic loading, lb BOD/1000 cu ft·d	40	Carter and Young, 1984
DMF	Compound loading, DMF/1000 cu ft	40	Carter and Young, 1984
DMF	Compound loading, MLSS, mg/L	14 800	Carter and Young, 1984
		2470–4200	Bell and Hardcastle, 184
		2500	Powell and Reitano, 1982
Full-scale, organic solvent; 1900–10 900 mg/L COD		5300–9650	Marston and Woodard, 1986
Refinery waste	SRT, days	31–105	Bell and Hardcastle, 1984
		57–380	Tsai et al., 1984
	HRT, days	4.4	Powell and Reitano, 1982
		2.6–7.1	Tsai et al., 1984
		0.4–1.5	Brown and Weintraub, 1982

[a]COD = chemical oxygen demand, MLSS = mixed liquor suspended solids, DMF = dimethylformamide, BOD = biochemical oxygen demand, SRT = solids retention time, and HRT = hydraulic retention time.

to remove such species would be necessary before attempting further biological treatment studies.

- The need for pH control. It should be noted that improper pH control in the bench or pilot study can lead to data that are unreliable for scale up or to false conclusions that stable operation is not possible.
- The need for nutrient supplementation. Nitrogen and phosphorus are the two most common nutrients that may be deficient in a waste. Sulfur may also be important in anaerobic systems. Metals (iron in particular) can be limiting nutrients as well. Determining the quantity of a nutrient to be added should be done in consideration of growth requirements, which will vary depending on the reactor being used and the manner in which it is operated. Overdosed nutrients can sometimes be evaluated by observing abnormally high concentrations in the effluent.
- Mass balances on the major constituents of interest. Nonbiological removal mechanisms must be taken into consideration.
- The reliability of process performance. Inconsistent performance in meeting final treatment objectives or the inability of the biological process alone to remove all of the target organics indicates that a polishing step will be needed after biological treatment.
- Kinetics and stoichiometry or organic removals.
- Organic and hydraulic loading criteria.

Standard engineering practice can be used to combine the above information for a feasibility evaluation and for comparison to other waste management alternatives. Several examples of scale up for full-scale treatment of hazardous wastes are discussed below.

Marston and Woodard (1984) conducted a pilot study for the treatment of high-strength solvent containing wastewater. Influent solvents included glycerin, methanol, ethanol, butanol, acetone, dimethylacetamide, dimethylsulfoxide, and formamide. The BOD concentration averaged 3150 to 3310 mg/L in the pilot study, when consistent removals of greater than 90% were achieved.

Important considerations in the design of the pretreatment facility are given below.

- Wastewater volume and strength varied throughout the day. A tank for flow and concentration equalization was required.
- The waste streams to be collected for pretreatment contained only solvents from the manufacturing processes. Chemical feed equipment to add nutrients was required.
- Cleaning compounds were in the wastewater that could cause foaming problems. A foam spray system in the aeration tank was required.
- Very little space for a pretreatment system was available. A deep aeration tank and common wall construction between the equalization tank, aeration tank, and clarifier were used to keep the system compact.

Table 5.10 Design criteria for a full-scale plant.

Parameter	Value
Flow	43 000 gal/d (average)
	50 000 gal/d (maximum)
Solvent load	800 lb/d (average)
	1 200 lb/d (maximum)
Solvent removal	90+%
Food:microorganism ratio	0.5/d (maximum)
Oxygen required	911 lb/d (average)
	1 352 lb/d (maximum)
Mixed liquor suspended solids	5 000 mg/L
Influent temperature	100 °F (maximum)

- The proximity of a residential area required minimization of the noise generated by the plant.
- There were no maximum limits for suspended solids; therefore, waste activated sludge could be discharged directly to the sewer. This eliminated the need for sludge handling facilities and minimized the clarifier size required.

Design criteria are presented in Table 5.10 for the full-scale plant. Performance and loading are shown in Tables 5.11 and 5.12 for October through December 1983.

As shown in Tables 5.11 and 5.12, the pretreatment system removed an average of 94% of the COD and 99% of the solvents contained in the con-

Table 5.11 Plant loading and performance for a full-scale, activated-sludge plant treating a hazardous waste.

Parameter	Value
Flow	15 000 gpd[a] (average)
	5 000–22 000 (range)
Influent chemical oxygen demand	6 700 mg/L (average)
	3 100–10 000 (range)
Influent solvents (Total)	4 230 mg/L (average)
	1 900–10 900 (range)
Effluent chemical oxygen demand	435 mg/L (average)
	70–1 000 (range)
Effluent solvents (Total)	<42 mg/L (average)
	<42–80 (range)
Food:microorganism ratio	0.2/d (average)
	0.04–0.4 (range)
Mixed liquor suspended solids	7 850 mg/L (average)
	5 300–9 652 mg/L (range)

[a]gpd $\times 3.785 \times 10^{-3}$ = m^3/d.

Table 5.12 Solvent removal.

Compound	Influent, mg/L	Effluent, mg/L
Methanol	1140 (average)	<4 (average)
	180–2900 (range)	<4–20 (range)
Ethanol	1458 (average)	<8 (average)
	240–4400 (range)	<8–17 (range)
Butanol	958 (average)	<2 (average)
	190–2000 (range)	
Acetone	571 (average)	<8 (average)
	49–2700 (range)	
DMAC[a]	80 (average)	<5 (average)
	5–220 (range)	<5–29 (range)
DMSO[b]	<10 (average)	<10 (average)
DMF[c]	23 (average)	<5 (average)
	<5–190 (range)	

[a]DMAC = dimethylacetamide.
[b]DMSO = dimethylsulfoxide.
[c]DMF = dimethylformamide.

centrated waste streams treated. The pretreatment system achieved consistently high removal rates in spite of significant variations in loading; however, on the average, the system is loaded at less than 50% of design capacity.

Herzbrun et al. (1985) operated sequencing batch reactor pilot units to treat a wastewater with more than 500 mg/L phenol. Results from the bench-scale studies were used to design a full-scale demonstration facility.

Construction of the 1900 m3 (67 000 CE) unit was completed and TOC and phenol degradation averaged 76 and 99%, respectively, during the first month of operation. Table 5.13 presents full-scale performance data. Data for phenol, TOX, effluent suspended solids, and sludge volume index (SVI)

Table 5.13 Performance data from a full-scale SBR.[a]

Parameter	Value
Influent TOC, g/m^3	1100
Effluent TOC, g/m^3	260
TOC degraded	76%
Influent phenol, g/m^3	39.6
Effluent phenol, g/m^3	0.4
Phenol degraded	99.0%
MLSS, g/m^3	1100
MLVSS, g/m^3	930
Effluent SS, g/m^3	80
SVI, mL/g	60

[a]TOC = total organic carbon, MLSS = mixed liquor suspended solids, SS = suspended solids, and SVI = sludge volume index.

indicate acceptable treatment as anticipated from the bench-scale reactors. In addition, TOC degradation decreased from 86% in week 2 to 72% in week 5. Such results were expected because lab reactors typically require 4 to 8 weeks for treatment to stabilize fully.

AEROBIC FIXED FILM. The U.S. EPA conducted a study of 40 POTWs (U.S. EPA, 1982) to determine the levels of VOCs, semivolatile organics, pesticides, and metals in the influent, effluent, and sludge. Seven of the POTWs used trickling filters and one used RBCs. The trickling filter plants were numbered 11, 15, 21, 24, 29, 39, and 40. Plant 33 was the RBC facility.

Plant 11 received approximately 4% of its average daily flow from food products and fabricated metal machinery industries. Volatile organics detected in the influent were effectively removed. The lowest removal observed was 77% for methylene chloride. The combined primary and secondary sludge generally showed higher concentrations of VOCs than were contained in the effluent. Chloroform, 1,1,1-trichloroethane, tetrachloroethylene, and trichloroethylene were the exceptions. The VOC concentration in the secondary sludge was only slightly higher than the influent and effluent concentrations. Eighty-eight percent of the oil and grease was removed and like the VOCs tended to concentrate in the combined sludge more than the secondary sludge.

The few semivolatiles detected in plant 11 were not significantly removed. (2-Ethylhexyl) phthalate concentrated primarily in the combined sludge and less so in the secondary sludge. Metal removal ranged from 11% for cadmium to 75% for copper and mercury. The metals were highly concentrated in the combined and secondary sludges.

Plant 15 averaged 6.6 mgd, 25% of which came from fabricated metal products, electroplating, and electrical component industries. Eighty-nine percent of the oil and grease was removed, but the primary and combined sludges contained more than 100 times the level of oil and grease. The primary volatile organic detected in the influent was 1,1,1-trichloroethane.

The treatment process removed 97% of the influent 1,1,1-1-trichloroethane with none of the volatile detected in the sludges. The other volatiles detected in the influent were also effectively removed.

As in plant 11, the primary semivolatile detected at plant 15 was (2-ethylhexyl) phthalate. Di-N-butyl phthalate was 60% removed by the system and was much less concentrated in the sludges as compared to *bis*-(2-ethylhexyl) phthalate. Metal removal ranged from 59% for copper to 85% for nickel. As discussed previously, the metals concentrated in the primary and combined sludges.

The data for plants 21, 24, 29, 33, 39 and 40 indicate similar trends as has been discussed for plants 11 and 15. In general, the volatile organics were effectively removed and did not concentrate in the sludges, indicating significant removal by biological degradation. The semivolatile compounds were not as efficiently removed and tended to concentrate in the sludge. These compounds may then be subject to biological degradation, but this possibility was not addressed in the report. The metals also concentrated in the sludges.

ANAEROBIC PROCESSES. Phenols and Coal Gasification Wastes. Phenolic compounds are often the major pollutant in aqueous effluents from coal gasification facilities, coke oven batteries, oil refineries, and some petrochemical plants. The liquid wastewater from industrial operations that use phenolic compounds as raw manufacturing materials also contain phenolic pollutants.

The use of anaerobic activated carbon filters to remove phenolic compounds has been studied by a number of researchers. Granular activated carbon anaerobic treatment processes combine the advantages of the energy-efficient anaerobic filter and the adsorptive capability of activated carbon for extended retention of toxic or less readily biodegradable compounds. Activated carbon protects microorganisms from shock loading through rapid initial adsorption into pores and slow subsequent release by desorption. This desorption phenomenon, accompanied by biodegradation of desorbed material, has been frequently referred to as bioregeneration.

O'Barsky et al. (1978) concluded that by pretreating polysulfide rubber wastewater, 80% sulfur removal could be achieved even when sulfate concentration exceeded 500 mg/L. The results obtained from running a GAC reactor (Kim et al., 1986) for approximately 200 days (starting with fresh GAC) indicated that contributions of biogas production, adsorption, and biomass production were all important in the removal of phenol. Among them, the biogas production was most important. Bioregeneration occurred as adsorbed phenol was desorbed because of reduced influent phenol concentration. This was also shown by increased biogas production during the same period.

Wang et al. (1986) concluded that the organic removal efficiency of an expanded bed anaerobic reactor using GAC as a biological attachment medium was very high when subjected to a wide range of feed phenol concentrations. During 588 days of continuous operation, four steady-state conditions were obtained under feed phenol concentrations of 358, 716, 1432, and 2864 mg/L. Steady-state phenol, COD, and DOC removal efficiencies exceeded 99.9, 92, and 93%, respectively, even for the highest loading conditions. By the end of this study, 87.5% of the feed COD was converted to methane gas; only 6% of the feed COD escaped in the aqueous effluent from the anaerobic reactor. The high conversion to methane and the low retention of COD within the reactor demonstrate that biological utilization was the primary removal mechanism and the net growth of bacteria in the reactor was low. The major conclusion of this study was that the GAC expanded bed reactor achieved nearly 100% removal of phenol when the surface loading ranged from 0.028 to 0.23 mg/cm2d COD.

The expanded bed reactor responded favorably to the sudden doubling of feed concentration. Gas production from the reactor increased rapidly immediately after the increase in organic load. Gas production doubled within 1 week after the feed phenol concentration was doubled, but increases in effluent phenol concentrations were small. The use of partially saturated GAC media and phenol-acclimated activated carbon inocula enabled biological utilization of phenol to start in 8 days.

In another study in Canada, Fedorak et al. (1986) diluted high-strength wastewater (7600 mg/L phenolics) from the H-coal liquefaction process and fed it to anaerobic methane-producing cultures. Total phenolic concentrations of 150 and 300 mg/L were added to 50 mL semicontinuous cultures with HRTs of 12.5, 16.7, and 25 days. The rates of methane production and effluent concentrations of three fermentable phenolics (phenol, *p*-cresol, and its presence) served as the first indicator of reduced phenolic removal capability. The effluent *m*-cresol concentrations from cultures receiving 300 mg/L total phenolics followed simple washout curves, suggesting that its degradation stopped abruptly. Later *p*-cresol and ultimately phenol appeared in the effluents from the cultures that received the highest phenolic mass loadings.

A mathematical model was developed by Mack et al. (1978) to represent specific phenol utilization by anaerobic process. This model indicated a maximum specific utilization rate by anaerobic process at a phenol level of 700 mg/L with process instability occurring at phenol levels in excess of this value. Methane was not produced during these experiments.

Chlorinated Solvents. Many environmentally important anthropogenic compounds are halogenated (Vargas and Ahlert, 1978). The list of these species includes pesticides, plasticizers, solvents, and trihalomethanes. Of the halogenated compounds, the best known and most studied are the chlorinated compounds (Kobayashi and Rittmann, 1982).

Anaerobic studies with halogenated compounds have shown that several of these compounds are biodegradable. It has been shown that transformations of trihalomethanes can occur under anaerobic conditions. The biotransformation of some haloaliphatic compounds suggests a series of reductive dechlorination reactions (Bouwer and McCarty, 1983).

In a study by Vargas and Ahlert (1978), the biodegradability of 1,1,1-trichloroethane (TCA), 1,1-dichloroethane, and dichloromethane under anaerobic conditions was investigated. The acclimation potential of a mixed anaerobic culture to these compounds was also accessed.

The buildup of acetic acid lowers pH, making conditions unfavorable for the methanogens. Also, increased hydrogen partial pressure inhibits the acetogens (Bouwer et al., 1981). If the acetogens were inhibited, recovery was more likely. During the lag phase, no gas production was observed because the acetogens were inhibited and ethanol was not broken down for use by the methanogens. After the methanogens degrade the toxicant to less than inhibition levels, acetogens start using ethanol, and gas production resumes. Studies with separate anaerobe populations indicate that the methanogens, known for extreme sensitivity, may be responsible for degradation of chlorinated compounds. There is some indirect evidence to support these results.

Johnson and Young (1983) conducted laboratory tests to study the inhibition of anaerobic cultures by toxic organic chemicals and to examine the conditions affecting recovery from this inhibition. Screening tests were used in phase 1 to isolate organic compounds that have severe inhibitory effects on anaerobic biological reactions. The nature of the recovery of anaerobic cultures from inhibitory effects was studied in phase 2, and phase 3 was designed to identify inhibition contributory factors. Concentrations of 100 mg/L of hexachloroethane, hexachlorocyclopentadiene, hexachloro-1,3-butadiene, 4-nitrophenol, nitrophenol, 2,4-dichlorophenol, and nitrobenzene inhibited anaerobic cultures at 37 °C. Acclimation adsorption and the concentration of toxic compounds were found to be effective factors in the reduction of toxicity.

Nitrogen-Substituted and Sulfonated Benzene Aquifer Contaminants. A literature survey of groundwater contaminants indicated that aquifers are repositories for hazardous wastes, including N′-and S-substituted benzene derivatives. They examined the susceptibility of several anilines, benzamides, benenesulfonic acids, and benenesulfonamides to anaerobic metabolism by aquifer microorganisms. Under sulfate-reducing and methanogenic conditions the carboxylated anilines were biotransformed within 3 months while unsubstituted or methylated anilines required longer incubation times. Benzamide as well as an arylmethyl and an N-methyl derivative were biodegraded under redox conditions. The anaerobic degradation of the N-methylated benzamide was favored in sulfate-reducing rather than methanogenic incubations. However, the addition of a second N-alkyl group rendered the resulting compounds resistant to anaerobic decay. Only one of seven benzenesulfonates and two of five benzenesulfonamides proved amenable to anaerobic metabolism (Kuhn and Suflita, 1986).

Scale Up Considerations. The steps involved in implementing a biotreatment program begin with feasibility determinations and end with engineering and economic determinations. Wick and Pierce have described an integrated approach to the development and implementation of bioremediation, which involves the careful examination of the site, estimation of the amount and type of material to be degraded, large-scale demonstration and cost analysis of the process, consideration of the regulatory issues, estimation of time required to complete the remediation process, and evaluation of whether any other procedures may be combined with biotreatment to speed up the cleanup or decrease the cost. A detailed examination of all of the above factors is essential for successful completion of a bioremediation project.

In general, the use of biological treatments enables the user to proceed with a cost-effective treatment process that can bring the level of toxic materials down below that required by federal and state agencies, whether it is in air, water, soil, sludge, or other media. Although the process occasionally takes longer than nonbiotic processes such as incineration, it typically results in complete mineralization with less possibility for exposure of human populations to potentially toxic and carcinogenic wastes.

EMERGING TECHNOLOGIES

GENERAL. A number of new technologies have recently emerged and have shown promise for the treatment of hazardous wastes. Although much of the experience is limited to laboratory and pilot reactors, further research and full-scale applications have proven these technologies have advantages over typical methods of conventional physical-chemical and biological hazardous waste treatment. These technologies involve the use of aerobic and anaerobic bacteria, as well as fungi, to degrade and treat numerous hazardous organic wastes once thought to be incapable of bioremediation. Bioremediation is being used at more sites than ever before and continues to increase in usage as an efficient and cost-effective treatment alternative. The U.S. EPA has determined that in 1994, bioremediation was chosen as the treatment technology for 33% of petroleum contaminated sites, 28% of wood preserving sites, 21% of solvent sites, 7% of pesticide sites, and 6% of all other types of sites. These percentages are expected to increase as the technology of using bacteria and fungi becomes more sophisticated, yet easier to apply.

BACTERIA. Immobilized bacterial cells have been used for the biodegradation of toxic wastes. The microorganisms are trapped on a media that is then fed a wastewater containing the target hazardous compounds. In this technique, bacterial cultures are centrifuged to concentrated pellets and immobilized on a gel of sodium alginate. Other larger cells, such as plant cells, animal cells, and high molecular weight enzymes, can also be immobilized.

Because of the low solubility of oxygen in water and the high local cell density, oxygen transfer is often the rate limiting factor in the performance of aerobic immobilized cell systems. Another disadvantage is the difficulty in assessing the biomass concentration (as is true in more conventional fixed film processes). Immobilization has advantages over the conventional free cell systems. Prevention of washout of biomass in continuous-flow reactors, easy separation, and a greater degree of operational flexibility have been reported (Lakhwala et al., 1988). Immobilized cells can also be much more resistant to high concentrations of toxic chemicals. The cell density of immobilized cells can be much higher than that of the free cells, resulting in higher rates of biodegradation per unit volume of the reactor. Moreover, the immobilized cells can also be dried and stored as a convenient and accessible source of reproducible biomass.

In a study by Lakhwala et al. (1988), activated sludge from a municipal wastewater treatment plant was acclimated in a nongrowth medium with phenol (100 mg/L) and with continuous aeration. The immobilized microorganisms could withstand concentrations of 5000 mg/L of 2-chlorophenol, while concentrations of 750 mg/L were fatal to acclimated free microorganisms. Thus, the immobilized cells were more resistant to higher concentrations of the toxic compound studied.

Engineers at Wright–Patterson Air Force Base in Ohio are developing bacteria and fungal cultures that attack polyurethanes used in airplane paints. Soil and groundwater at the Hanford Nuclear reservation in Washington contaminated with carbon tetrachloride are being studied for degradation with an isolated organism that feeds on acetate and that will also convert toxic nitrates to nitrogen gas. Research, pilot testing, and full-scale processes are still being developed for the use of aerobic and anaerobic bacteria to degrade chlorinated compounds and other organics originally deemed nonbiodegradable.

FUNGI. Aitken and Irvine (1987) presented an overview of the use of fungi for biodegrading xenobiolics (meaning foreign to the biological system, without regard to the biological response). Aitken and Irvine list more than 30 citations conducted mostly in the laboratory for which fungi-degraded aromatics, polynuclear aromatics, dyes, chlorinated lignins, cyanides, PCBs, PCP, 2,3,7,8,-TCDD, and lindane.

An example is a study by Lewandowski et al. (1988) in which white rot fungus was used to degrade 2-chlorophenol. They combined the fungi with immobilized bed technologies using both a packed bed reactor using a silica-based porous support for the fungus and a well-mixed reactor using alginate beads as the immobilizing medium.

In addition, balsa wood chips were also used as a support for the fungi. Influent concentrations of 520 mg/L of 2-chlorophenol were reduced to 0.1 to 170 mg/L in the immobilized packed bed and were reduced to 0.1 to 210 mg/L in well-mixed reactors with immobilized cells, respectively. The degree of removal was dependent on the influent flow rate to the bench-scale reactors, with better removals at lower flow rates. The most important external factors affecting the activity of the fungi were reportedly temperature, pH, DO concentration, and fixed nitrogen concentration.

Although most biological treatment systems use bacteria, fungus is becoming better known because of its promising results in lab-, pilot-, and full-scale applications. White rot fungi, which decomposes wood, also produces enzymes that have been shown to breakdown PCBs, pesticides, dioxins, chlorinated solvents, TNT, creosote, pentachlorophenol, cyanide, and other difficult to treat compounds. Several research grants issued to universities has gone toward the exclusive research of white rot fungus (WRF) bioremediation. Utah State University is one such university, which under the direction of Dr. Steven Aust has performed extensive laboratory testing of WRF. This research and development work, which has taken place over the past 8 years, is being transformed into pilot-scale testing, as well as full-scale treatment applications. This WRF technology is being tested throughout the world for wastewater, groundwater, soil, sediment, and other media, and it holds promise for bioremediation of hazardous wastes.

Aitken and Irvine (1987) reviewed the use of algae for removal of xenobiotics. However, means other than biological degradation were reportedly responsible for the removals observed. Sloan et al. (1984) identified a sequestering phenomenon that produced the removal of metal ions by the green alga, *Chlorella pyrenordosa.* The uptake of metals by the alga depended on the ionic strength of the solution and the metal ion concentration. Their results suggested the rapid uptake of metal ions (cadmium, copper, and lead) resulted from ion exchange, independent of the alga cell's metabolism. The algae cell's uptake was found to be best modeled as a weakly acidic cation exchange resin.

OVERALL TECHNOLOGICAL OUTLOOK

PROMISE OF BIOREMEDIATION. More than a decade after the passage of federal laws intended to readdress the problem of treatment and disposal of hazardous waste, the nation still faces a problem of enormous magnitude. Approximately 1200 hazardous waste sites are listed on the National Priorities List (NPL) established under Superfund. The U.S. EPA also estimates that an additional 4000 sites may need corrective action under RCRA.

Faced with a task of these proportions, bioremediation offers some clear advantages over conventional hazardous waste treatment technologies. One leading advantage is cost. Bioremediation offers significant savings over the leading hazardous waste treatment technologies. The costs of incineration, for example, range from $300 to $1000 per ton of soil, while land disposal costs between $200 and $300 a ton. A second major advantage is that bioremediation can be accomplished with less environmental insult to the affected site and quite possibly to the environment in general. In situ bioremediation, for example, would ordinarily involve minimal disruption of surface soils. In addition, bioremediation avoids some of the significant environmental drawbacks associated with the leading treatment technologies.

Perhaps the greatest advantage of bioremediation, however, is as yet untapped. That is the use of bioremediation technologies as an instrument in pollution prevention. The U.S. EPA is well aware of this potential. U.S. EPA Administrator William Reilly, speaking in June 1991 at a meeting between representatives of the bioremediation industry and the agency, stated that the various applications of biotechnology "need to become an integral part of our future discussions and actions in pollution prevention."

Although biological treatment of waste organic matter has been studied for approximately a century, new wastewater treatment challenges, in the form of the biological treatment of hazardous chemicals and of more stringent effluent regulations, continue to motivate the development of new and innovative treatment processes. Many of these new processes have been described in this chapter. Predictions of trends must be based on events and discoveries of the immediate past and the present. The primary areas in which current research is moving forward rapidly are the application of molecular genetics to develop new bacterial strains capable either of more complete degradation of compounds or of metabolizing compounds otherwise resistant to microbial metabolism and the development of new engineering technologies to enhance rate and extent of degradation. Similar observations have been made in which examples of specific applications combining microbial and engineering approaches were discussed. We note at the outset that the regulatory infrastructure must be addressed in parallel with innovations in the science and engineering of biodegradation because new strains cannot be used except within a regulatory framework.

Much progress is being made in advancing the science and technology of biodegradation and bioremediation. In 1991, the market for bioremediation of hazardous wastes was approximately $60 million, and it is was expected to grow to between $125 million and $300 million by 1995, and more than $500 million by 2000. A growth rate of 65% per year has been forecast. Another important indication of the industry's maturation is seen in the efforts of the members of the Applied Biotechnology Association, who are using, to date, only nonengineered microorganisms. The association was formed in 1988 to promote the interests of firms involved in biotreatment. It has prepared a compendium of successes and is currently devising a means of accreditation for waste treatment providers using nonengineered microorganisms to treat a wide variety of wastes.

The significance of the regulatory infrastructure is becoming more evident. For example, specific regulatory factors driving the development of the entire biotreatment industry are (1) federal and state cleanup regulations and (2) real estate transfer issues. Closely related is the need to demonstrate unequivocally that microbial degradation is the primary factor in the disappearance of pollutants and that the process results in mineralization or production of innocuous products.

Unfortunately, in some areas progress is less rapid. After conducting a literature survey for evidence of successful biotreatment, only seven cases were listed in which biodegradation attributed to microbial action could be documented without equivocation. These included three soil, two aquatic, and two marine field applications. Only one of the soil treatments included addition of microorganisms; the others used only addition of nutrients in the form of fertilizer. None of the marine or aquatic trials employed addition of microorganisms. In two cases, indigenous microorganisms were enhanced by addition of fertilizer. A number of other studies were identified in which laboratory work predicted the ability of indigenous microorganisms to degrade particular wastes, but the associated field studies did not generate conclusive data to support the conclusion that biodegradation was a result of microbial action. Clearly, additional data detailing in situ effects of microbial action are required before regulators, the public, and prospective clients accept biotreatment as a viable, safe, and cost-effective alternative.

Areas for high-priority research include (1) increasing bioavailability of pollutants, (2) improving reactor/in situ remediation designs, (3) overcoming scale-up problems, and (4) finding improved bioremediation processes. Given enough time, many, if not most, compounds are degraded by microorganisms, but the thrust for the next decade must be to increase efficiency. The importance of demonstrating that a process can be scaled up and operated under field conditions cannot be overemphasized. In fact, scale up feasibility (and actuality) is a primary obstacle in the development of biotechnology and particularly of bioremediation.

Specific to the use of genetically engineered microorganisms (GEMs) in bioremediation is the question of safety because of possible distribution of either the GEM itself or its genetically engineered DNA sequences into the environment, with the potential for transfer among naturally occurring microorganisms. One approach that has gained some support is the incorporation of a "suicide" gene into a bacterium proposed for deliberate release. The killing function can be combined with the engineered degradative functions so that the bacteria die as soon as the hazardous waste is depleted, that is, as soon as the substrate is metabolized. In one system, a protein (the GEF protein) is inserted into the cell membrane. When degradation ceases, unrestricted flow into and from the cell is facilitated, resulting in cell death. The GEF-based system has been demonstrated to function in several species suggested for deliberate release.

New methods of biotreatment being developed include technologies for treating volatile chlorinated and light aromatic hydrocarbons in groundwater and in air streams. There are a variety of innovative processes for reclamation of contaminated subsurface environments, particularly streams containing chlorinated aliphatic hydrocarbons and the light aromatic constituents of petroleum products. They concluded that 60 to 90% removal was achievable using existing techniques. The extent of removal depended on the type of material, type of reactor substrate used, flow rate, temperature, and related environmental parameters. Obviously, further research will yield improved performance.

For methods for biofiltration of air streams, carbon adsorption of contaminants, such as trichloroethylene (TCE), requires 5 to 10 times as much carbon as the weight of the amount of TCE and that the material is not changed, but simply adsorbed. In contrast, biofiltration results in mineralization of the contaminant, without byproducts (such as vinyl chloride), which may have adverse properties. The technique has already proved effective for treatment of air streams containing low concentrations of organic compounds.

Bioremediation techniques for highly concentrated process flows are in active development. Biofiltration can also be used effectively when venting is used as a remediation process. A recently completed study for the U.S. Air Force showed that 50 g/h of jet fuel was removed using a biofilter. Capital and operating costs for biofiltration were estimated to be approximately the same as for catalytic oxidation systems. Operating costs were about one-third of those for carbon adsorption systems.

Treatment of landfill leachate is an increasingly common biotreatment application. With increased (and mandated) use of lined landfills, collection and treatment of leachates pose difficult problems. Treatment in a semicontinuous mode makes it possible to maintain treatment for prolonged periods of time. Semicontinuous biotreatment of a landfill leachate containing phenoxy herbicide chemical allowed accommodation of an average daily loading of 1.6 kg of phenoxies and 0.5 kg of phenols, with greater than 98% degradation of both.

Use of reductive conditions in degradation has allowed new ground to be broken. Currently, many halogenated compounds are dehalogenated using anaerobes, after which ring cleavage occurs rapidly under aerobic conditions. Fluidized bed systems can be used for anaerobic waste treatment. In fact, development of various anaerobic reactors, such as the upflow biomass bed and filter reactor for treating synthetic sugar wastes, is moving ahead. Incidentally, anaerobic conditions also aid degradation of cyanide, the biodegradation of which was discussed previously.

Fluidized bed reactors are also used extensively for aerobically degraded wastes, including degradation of S-triazine-containing industrial wastewater. Up to 18 g/L volatile suspended solids can be removed with the S-triazines as the sole nitrogen source for microorganisms carrying out degradation. A maximum removal efficiency of 80% has been reported. Mixed cultures of microorganisms have been shown to degrade 3,4-dichloroaniline in a three-phase draft tube fluidized bioreactor at efficiencies of 95%.

Although many degradative reactions occur when suspended cultures are used, it may be advantageous to immobilize cells before starting treatment. Bioprocess engineering has shown that immobilization leads to enhanced reaction stability, perhaps by preventing degradation of the biocatalyzing organisms. An immobilized Arthrobacter species (NCIB 11075) has been shown to catalyze the rate-limiting step of trimethyl lead degradation to dialkyl lead. Contaminated groundwater has been treated for more than 90 days at a flow rate of 80 gal/d using an immobilized-cell packed-bed reactor. Also, immobilized cells of the fungus *Phanerochaete chrysosporium* have been reported to degrade pentachlorophenol.

Although certainly not new, some rather innovative uses of traditional methods of degradation, such as composting and land farming, are proving very useful in achieving degradation of hazardous wastes. Improvement in efficiency and enhancement of efficacy can be achieved by continuing refinement of these methods. Composting has been used to decontaminate sludge containing the explosive trinitrotoluene (TNT), thus cost effectively reducing hazard.

Major changes have occurred in the regulatory status of biotreatment procedures; many of these have taken place at the federal level. The changes have spurred development of new techniques and expansion of field application of existing methods. Acceptance of bioengineering at the federal level has placed greater attention on biodegradation. The pressure of this attention and the costs associated with physical-chemical methods comprise driving forces that create significant interest in biotreatment projects.

References

Aitken, M.D., and Irvine, R.L. (1987) Biological Hazardous Waste Treatment. Paper presented at Haz. Wastes Treatment Process Design Workshop, 60th Annu. Conf. Water Pollut. Control Fed., Philadelphia, Pa.

Blackburn, J.W., et al. (1987) Molecular Ecology of a Napthalene Degrading Genotype in Activated Sludge. *Environ. Sci. Technol.,* **21,** 884.

Bouwer, E.J., and McCarty, P.L. (1983) Transformations of 1- and 2-Carbon Halogenated Aliphatic Organic Compounds Under Methanogenic Conditions. *Appl. Environ. Microbiol.,* **45,** 1286.

Bouwer, E.J., et al. (1981) Anaerobic Degradation of Halogenated 1- and 2-Carbon Organic Compounds. *Environ. Sci. Technol.,* **15,** 596.

Calmbacher, C.W. (1991) Biological Treatment Gaining Acceptance. *Environm. Protection.*

Chang, S.Y., et al. (1986) Effect of Cd(II) and Cu(II) on a Biofilm System. *J. Environ. Eng.,* **112,** 94.

Cormack, J.W., et al. (1983) A Pilot Study for the Removal of Priority Pollutants by the PACT Process. *Proc. 38th Ind. Waste Conf.,* Purdue Univ., West Lafayette, Ind.

Dobbs, R.A., and Cheng, K.-Y. (1988) Partitioning of Toxic Organic Compounds on Municipal Wastewater Treatment Plant Solids. In *Biotechnology for Degradation of Toxic Chemicals.* R.J. Scholze, Jr., et al. (Eds.), Noyes Date Corp., Park Ridge, N.J.

Elenbogen, G., et al. (1985) Studies of the Uptake of Heavy Metals by Activated Sludge. *Proc. 40th Ind. Waste Conf.,* Purdue Univ., West Lafayette, Ind.

Fedorak, P.M., et al. (1986) Anaerobic Treatment of Phenolic Coal Conversion Wastewater in Semicontinuous Cultures. *Water Res.* (G.B.), **20,** 113.

Ford, D.L., et al. (1971) Analytical Parameters of Petrochemical and Refinery Wastewaters. *J. Water Pollut. Control Fed.,* **43,** 1712.

Freeman, H.M. (Ed.) (1988) *Standard Handbook of Hazardous Waste Treatment and Disposal.* McGraw-Hill, New York.

Grady, C.P.L., Jr. (1988) Biodegradation of Hazardous Wastes by Conventional Biological Treatment. *Haz. Waste Haz. Mater.,* **3,** 333.

Grady, C.P.L, Jr., and Kim, H.C. (1980) *Biological Wastewater Treatment.* Marcel Dekker, Inc., New York.

Hannah, S.A., et al. (1986) Comparative Removal of Toxic Pollutants by Six Wastewater Treatment Processes. *J. Water Pollut. Control Fed.,* **58,** 27.

Herzbrun, P.A., et al. (1985) Biological Treatment of Hazardous Waste in Sequencing Batch Reactors. *J. Water Pollut. Control Fed.,* **57,** 1163.

Hickman, G.T., and Novak, J.T. (1984) Acclimation of Activated Sludge to Pentachlorophenol. *J. Water Pollut. Control Fed.,* **56,** 364.

Hill, N.P., et al. (1986) Behavior of Chlorophenoxy Herbicides During the Activated Sludge Treatment of Municipal Wastewater. *Water Resour.,* **20,** 45.

Horvath, R.C. (1972) *Bacterial Rev.,* **36,** 2, 146.

Johnson, L.D., and Young, J.C. (1983) Inhibition of Anaerobic Digestion by Organic Priority Pollutants. *J. Water Pollut. Control Fed.,* **55,** 1441.

Kim, B., et al. (1986) Adsorption, Desorption, and Bioregeneration in an Anaerobic, Granular Activated Carbon Reactor for the Removal of Phenol. *J. Water Pollut. Control Fed.,* **58,** 35.

Kobayashi, H., and Rittmann, B.E. (1982) Microbial Removal of Hazardous Organic Compounds. *Environ. Sci. Technol.,* **16,** 170.

Kodukula, P.S., and Patterson, J.W. (1983) Distribution of Cadmium and Nickel in Activated Sludge Systems. *Proc. 38th Ind. Waste Conf.,* Purdue Univ., West Lafayette, Ind.

Kugelman et al. (1982) Sludge Treatment. *Pollut. Eng. Technol.,* 14.

Kuhn, E.P., and Suflita, J.M. (1986) Anaerobic Biodegradation of Nitrogen-Substituted and Sulfonated Benzene Aquifer Contaminants. *Haz. Wastes Haz. Mater.,* **6.**

Lakhwala, E.S., et al. (1988) Design of Toxic Waste Treatment Bioreactor: Viability Studies of Microorganisms Entrapped in Alginate Gel. Paper presented at Int. Conf. Physiochem. Biol. Detox. Haz. Wastes, Atlantic City, N.J.

Lange, C.R., et al. (1987) Constraints of Bioaugmentation in Enhancing Biological Treatment Process Performance. *Proc. 42nd Ind. Waste Conf.,* Purdue Univ., West Lafayette, Ind.

Lawrence, A.W., and McCarty, P.L. (1969) Kinetics of Methane Fermentation in Anaerobic Treatment. *J. Water Pollut. Control Fed.,* **41,** 2.

Levin, M.A., and Gealt, M.A. (Eds.) (1993) Biotreatment of Industrial and Hazardous Waste. McGraw-Hill, New York.

Lewandowski, G.A., et al. (1988) Reactor Design for Hazardous Waste Treatment Using a White Rot Fungus. Paper presented at Spring Am. Inst. Chem. Eng. Meeting, New Orleans, La.

Lowe, W.L., and Gaudy, A.F., Jr. (1985) Removal of Cadmium at High and Low Dosages by an Extended Aeration Process. *Proc. 42nd Ind. Waste Conf.,* Purdue Univ., West Lafayette, Ind.

Mack, J.D., et al. (1978) Kinetics of Anaerobic Decomposition of Phenol. Am. Soc. Civ. Eng. Spring Meeting, Pittsburgh, Pa.

Marshall, B.R., et al. (1982) Some Problems and Alternatives in Applying Biological Treatment Models to a Coal Conversion Wastewater. *Proc. 36th Ind. Waste Conf.,* Purdue Univ., West Lafayette, Ind.

Marston, K.R., and Woodward, F.E. (1984) Treatment of High-Strength Wastewater Containing Organic Solvents. *Proc. 39th Ind. Waste Conf.,* Purdue Univ., West Lafayette, Ind.

Masunaga, Y., and Yonezawa, Y. (1986) Biodegradation Pathway of Cresol by Heterogeneous Culture: Phenol Acclimated Activated Sludge. *Water Resour.,* **20,** 477.

McCarty, P.L. (1964) Anaerobic Waste Treatment Fundamentals. *Public Works,* **95,** 9, 107; 10, 123; 11, 91; 12, 95.

McDermott, J.B., et al. (1989) Two Strategies for PCB Soil Remediation: Biodegradation and Surfactant Extraction. *Environ. Progress,* **8,** 46.

Melcer (1986) Control of Toxic Trace Contaminants in Municipal Wastewater Treatment Plants. Technol. Transfer Conf., Toronto, Ont., Canada.

Melcer (1987) Biological Treatment of Industrial Process Wastewaters Containing Hazardous and Toxic Contaminants. Int. Congress Recent Adv. Manage. Haz. Toxic Wastes Process Ind., Vienna, Austria.

Metcalf and Eddy, Inc. (1979) *Wastewater Engineering Treatment, Disposal, and Reuse.* McGraw-Hill, New York.

Middleton, A.C.; Nakles, D.V.; and Linz, D.G. (1991) The Influence of Soil Composition on Bioremediation of PAH-Contaminated Soils. Remediation.

Nyar, E.K. (1992) *Groundwater Treatment Technology.* 2nd Ed., Van Nostrand Reinhold, New York.

O'Barsky, B.J., et al. (1978) Sulfur Removal of Polysulfide Rubber Manufacturing Wastewaters by Anaerobic Treatment. *Proc. 33rd Ind. Waste Conf.,* Purdue Univ., West Lafayette, Ind.

Philbrook, M.J., and Grady, C.P.L., Jr. (1985) Evaluation of Biodegradation Kinetics for Priority Pollutants. *Proc. 40th Ind. Waste Conf.,* Purdue Univ., West Lafayette, Ind.

Powell, R.W., and Reitano, A.J., Jr. (1976) Screening the Effects of Inhibitor Contaminants in Activated Sludge Feed. Exxon Proprietary Rep.

Reitano, A.J., Jr. (1981) Startup and Operation of a Refinery Activated Sludge Plant. *Proc. 36th Ind. Waste Conf.,* Purdue Univ., West Lafayette, Ind.

Rogers, C.J. (1979) *Selected Biodegradation Techniques for Treatment and/or Ultimate Disposal of Organic Material.* EPA-600/2-79-006, U.S. EPA, Cincinnati, Ohio.

Roy, K. (1992) Government Scientists Discover Microbes Hungry for Metals. *Hazmat World.*

Rozich, A.F., and Gaudy, A.F., Jr. (1985) Response of Phenol Acclimated Activated Sludge Process to Quantitative Shock Loading. *J. Water Pollut. Control Fed.,* **57,** 795.

Sloan, F.J., et al. (1984) Algae, Ion Exchange, and Metal Finishing Wastes. *Proc. 39th Ind. Waste Conf.,* Purdue Univ., West Lafayette, Ind.

Stephenson, T., et al. (1987) Mechanism of Metal Removal in Activated Sludge. *J. Sanit. Eng. Div., Proc. Am. Soc. Civ. Eng.,* **113,** 1074.

Stover, E.L., and Kincannon, D.F. (1981) Biological Treatability of Specific Organic Compounds Found in Chemical Industry Wastewaters. *Proc. 36th Ind. Waste Conf.,* Purdue Univ., West Lafayette, Ind.

Tabak, H.H., et al. (1981) Biodegradability Studies with Organic Priority Pollutant Compounds. *J. Water Pollut. Control Fed.,* **53,** 1503.

Unger, M.T., and Claff, R.E. (1985) Evaluation of Percent Removal Variability for Priority Pollutants in POTWs. *Proc. 40th Ind. Waste Conf.*, Purdue Univ., West Lafayette, Ind.

U.K. Department of the Environment (1971) Inhibition in the Anaerobic Digestion Process for Sewage Sludge. *Notes Water Pollut.*

U.S. Code of Federal Regulations (1989) Title 40, 40 CFR Parts 261 to 170. U.S. Government Printing Office, Washington, D.C.

U.S. Environmental Protection Agency (1982) *Fate of Priority Pollutants in Publicly Owned Treatment Works.* EPA-440/1-82-303, U.S. EPA, Cincinnati, Ohio.

U.S. Environmental Protection Agency (1988a) *Technology Screening Guide for Treatment of CERCLA Soils and Sludges.* EPA-540/2-88-004, Cincinnati, Ohio.

U.S. Environmental Protection Agency (1988b) *Treatment Potential for 56 EPA-Listed Hazardous Chemicals in Soil.* EPA-600/6-88-001, U.S. EPA, Cincinnati, Ohio.

Vargas, C., and Ahlert, R.C. (1978) Anaerobic Degradation of Chlorinated Solvents. *J. Water Pollut. Control Fed.,* **59,** 964.

Wang, Y.T., et al. (1986) Anaerobic Treatment of Phenol by an Expanded-Bed Reactor. *J. Water Pollut. Control Fed.,* **58,** 227.

Williamson, K.J., and McCarty, P.L. (1975) Rapid Measurement of Monod Half-Velocity Coefficients for Bacterial Kinetics. *Biotechnol. Bioeng.,* **17,** 915.

SUGGESTED READINGS

Anderson, W.C. (Ed.) (1993) *Innovative Techology—Bioremediation.* WASTECH, Annapolis, Md.

Brown, J.A., Jr., and Weintraub, M. (1982) Biooxidation of Paint Process Wastewater. *J. Water Pollut. Control Fed.,* **54,** 1127.

Carter, J.L., and Young, D.A. (1983) Biodegradation of Chemical Plant Wastewater Containing Demethylformamide. *Proc. 38th Ind. Waste Conf.,* Purdue Univ., West Lafayette, Ind.

Cookson, J.T. (1995) *Bioremediation Engineering: Design and Application.* McGraw-Hill, New York.

47 Federal Register FR 42898, Sept. 28, 1982.

Freeman, H.M. (Ed.) (1988) *Standard Handbook of Hazardous Waste Treatment and Disposal.* McGraw-Hill, New York.

Patterson, J.W., and Kodukula, P.S. (1984) Metals Distributions in Activated Sludge Systems. *J. Water Pollut. Control Fed.,* **56,** 432.

Tsai, K.C., et al. (1983) Biotoxicity of Coal Liquidification Waste in the Activated Sludge Process. *Proc. 38th Ind. Waste Conf.,* Purdue Univ., West Lafayette, Ind.

Verschueren, K. (Ed.) (1983) *Handbook of Environmental Data on Organic Chemicals.* 2nd Ed., Van Nostrand Reinhold Co., New York.

Chapter 6
Physical Treatment of Hazardous Wastes

INTRODUCTION

The physical treatment of hazardous wastes is distinct from chemical treatment. However, confusion arises on this point because chemical treatment is used to induce changes of state in the wastes, and physical treatment is used to remove hazardous wastes from the liquid. Physical treatment processes involve a physical removal of the waste from its carrier. Chemical treatment transforms the hazardous waste into a different form or compound. Physical treatment processes include coagulation, sedimentation, absorption, stripping, flotation, centrifugation (high-gravity sedimentation), ion exchange and evaporation, and encapsulation or stabilization. The overlap between physical and chemical treatment can be understood by the following example. Precipitation of hazardous waste constituents from solution occurs by a chemically induced change, such as when soluble mercuric and nickel compounds are precipitated by the addition of sodium sulfide or hydrogen sulfide to the waste stream. The insoluble precipitate formed is coagulated to form floc and removed by gravity separation, centrifugation, or filtration. The addition of coagulant compounds and the addition of acids or bases to change the pH may also be necessary to form the precipitate or to increase the effectiveness of the removal. Chemical addition has been discussed in a previous chapter.

Before embarking on an waste treatment program, it is important to define a hazardous waste and answer the question of whether the compound is amenable to physical or chemical treatment.

CHEMICAL CHARACTERIZATION AND TESTING

According to hazardous waste regulations (40 CFR Part 261), a waste is hazardous if it is ignitable, explosive, toxic, corrosive, or reactive or if it is from specific sources or contains specific compounds as listed in the tables in 40 CFR Part 261. It is important to distinguish that a hazardous waste must also be handled in a "waste-like manner," as opposed to a material that is for recycling or reuse. The rules are complex and beyond the scope of this chapter. If the material is handled in a waste-like manner and if it contains any of the compounds in one of several lists, it is probably a hazardous waste. The determination of whether one has a hazardous waste should be made carefully because inadvertent mischaracterization can result in heavy fines.

Hazardous wastes will be found in either liquid or solid states. Sludges are a mixture of waste materials in both of these states. Gases are a separate problem, again beyond the scope of this chapter. Hazardous waste gases are often an air pollution emission problem. The treatment options are limited unless the gas stream contains high levels of solvents or particulates that can be incinerated, filtered, condensed, or otherwise removed from the gas stream.

It is important to determine which hazardous waste classification is being used to categorize a waste because many waste streams can be characterized differently. A soil contaminated with benzene can be analyzed and reported as a hazardous waste, or it can be analyzed and reported as a soil containing total petroleum hydrocarbons (TPH). The classification depends on whether one has foreknowledge of the presence of the constituent material that makes the soil a hazardous waste. If the soil is reported as containing TPH, it may not be a hazardous waste, and the difference in cost of disposal can be substantial.

FIRST PRINCIPLE

The objective of any physical treatment process is to minimize the cost of the wastes to be disposed. Volume reduction, process modification and waste elimination are all important steps in this process. Before developing an overall waste disposal strategy, the limitations of the toxic characteristic leaching procedure (TCLP) protocols must be understood and addressed because the TCLP will help determine the effectiveness of the treatment of the hazardous wastes and the limitations of its disposal.

TOXIC CHARACTERISTIC LEACHING PROCEDURE

The TCLP is an extraction test that was developed by the U.S. EPA to determine whether a waste is hazardous and whether a waste can be disposed on land. If the waste exhibits the characteristics of a hazardous waste after being subjected to the TCLP extraction or if the chemicals in the extraction fluid exceed the values set forth in 40 CFR Part 261.24, Table 1, the waste is a hazardous waste. If the waste, even after treatment, exceeds the values listed in 40 CFR Part 268.41 and following paragraphs and tables, the waste is prohibited from being disposed on land. In specific instances, there may be conflicts in values between the limits expressed in the Landban Disposal Restrictions (LDR) and the numerical values in 40 CFR Part 261.24. In those instances, the LDRs govern disposal.

It is important to distinguish between the analyses of a waste and the TCLP analysis of a waste stream. The TCLP analysis determines when the waste is hazardous and when the treatment is finished. The TCLP test (described in SW-846) includes a size reduction for solids to approximately 1 cm diameter, followed by extraction with a dilute solution of acetic acid in mixing reactor.

If a semisolid paint sludge contains 60 ppm of lead and 180 ppm of chromium, it might be a hazardous waste. Because it is a semisolid, it must be directly analyzed using the TCLP. The results of the extraction are analyzed and compared with the values in 40 CFR Part 261, Table 1, for the determination of whether the waste is hazardous. Note that in the TCLP the volume of dilute acid used is 20 times the volume of the waste. This provides an immediate 20-fold volumetric dilution of the waste. If the acid solution left after the 24-hour extraction does not contain more than the values in Table 1, the waste is not hazardous.

Because we have provided a 20-fold dilution of the waste, we can, theoretically at least, divide the concentrations in the paint sludge by 20 and see if the values are greater than those in Table 1. The chromium level is going to be less than 9 mg/L (below the regulated value of 10 mg/L), and the theoretical maximum lead level will be 3 mg/L (which is below the 5.0 mg/L regulatory limit). On subsequent samples and after one is more comfortable with the TCLP, it might be possible to declare the waste stream nonhazardous from inspection of the lead and chromium levels in the waste directly, without performing the TCLP. However, that should be done only after one is familiar with the regulatory framework and the TCLP results on the wastes.

If the concentration of the lead and chromium in the paint sludge was 1000 mg/L or greater, the answers are less clear because the paint is resistant to dilute acids and the values will be less than the theoretical maximum but above the Table 1 waste concentrations. If one analyzed the paint waste by conventional means and then performed the TCLP, what first appeared to be a hazardous waste would not be hazardous at all. However, check and test; do not assume. If you guess and are wrong, you can be heavily fined.

WASTE CHARACTERIZATION AND TREATABILITY STUDIES

Consider the regulatory framework before initiating the design of a waste treatment system. If the waste is a liquid and if one can obtain discharge permits for the wastewater from the process, a Resource Conservation and Recovery Act (RCRA) Part B permit may not be necessary as long as the requisite pretreatment standards and permit conditions are met with the plant. However, if a hazardous waste is treated and if there is no plant or National Pollutant Discharge Elimination System (NPDES) permit, the waste treatment works may require a RCRA Part B permit.

Volume reduction is essential in the waste treatment process. The treatment of dilute waste streams is costly and inefficient. To that end, treatability studies and in-process reviews of waste generation practices should be considered. Washdown or cleaning practices within the facility often generate a highly variable waste stream that is difficult (and expensive) to treat.

Waste treatability studies are often valuable as an inexpensive way to identify potential problems with a waste treatment system before it is built in full scale. However, if the study is not designed and supported with a good sampling program, it may be of little use. The objectives of the sampling study and the treatability study are to identify the extreme values of the waste stream and design a treatment regimen that will produce a consistent effluent under those conditions. One should know what to do with the data that will be collected before they are obtained, and if the data show a permit violation, they probably will have to be reported to the appropriate regulatory agency.

The operation of a larger-scale pilot system may help identify some operation problems that laboratory studies cannot identify. A larger-scale pilot plant makes it difficult to overlook or hide bad results or operation problems. If the waste stream permits, a pilot system should be set up and operated at between 18 and 20 L/min (5 and 10 gal/min) or at an absolute minimum of at least 3 L/min. All pilot systems, and especially the smaller systems, should be operated until the potential problems of full-scale operations have been identified. For batch processes or semisolid systems a waste stream of between 5 and 10% may be adequate for design purposes. The results from small-scale pilot studies should be examined closely when the results are scaled up by factors of 10 or more.

ENCAPSULATION AND STABILIZATION

Encapsulation and stabilization are discrete but related processes. Both immobilize the waste, sometimes treating and at other times not treating the waste. Encapsulation forms a nonreactive barrier around the waste stream. Stabilization is a chemical process that often reacts with the waste and tie it up so that it is unavailable to the environment or to the acids in the TCLP test.

Examples of encapsulation include mixing various types of metallic sludges in with concrete or asphalts or plastics. A simple example of encapsulation would be placing the wastes in a Tupperware container. Encapsulation has been used successfully in the nuclear industry when radioactive wastes have been treated by surrounding them with molten glass, polyethylene, and asphalts. The molten glass is a high-temperature process that is currently being used for encapsulation of the most radioactive substances.

Stabilization involves mixing the wastes with reacting substances (typically inorganic). An example of a stabilized sludge is that of chromium, which is reduced and precipitated with the excess lime process. In this latter instance, the calcium hydroxide/calcium carbonate mixture in the lime is so strongly basic that it causes the waste to pass the TCLP test almost every time. Encapsulation and stabilizaton will add between 25 and 50% to the volume of wastes treated.

DISCUSSION OF SPECIFIC TREATMENT TECHNOLOGIES

COAGULATION AND GRAVITY SEPARATION. Coagulation is a transformation of a liquid into a semisolid. In practice, coagulation includes the process of chemical addition and low shear mixing, which causes the agglomeration of the precipitates into a larger floc particle that is then removed by sedimentation or filtration. Physically, coagulation changes the zeta potential (strength of the surface charge) of a particle, reducing it to the point where two particles stick together when they contact each other. There are commercial devices for measuring zeta potential, but often the cost of using these devices often exceeds their value. Good results can also be obtained by running properly set up jar tests, with the aid of an experienced polymer representative or with the aid of a experienced water treatment consultant.

Most of the particles in nature come with a net negative surface charge. This is true for most hydroxides, emulsions, colloids, and clays. Most emulsions are broken first by adding highly positive valence metal salt solutions (such as alum or ferric chloride) by themselves or in combination with or strongly cationic polymers and adjusting the pH of the solution to an optimum range. The optimum range will depend on the emulsion and its condition. For some emulsions, a variant on the lime softening technique can be used to break the emulsion and promote coagulation.

Adjustment of the pH downward with sulfuric or other acids and the addition of aluminum sulfate or ferric salts will break many emulsions. After the emulsion breaks, the pH can be raised to neutral by using lime. This final pH adjustment forms a very large and heavy aluminum hydroxide floc and a calcium sulfate floc, precipitating most of the solids added to the system. The floc also tends to scrub many of the remaining suspended materials from solution. The principal remaining dissolved salt still in solution is calcium sulfate, which is soluble to approximately 2500 mg/L and does not make the water too saline for most discharges.

Table 6.1 and Figure 6.1 present a list of flocculant materials that are typically used in precipitation and emulsion breaking and an idealized picture of two- and three-stage flocculation systems. The three-stage system is preferred because it provides the floc with a chance to grow. The basic elements in a precipitation system include a chemical addition and flash mixing, followed by a low shear mixing to permit particle agglomeration.

Other things that enhance floc growth are the polymers and the sols. Polymers are purchased in solution and crystal forms, and are made from high molecular weight water soluble proteins and carbohydrates. The molecular weights of polymers range from several thousand to almost 1 million. Polymers are available in liquids, powders, and suspensions. The powders tend to require careful handling when they are hydrolyzed, and the mixing conditions must be observed closely, including the hydrolyzation time and the initial mixing to break up lumps of undissolved polymer.

Polymers are manufactured with a variety of strengths of electrical charges, from strongly cationic to nonionic to strongly anionic, with all degrees of charge and molecular weights. Polymers are typically applied at a rate of a few parts per million of polymer, largely because of the expense. Polymers are also used to condition sludges to provide desirable handling characteristics. Those polymer addition rates are in the percent ranges.

Table 6.1 List of inorganic coagulants and flocculants.

Compounds	Formula	Commercial strength	Grades available	Weight, g/m^3	Suitable handling materials	Remarks
Coagulants						
Aluminum sulfate	$Al_2(SO_4)_3 \cdot 18H_2O$	17% Al_2O_3	Lump, powder, granules	Powder: 0.61–0.72 Other: 0.91–1.07	Lead, rubber, silicon, iron	Coagulation and sedimentation systems; before pressure, filters for removal of suspended matter and oil
Sodium aluminate	$Na_2Al_2O_4$	55% Al_2O_3	Crystals	0.80–0.96	Iron, steel, rubber, plastics	Usually added with soda ash to softeners
Ammonium alum	$Al_2(SO_4)_3(NH_4)_2SO_4 \cdot 24H_2O$	11% Al_2O_3	Lump, powder	0.96–1.09	Lead, rubber, silicon, iron, stoneware	Coagulation systems—not widely used
Potash alum	$Al_2(SO_4)_3 \cdot K_2SO_4 \cdot 24H_2O$	11% Al_2O_3	Lump, powder	1.02–1.09		Coagulation systems—not widely used
Copperas	$FeSO_4 \cdot 7H_2O$	55% $FeSO_4$	Crystals, granules	1.01–1.06	Lead, tin, wood	Suitable coagulant only in pH range of 8.5 to 11.0
Chlorinated copperas	$FeSO_4 \cdot 7H_2O + 1/2\ Cl_2$	48% $FeSO_4$				Ferrous sulfate and chlorine are fed separately

(Continued)

Table 6.1 List of inorganic coagulants and flocculants (continued).

Compounds	Formula	Commercial strength	Grades available	Weight, g/m^3	Suitable handling materials	Remarks
Ferric sulfate	$Fe_2(SO_4)_3$	90% $Fe_2(SO)_3$	Powder, granules	0.96–1.12	Lead, rubber, stainless steel, plastics	Coagulation—effective over wide range of pH, 4.0 to 11.0
Ferric chloride hydrate	$FeCl_3 \cdot 6H_2O$	60% $FeCl_3$	Crystals		Rubber, glassware, stoneware	Coagulation—effective over wide range of pH, 4.0 to 11.0
Magnesium oxide	MgO	95% MgO	Powder	0.40–0.56	Iron, steel	Essentially insoluble—fed in slurry form
Coagulant aids[a]						
Bentonite			Powder	0.96	Iron, steel	Essentially insoluble—fed in slurry form
Sodium silicate	$Na_2O(SiO_2)_{3-25}$	40 Bé solution	Solution	1.37	Iron, steel, rubber	
pH adjusters						
Lime, hydrated	$Ca(OH)_2$	93% $Ca(OH)_2$	Powder	0.40–0.80		pH adjustment and softening
Soda ash	Na_2CO_3	99% Na_2CO_3	Powder	0.54–0.83		pH adjustment and softening
Caustic soda	NaOH	98% NaOH	Flake, solid, ground, solution			pH adjustment, softening, oil removal systems
Sulfuric acid	H_2SO_4	100% H_2SO_4	Liquid			pH adjustment

[a]Other compounds, for which no information is available, suitable as coagulant aids are activated silica, clay, activated carbon, causticized starches, and ethyl cellulose.

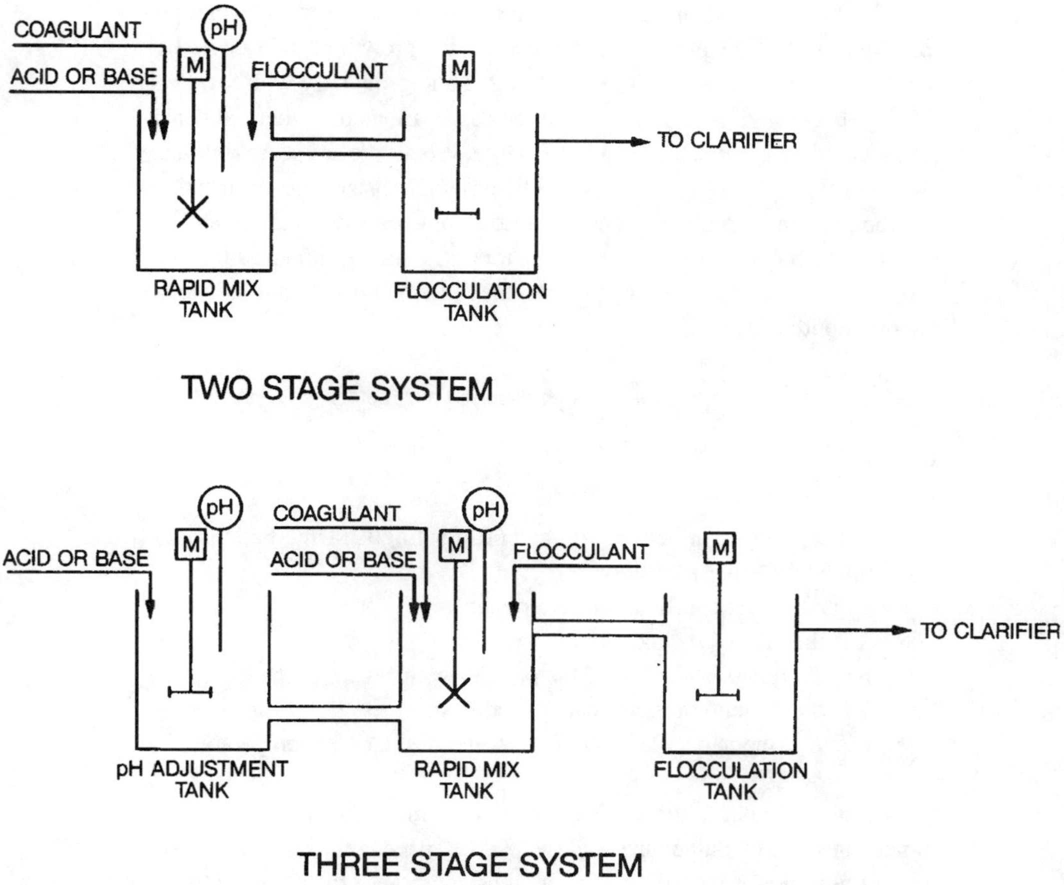

Figure 6.1 Two- and three-stage coagulation/flocculation systems.

Polymer usage should be minimized because of the expense. Some polymers cost more than $5/lb ($2.40/kg).

Sols are an old technology, principally now discontinued but often useful in hazardous waste treatment. Sols are typically manufactured from waterglass (sodium silicate) but can be manufactured from metal salts, acids, and many other compounds. The sol is formed when the strongly acidic or basic material is back titrated with something that will react and form a precipitate. The reaction is taken to a point just short of precipitate formation. In the case of waterglass, the silicate is titrated back to a near neutral pH with hydrochloric acid, hypochlorous acid, or other acids. The sols provide a nucleation center and work much the same as polymers, including having a positive or negative charge. Sols were the predominant technology before the development of polymers but have largely gone into disuse. In aqueous solution they can act as a weak acid or base, and they will help combine with various colloids and anions or cations in solution, aid in their precipitation, and form nucleation centers. The commercial literature on water treatment can provide a good source for manufacture of sols.

After the chemicals have been added to the waste stream to form a precipitate or a floc, the next task is to promote the growth of the floc to the size that it can settle easily. Flocs are fragile and grow only with low shear mixing. If the floc is too small, it will be easily affected by thermal density currents and will not settle. If it is too large, it may break up. A "right-sized" floc is one that settles relatively quickly, has a large size, and compacts readily upon settling. To obtain right-sized floc, some experimentation and jar testing are necessary. The overall measure of the mixing required for floc growth is related to the energy put into the system over a period of time. The energy input is expressed as:

$$G \text{ (feet per second per foot or meters/second per meter)} = [C_d A v^3/(2 \nu V)]^{1/2}$$

Where

- G is optimized at between 30 and 60 (for English Units) and is determined by experimentation;
- C_d is the drag coefficient of the mixer;
- A is the area of the paddle;
- v is the relative velocity of the paddles with regard to the liquid;
- V is the volume of the flocculate, cu ft; and
- ν is the kinematic viscosity of the water at the given temperature.

The energy input to the jar tests can be determined by noting the paddle dimensions, the container size and the paddle speed, and liquid temperature. These observations will allow the rough calculation of the G factor for pilot plant design. The pilot plant coagulation system should be flexible enough to permit a wide range of G factors to be investigated.

GRAVITY SEPARATION OR SEDIMENTATION. The Navier–Stokes laws govern settling of particles through a liquid. The best settling data is obtained from a pilot test or from laboratory settling studies. A large-diameter graduated cylinder should be used for measuring settling rates. With small-diameter cylinders, the edge effects of the wall can produce results that are too optimistic. It is virtually impossible to determine the settlability of a flocculant solution by straight calculation because there are too many unknowns.

There are a number of empirical formulas and rules that govern the selection of a sedimentation system. These rules of thumb like overflow rate (volume per day per unit area of clarifier) and weir flow rate (volume per day per length of weir) have been developed from municipal and industrial water and wastewater treatment plants and typically provide excellent information and practical assistance for developing the design of the sedimentation tank. Almost any good book on environmental engineering will provide sufficient information to rough size a clarifier. Table 6.2 provides some operational parameters for clarifier design.

The final design of a clarifier system should be made in conjunction with a consultant and preferably not a vendor who has a vested interest in making a sale. Clarifiers can be rectangular, square, and circular. The relative advantages to each type depend on the capital investment, the ease of operation of the sludge collection equipment, and the maintenance requirements. Table 6.3 provides a good discussion on the positives and negatives of various types of clarifiers and some of the problems that can be encountered with their selection and operation. Figure 6.2 shows cross sections of various types of clarifiers.

Circular clarifiers are easier to maintain and operate because they do not have a drag-out chain or many moving parts beneath the water; the sludge collection mechanism is hung from a center post in the clarifier. This apparent ease of operation is offset by the difficulty that can occur when or if the sludge or chemical deposits plug the collection well or the elbow under the center of the clarifier. The maintenance requirements and characteristics of the sludge should be considered. High solids concentrations and lime sludges tend to precipitate on the pipes and surfaces inside all clarifiers. Ease of maintenance should be considered in the clarifier design and equipment selection.

Wastes removed by clarification settle very slowly. A very slow settling floc will settle at the rate of approximately 0.3 m/h, while a rapidly settling floc will settle at a rate approaching 3 m/h. At low settling rates, differences in hydraulic flow rate, the movement of sludge collection equipment, and density currents can affect the settling of the particles. An increase in hydraulic flow rate in the clarifier can cause serious problems for slow settling sludge and cause it to wash out of the clarifier. Another thing that can hamper clarifier performance is biologic activity in the clarifier. If the waste has a significant organic component, biological degradation may occur in the

Table 6.2 Typical design information for clarifiers.[a]

	Overflow rate, $m^3/m^2 \cdot d$		Loading, $kg/m^2 \cdot h$		
Type of treatment	**Average**	**Peak**	**Average**	**Peak**	**Depth, m**
Settling following trickling filtration	16–24	40–48	3.0–5.0	8.0	3–4
Settling following air-activated sludge (excluding extended aeration)	16–32	40–48	3.0–6.0	9.0	3.5–5
Settling following extended aeration	8–16	24–32	1.0–5.0	7.0	3.5–5

[a]Metcalf and Eddy, Inc., 1979.

Table 6.3 Clarifier types—advantages and disadvantages.

Clarifier type	Advantages	Disadvantages
Rectangular clarifier—chain drive scraper	Requires less land area than circular clarifiers when multiple units are used Common wall for meeting with other clarifiers or treatment units Easier to cover than circular clarifiers Small units have less short circuiting than comparable circular clarifiers Lower inlet–outlet losses Power consumption of endless chain sludge removal mechanism less than that of other mechanisms	Adversely affected by flow surges Restricted in width by collection equipment Requires multiple weirs to maintain low overflow rates High maintenance costs of the chain and flight sludge removal mechanism Tank must be dewatered for gear or chain repair
Rectangular clarifier—bridge drive scraper	Requires less land area than circular clarifiers when multiple units are required Common-wall construction is possible Low inlet–outlet losses Good hydraulic performance Small units have less short circuiting than comparable circular clarifiers Lower maintenance cost than chain-type designs No width restrictions as with chain-type designs Tank dewatering not required for service or replacement Unit will operate with "light" icing conditions Longer mechanical life with no underwater bearings or gears Standard design permits scraper repair or replacement without tank dewatering	These units will not operate with heavy ice-covered tank
Rectangular clarifier—bridge drive suction	Requires less land area than circular clarifiers when multiple units are required Common-wall construction is possible Low inlet–outlet losses Good hydraulic performance Small units have less short circuiting than comparable circular clarifiers Lower maintenance cost than chain-type designs No width restrictions as with chain-type designs Tank dewatering not required for service or replacement Optimum control of sludge residence time is possible Better pickup of light sludges than scraper-type mechanisms Lower cost tank construction	These units will not operate with ice-covered tank

Center feed peripheral overflow circular clarifier	Low weir overflow velocities Good flocculation of solid particles Ease of design and construction Standardization of sizes Hydraulic performance better than other types of circular clarifiers for large diameter (100 ft) clarifiers	Excessive short circuiting for small diameter (<20 ft) clarifiers Sensitive to density currents Requires more land area than rectangular clarifiers for multiple units
Peripheral feed peripheral overflow circular clarifier	Intermediate size (50 to 100 ft diameter) clarifiers exhibit good hydraulic performance Intermediate size clarifiers produce better solids removals than other types of clarifiers at higher overflow rates Minimizes density currents by introducing flow near the bottom of the clarifier Lower weir overflow velocities	Large diameter (100 ft) clarifiers do not use the entire surface area of the clarifier for settling Requires large peripheral skirt for flow distribution Requires deep tanks so deposited sludge is not resuspended
Peripheral feed center overflow circular clarifier	Intermediate size clarifiers exhibit good hydraulic performance Minimizes density currents by introducing flow near the bottom of the clarifier	High weir overflow velocities Requires large peripheral skirt for flow distribution Requires deep tank so deposited sludge is not resuspended
Solids contact clarifier	Better hydraulic performance than other types of circular or rectangular clarifiers Sludge blanket acts as a filter mat for capturing small particles Not upset by flow surges Good floc formation Reduced detention time for equivalent solids removal	Long sludge detention tine makes it impractical for wastewater treatment applications
Tube settler clarifier	Provides greater surface area, thereby reducing clarifier size Provides laminar flow and quiescent settling conditions within the tubes	Long sludge detention time and sludge deposits on tube surfaces may cause septic conditions Shorter life of plastic modules Short circuiting when influent warmer than basin temperature
Parallel plate clarifier	Low space requirements because of small size Not affected by wind currents Does not require sludge removal mechanisms in smaller units	Sensitive to loading fluctuations Large units use chain- and flight-type of sludge removal mechanisms with their attendant maintenance problems Long sludge detention time and sludge deposits on plates may cause septic conditions Narrow flow distribution channels susceptible to clogging

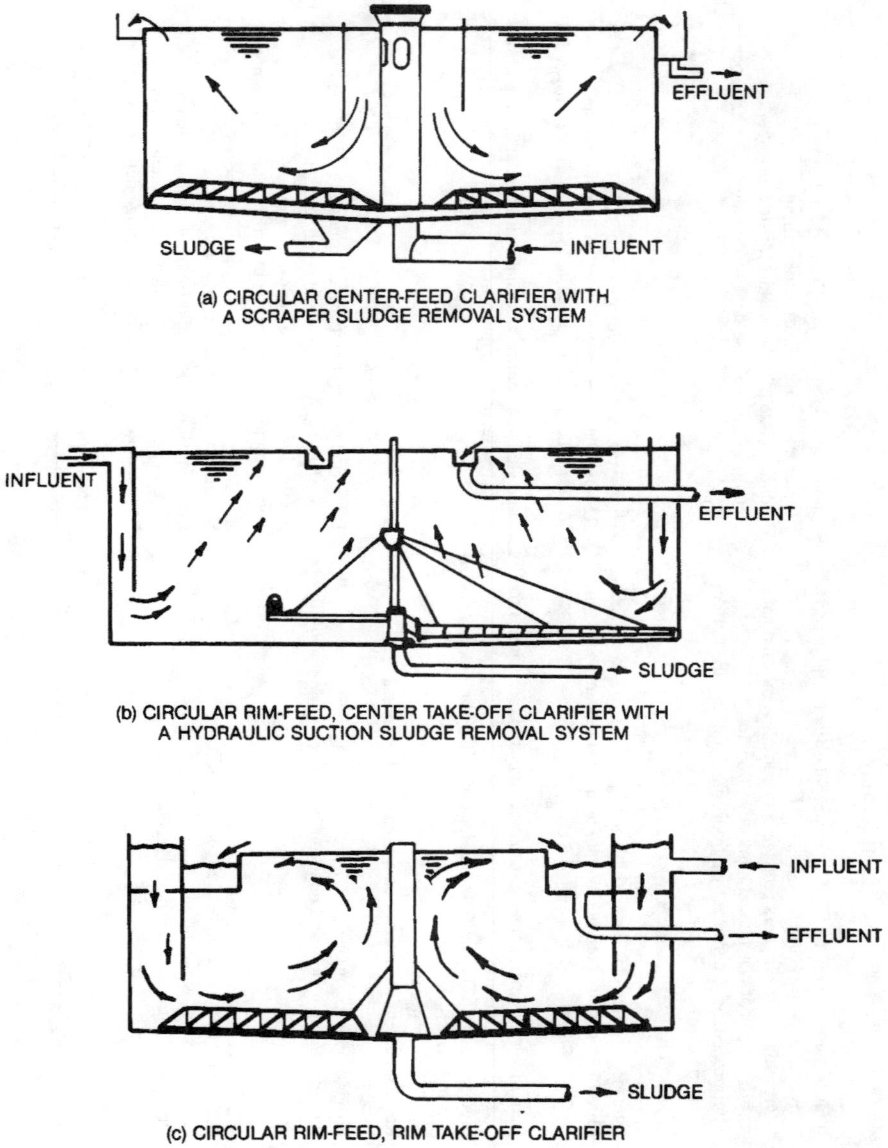

Figure 6.2 Typical clarifier configurations.

clarifier. This can result in the formation of gas bubbles that will disturb the flow and reduce settling. In many wastewater treatment plants, clarifier residence times are deliberately held below 3 hours to minimize development of anaerobic conditions, odors, and sludge gas generation.

ENHANCED GRAVITY SEPARATION. Clarifier performance can be enhanced by installing settling tubes. These tubes, which are installed at a slant, are used to fill the space in the clarifier and shorten the distance a particle or floc has to travel before it settles out on the surface of the tube and flows toward the clarifier collection area. The tubes typically have a square cross section and vary in size from approximately 5 cm to 15 cm (2 to 6 in.) or are a series of parallel plates with the same approximate spacing.

An alternative to a clarifier that uses a modification of the tube settling principle is a lamella. This device, shown in Figures 6.3 and 6.4, uses a set of parallel plates to minimize the vertical distance of particle travel before it is removed from the liquid. Lamellas are quite successful in many industrial

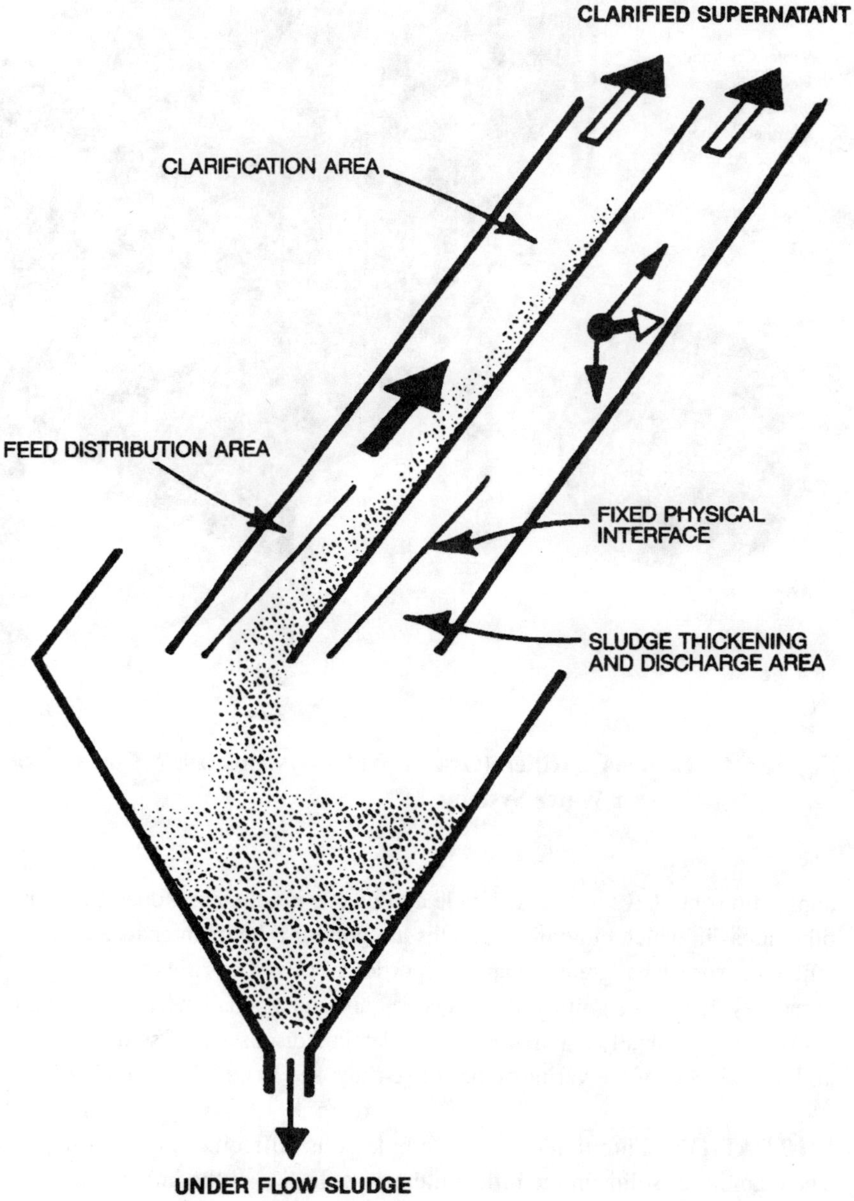

Figure 6.3 Inclined surface clarifier.

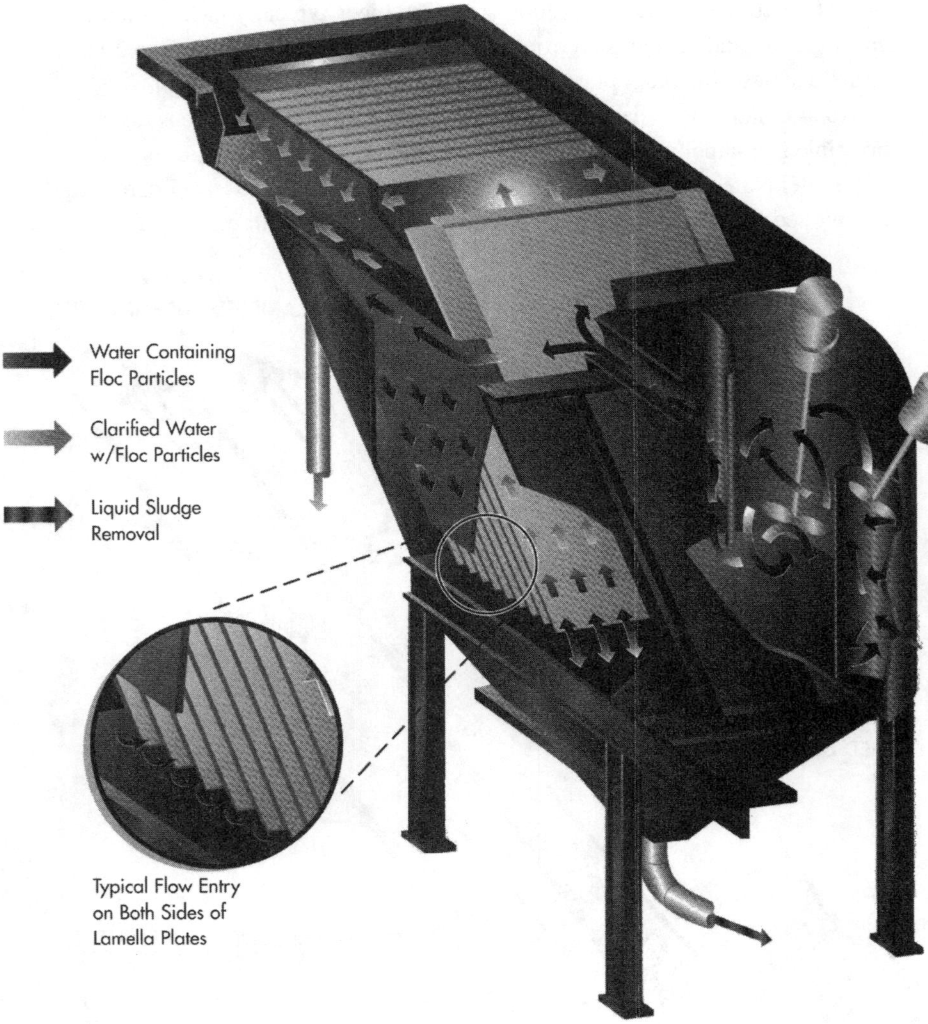

Figure 6.4 Lamella clarifier. (Graver Water Systems, 1999. Courtesy of Graver Water Systems.)

applications but have not gained wide acceptance in municipal operations and other areas in which biological growths are present. The maintenance of a lamella is somewhat greater than a proportionally sized clarifier, but the lamella typically is significantly less costly and much smaller. The throughput is often proportionally smaller. For more detailed discussion of settling theory and the subjects of the various types of settling, see Benefield et al. (1982).

FILTRATION. Filtration can be as simple or as difficult as one chooses. The quantity of solids in the influent, the particle size of the influent solids, and the quantity and size of particles in the filter effluent will partially dictate the selection of the filter system. The types of filtration systems currently fall into three broad classifications: depth filters, skin filters, and combinations. Depth filtration is characterized by a sand or granular media filter system. A membrane filter is an excellent example of a skin filter. A combination filter is one that uses both depth and skin filtration by application of a precoat system to remove the particles from the liquid. This latter type of filter is also characterized by a swimming pool filter or filter press that uses a precoat system.

Filtration is further classified into microfiltration and macrofiltration. Macrofiltration systems are the most common and most widely used, although microfiltration systems are being used more frequently in applications in which there is a high demand for molecular separation or for extremely high purity in the effluent and in desalination applications. In most instances, macrofiltration is used to remove more than 90% of the suspended and particulate matter in a system, and the microfiltration system removes the remainder. Most microfiltration systems require protection from a macrofiltration system to keep them from fouling and to reduce overall operation costs. Depending on the microfiltration application, the prefilter may be another microfiltration system that has larger pore openings.

The most typical types of macrofiltration systems currently in use are pressure or gravity sand and mixed media filters. The differences between pressure sand filtration and gravity sand filtration are the solids loading rate on the filter and the pressure drop allowed across the media before the unit is backwashed. The difference between sand filtration and mixed media filters is the size gradient of the media.

Sand filters have the finest particles at the top of the filter and the largest coarser particles at the bottom. Sand filter bed sizing is controlled by the backwash of the media, and because the larger particles settle more quickly, they are on the bottom of the filter. This also sets the smallest diameter particles at the top of the filter.

Mixed media filters use a variety of different density materials in the construction of the filter. As a result, the filter media has a reverse gradient; the larger sized (less dense) materials are on the top of the filter bed, and the smaller more dense particles are on the bottom. Because of this reverse gradient, the filter is able to use a higher hydraulic loading rate, and the solids penetrate deeper into the filter bed.

The filters indicate their saturation or failure in different ways. In the mixed media filter, the solids will, at the backwash point, fully penetrate the filter bed and break through into the filtrate. In the sand filter, the initial filtrate is often cloudy and sent into the backwash cycle waters until the filter has achieved its satisfactory level of performance. After the sand filter has achieved an initial bed coating, the solids removal capabilities improve with time, and the backwash point is described by unacceptable head losses rather than solids breakthrough.

Swimming pool filtration systems and filter presses are examples of combination filtration systems that use a cloth filter media supplemented by a "filter aide" (typically diatomaceous earth). The diatomaceous earth is used as a filter precoat material to form the filter cake on the filter cloth or element and is fed to the filter during the run to maintain the porosity of the filter cake and the flow through the filter. The pressure drop across these systems is often on the order of 104 kPa (15 psi) or more depending on the filtration element and the type and kind of precoat system being used. For some filter presses, the pressure drop may be 700 kPa (100 psi), or higher, but these are extreme cases. Figure 6.5 and Table 6.4 illustrate and tabulate data on granular media filtration systems.

FILTER BACKWASH VOLUMES. A filter is backwashed to remove the material it has removed from the waste stream. Most granular media filters are backwashed by reversing the hydraulic flow through the system. This flow reversal lifts the granules in the media and expands the filter bed. The bed is then mixed under high shear conditions that are often augmented by air or mechanical stirring. Mechanical abrasion removes the dirt from the sand. When no further dirt appears in the backwash cycle, the hydraulic flow

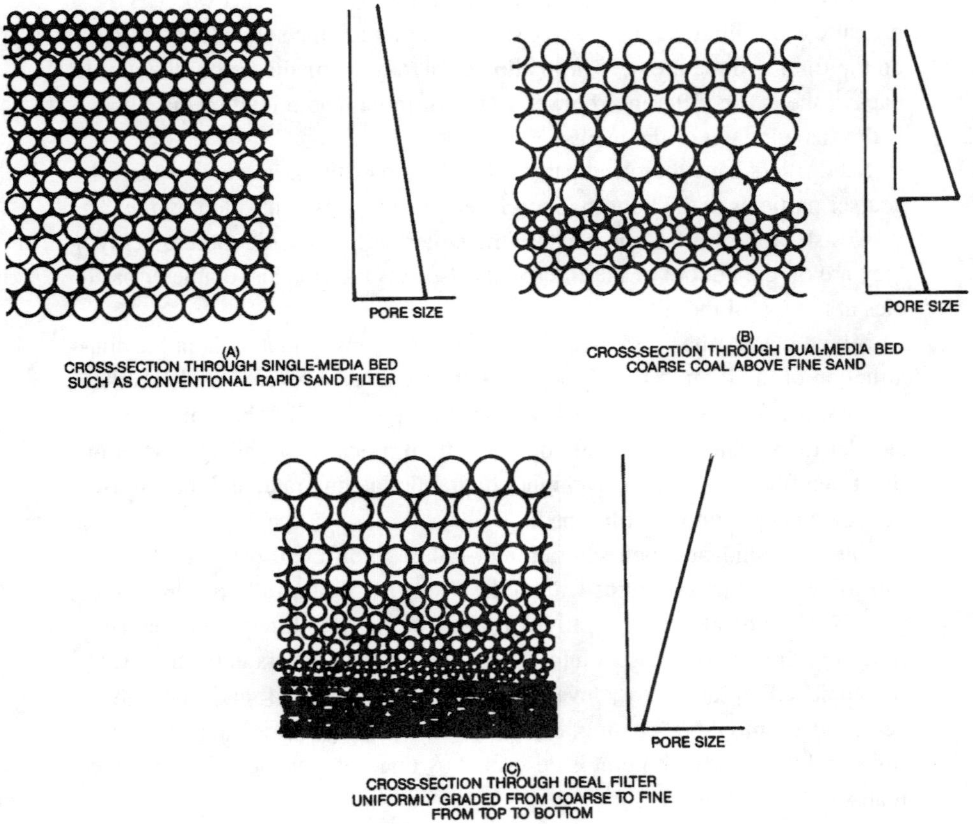

Figure 6.5 Various media designs.

Table 6.4 Typical design data for dual-media and multimedia filters.

Characteristic	Value	
	Range	Typical
Dual-media		
Anthracite		
Depth, mm	300–600	450
Effective size, mm	0.8–2.0	1.2
Uniformity coefficient	1.3–1.8	1.6
Sand		
Depth, mm	150–300	300
Effective size, mm	0.4–0.8	0.55
Uniformity coefficient	1.2–1.6	1.5
Filtration rate, L/m^2·min	80–400	200
Multimedia		
Anthracite (top layer of quadmedia filter)		
Depth, mm	200–400	200
Effective size, mm	1.3–2.0	1.6
Uniformity coefficient	1.5–1.8	1.6
Anthracite (second layer of quadmedia filter)		
Depth, mm	100–400	200
Effective size, mm	1.0–1.6	1.2
Uniformity coefficient	1.5–1.8	1.6
Anthracite (top layer of trimedia filter)		
Depth, mm	200–500	400
Effective size, mm	1.0–2.0	1.4
Uniformity coefficient	1.4–1.8	1.6
Sand		
Depth, mm	200–400	250
Effective size, mm	0.4–0.8	0.5
Uniformity coefficient	1.3–1.8	1.6
Garnet or ilmenite		
Depth, mm	50–150	100
Effective size, mm	0.2–0.6	0.3
Uniformity coefficient	1.5–1.8	1.6
Filtration rate, L/m^2·min	80–400	200

gradually is reduced, allowing the bed to sort itself out by particle density and size. The backwash restores the hydraulic performance of the filter.

Cloth or membrane type filters are washed by hydraulic scrubbing of the filter face to remove the accumulated solids. In some cases, such as that of the filter press, the solids are scraped off the surface with a squeegee, and that cleaning is followed with a high pressure/low volume spray wash to remove the remaining dirt. Other types of diatomaceous earth filtration systems are backwashed by stopping the flow and allowing the filter cake to slough off the filter. This is followed by a hydraulic or mechanical brushing of the filter face to remove the accumulated dirt and filter aid.

Mixed media or sand filtration systems have a backwash system that is between 2 and 8% of the throughput volume, and the upper value is more representative than the lower. Even 2% of the throughput volume may be a substantial amount when it contains a hazardous waste that needs treatment and disposal.

A filtration system using diatomaceous earth will have a backwash volume less than 10% of the backwash volume of granular media filtration systems. This is especially true of a filter press that uses high pressure to help dewater the filter cake. Filter press solids cakes often run 15% solids and higher. Vacuum and pressure diatomaceous earth filtration systems can often achieve good solids separations, but the potential for the same type of problems are possible.

MICROPOROUS FILTRATION SYSTEMS. There are three types of microporous filtration systems in use: depth, screen, and surface. Depth filters are matted fibers or materials compressed to form a matrix that retains particles by random adsorption or entrapment. Screen filters are uniform structures which, like a sieve, retain all particles larger than the precisely controlled pore size on their surface. Surface filters are made from multiple layers of media. When fluid passes through the filter, particles larger than the spaces within the filter matrix are retained, accumulating primarily on the surface of the filter.

The distinction between the various types of microporous filters is important because the three serve very different functions. Depth type microfilters are typically used as prefilters because they are an economical way to remove 98% of suspended solids and protect finer microfilter elements downstream from fouling or clogging.

Surface microfilters remove 99.99+% of suspended solids and may be used as either prefilters or clarifying filters. Microporous membrane (screen) filters are placed at the last possible point in a system to remove any remaining traces of suspended solids and other colloidal sized particles and microorganisms. For example, some membrane filters are used to sterilize intravenous solutions, serums, and antibiotics because the filter openings are so small that bacteria and viruses cannot pass through them.

Table 6.5 shows the relative advantages and disadvantages of membrane and microporous filters. Other properties of filtration systems are shown in Table 6.6. The relative performance of the various types of filtration systems is compared to gravity separation and other devices in Figure 6.6.

Membrane filtration systems use an ultrafine pore spacing in an acrylate or other organically prepared medium. The pore sizes are so small that in some cases they can operate as a molecular sieve and will remove even dissolved solids from the liquid being filtered. (These applications are reverse osmosis and ultrafiltration systems.) The various types of membrane filtra-

Table 6.5 Microporous and membrane filtration systems.

Advantages	Disadvantages
Removal of all particles and microorganisms greater than the filter pore size	Will not remove dissolved inorganic compounds and salts (dissolved inorganics are removed by reverse osmosis filtration systems)
Requires minimal maintenance	Potentially high costs for cartridge replacement
May permit reclamation and recycling of the liquid	Most cannot be regenerated, but reverse osmosis filter systems are continually regenerated by diverting a portion of the flow through the filter to a reject stream

tion systems include ultrafiltration (UF), reverse osmosis (RO), electrodialysis (ED), microfiltration (MF), and electrodialysis reversal (EDR). The electrodialysis treatment is not widely used in hazardous waste treatment because of the equipment capital cost and relatively low throughput volume.

All of these are distinguished from the other types of filtration systems mentioned above by two distinct features. They have an excellent water quality for their effluent, and they have a reject stream that composes a substantial fraction of the throughput volume. Reverse osmosis, ED, and EDR are used as process treatment steps in electroplating baths to maintain the purity of the bath and to minimize the need for bath dumping.

Table 6.6 Summary of key cross-flow membrane separation processes.[a]

Process	MF	UF	RO	ED/EDR
Pore size, Å	1 000	10	1	
	100 000	200	10	
Pressure, psid	10	10	100	
		100	1 000	
Material separated				
Dissolved salts			Y	Y
TOC		Y	Y	
Colloids (0.2)	Y	Y	Y	
Dissolved silica			Y	
CO_2 gas				
Bacteria	Y	Y	Y	
Viruses		Y	Y	
Pyrogens		Y	Y	
Oil emulsions	Y	Y		

[a]MF = microfiltration, UF = ultrafiltration, RO = reverse osmosis, ED = electrodialysis, and EDR = electrodialysis reversal.

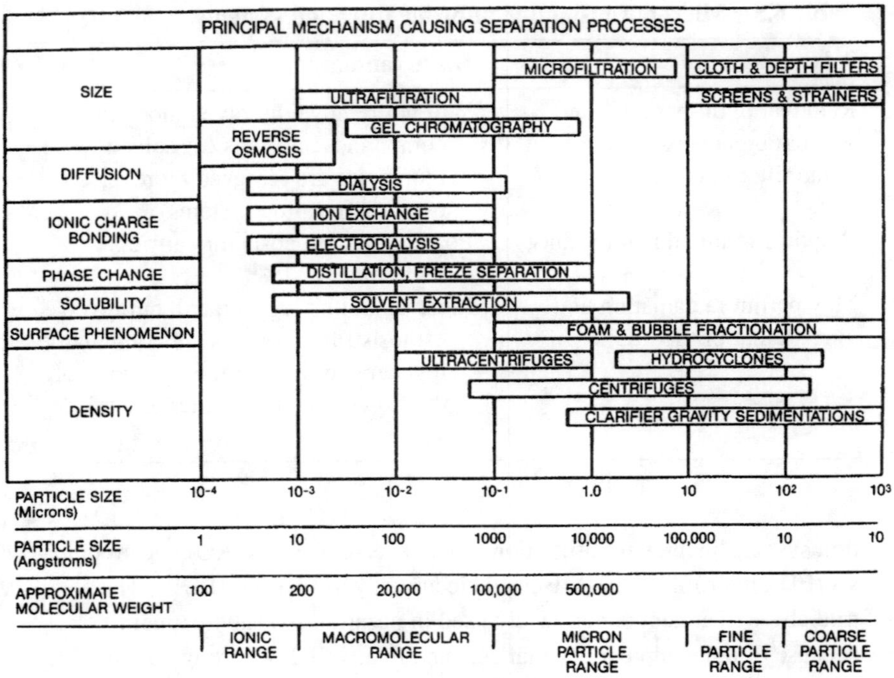

Figure 6.6 Separation processes and mechanisms.

Several water foundations and manufacturers have excellent sites on the Word Wide Web that can provide information about the equipment and the processes.

Most of the membranes used for RO, ED, UF, and similar processes are made up from poly-acetate, poly-acrylate, and other polyamid materials that are cast into thin membranes or fine hollow fibers. The membranes are constructed into a cartridge, which is reinforced with a stronger outer backing. The membranes for RO operate at pressures of between 1380 and 2760 kPa (200 and 400 psi) depending on the model. The effluent quality produced is extremely high. Typical removals are up to 98% of dissolved minerals and 99.99+% of colloidal and suspended matter.

Figure 6.7 shows a highly oversimplified and greatly enlarged cross section of the construction of a membrane filter. For these systems it is important to precondition the water to eliminate the membrane fouling and membrane attack. The acetates are particularly susceptible to attack from slimes and bacteria that are often in the water supply. Membrane lift is maintained by preconditioning the water with either prefiltration or carbon adsorption. Approximately 35% of the material fed can be recovered as permeate. (Ultra clean water is designed for reuse.) The balance is reject water used to keep the solids from plugging the micro-fine tubes. On large commercial systems, up to 80% of the feed solution can be recovered as permeate.

One recent report of electrodialysis technology for anodizing bath cleaning prepared by the U.S. Navy indicated that the installation of an electro-

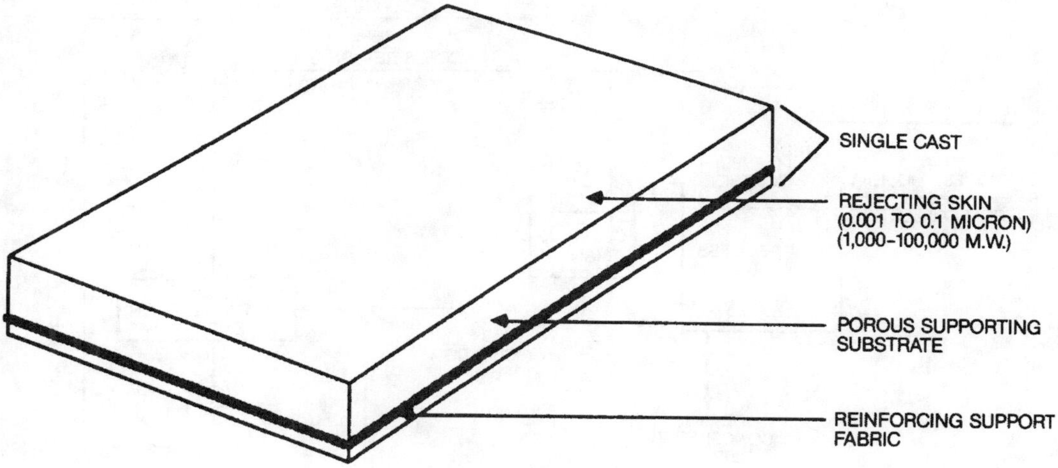

Figure 6.7 Spiral wound UF asymmetrical membrane in profile (micron × 1.000 = μm).

dialysis unit on a 18 925 L (5000 gal) ship that was changed out annually would reduce the amount of bath dumping and total volume of hazardous waste generated by more than one-half the first year and almost one-half again in subsequent years. The cost of this type system was approximately $30 000 for a relatively small system.

Electrodialysis consists of both anionic and cationic membranes assembled in an alternative fashion and separated by a proprietary gasket spacer design that directs the flow of solution into the proper chambers and promotes flow distribution. The driving force in an electrodialysis stack is the electricity applied as a direct current flowing from an anode located at one end of the stack through the stack to the cathode located at the other end of the stack. The ion exchange membrane, the current density, and the electrolyte to be processed affect the performance of the electordialytic system.

Membranes can become fouled or scaled from salts within the liquid. Ionics claims to have developed the EDR process, which prevents fouling by changing both the direction of the flow through the membranes and the polarity of the electrical charge several times per hour. Information provided by some of the manufacturers indicates that the waste reductions can be up to 90% and suggests that the ideal applications include desalting and purification of organic acids and solvents, including glycol and amines; concentration of dilute acid; base and salt streams; drinking water nitrate reduction; potable water production from brackish water; and wastewater minimization. Note that the applications are primarily for concentrating and not necessarily for final treatment before disposal. Additional processing may be required to achieve volume reductions for final waste disposal.

Depending on the type of media and the particular process selected, UF/RO systems can be used with caution to treat oily solutions of wastewater to concentrate the oils and the permeate can be discharged with mini-

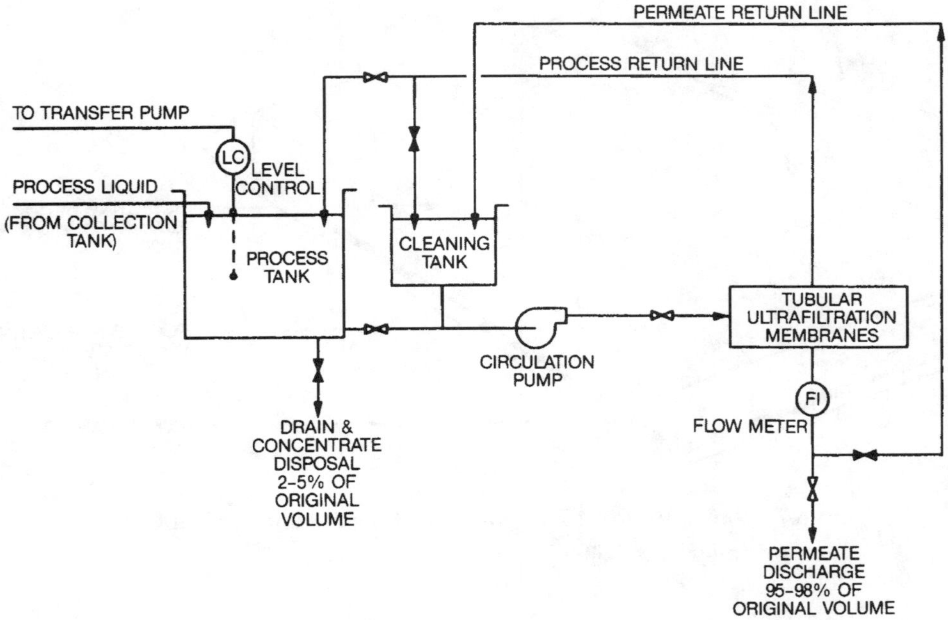

Figure 6.8 Ultrafiltration process diagram.

mal additional treatment (Davies et al., 1985). The use of ED/EDR may not be applicable to this type waste because of the compatibility of the wastes to the membrane and because the "oils" have a tendency to coat or adhere to the surfaces of the vessels and pipes they contact. Figures 6.8 through 6.11

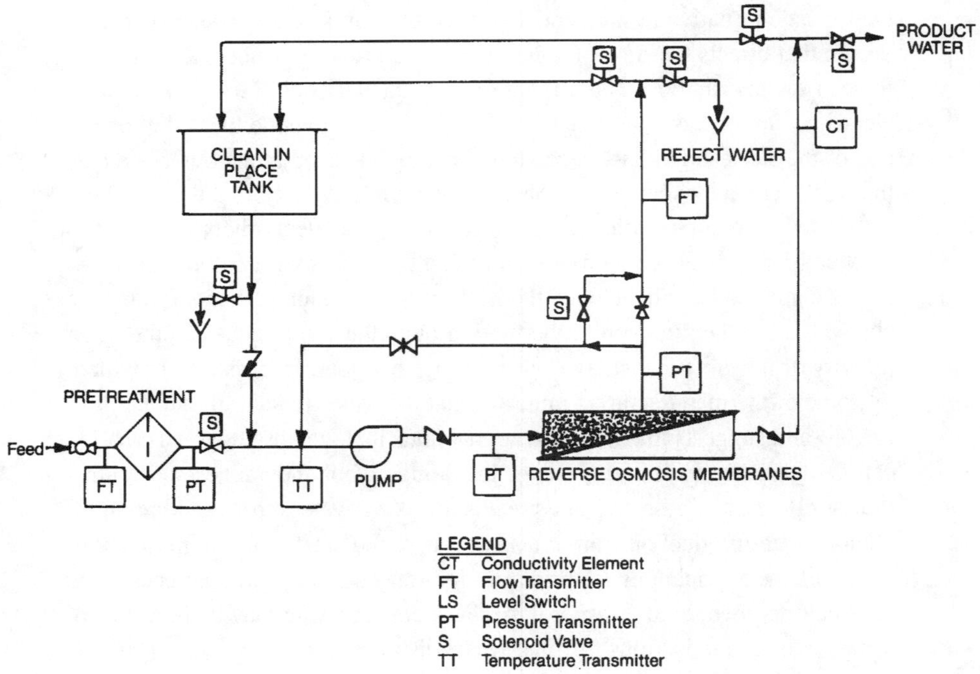

Figure 6.9 Single-stage reverse osmosis system.

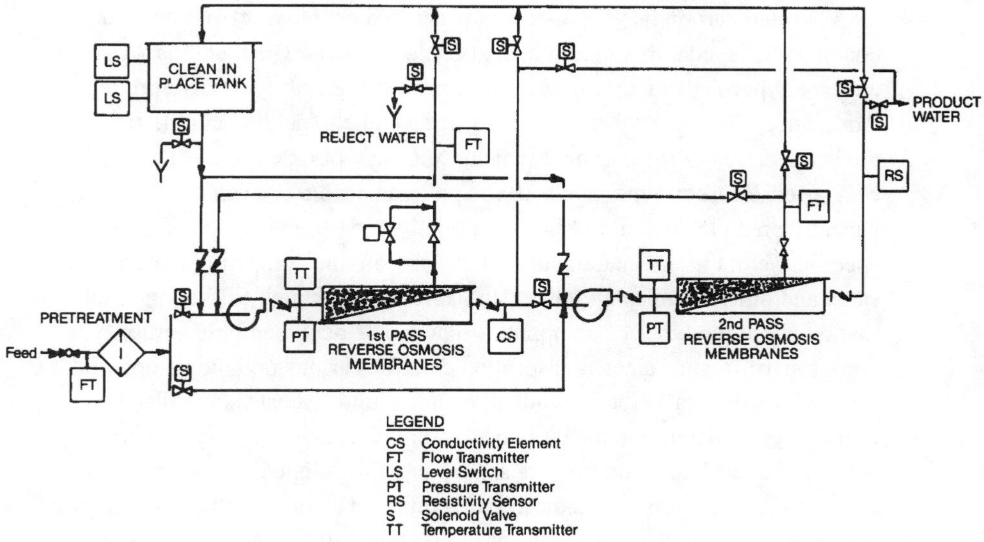

Figure 6.10 Two-stage reverse osmosis system.

provide typical flow diagrams for various types of microporous filtration systems.

ACTIVATED CARBON ADSORPTION. Activated carbon is used to treat hazardous wastes by adsorbing organics and some salts from relatively dilute contaminated liquids, and to transfer those contaminants on to the carbon. The carbon has a definite capacity, and when it is exhausted, the carbon can be separated from the liquid, regenerated, and restored to a substantial fraction (sometimes as high as 90%) of its initial adsorption capacity. The regeneration process typically includes thermal treatment, solvent regeneration, or steam regeneration. Thermal treatment is typically performed off-site. The purpose of the regeneration process is the removal of the contaminant, its transfer to a smaller volume or its destruction, and recovery of the carbon adsorption capacity.

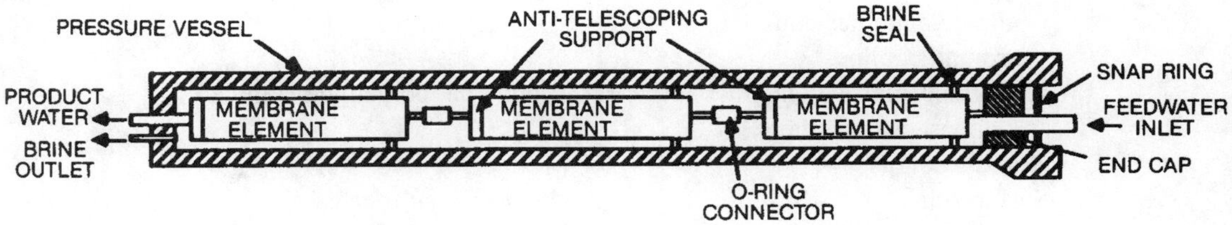

Figure 6.11 Cross section of spiral wound reverse osmosis pressure vessel (ppm × 1.000 = mg/L).

Activated carbon itself is specially processed charcoal often made from coconut shells, coal, petroleum coke, or select woods. The media is extremely porous and has an extremely large surface area for adsorption. Adsorption is an internal rather than a surface phenomenon, and the larger the surface area of the carbon, the better the adsorption capacity. The ability of a specific carbon type to adsorb the pollutants from a given waste is only partially predictable and depends on a number of variables, including the specific factors in the manufacture of the carbon, the competing materials or salts, and other chemicals in the water. Tables 6.7 through 6.13 contain information about the types of compounds that can be adsorbed onto carbon, carbon properties, and selected adsorption data for various organic compounds. Figure 6.12 illustrates the Freundlich equation and indicates how the equation is used to determine the carbon adsorption capacity for a particular waste material. The isotherm data are typically developed on a specific carbon for a specific waste stream. Manufacturers' data are helpful in predicting the carbon life and the economics of the system, but in the end there is no substitute for the actual pilot test on a particular carbon.

Table 6.7 Classes of organic compounds adsorbed on carbon.

Organic chemical class	Examples of chemical class
Aromatic hydrocarbons	Benzene, toluene, xylene
Polynuclear aromatics	Naphthalene, anthracenes, biphenyls
Chlorinated aromatics	Chlorobenzene, polychlorinated biphenyls, aldrin, endrin, toxaphene, DDT
Phenolics	Phenol, cresol, resorcinol, polyphenyls
Chlorinated phenolics	Trichlorophenol, pentachlorophenol
High-molecular-weight aliphatic and branch chain hydrocarbons[a]	Gasoline, kerosene
Chlorinated aliphatic hydrocarbons	1,1,1-Trichloroethane, trichloroethylene, carbon stetrachloride, perchloroethylene
High-molecular-weight aliphatic acids and aromatic acids[a]	Tar acids, benzoic acid
High-molecular-weight aliphatic amines and aromatic amines[a]	Aniline, toluene diamine
High-molecular-weight ketones, esters, ethers, and alcohols[a]	Hydroquinone, polyethylene glycol
Surfactants	Alkyl benzene sulfonates
Soluble organic dyes	Methylene blue, indigo carmine

[a]High-molecular-weight compounds include those in the range of 4 to 20 carbon atoms.

Table 6.8 Amenability to adsorption of selected hydrocarbons.

Compound	Molecular weight	Adsorbability, g compound/ g carbon	Remarks
Alcohols			
Methanol	32.0	0.007	Compounds highly polar and highly soluble, resulting in low amenability of compounds to adsorption; polarity decreases as molecular weight increases, resulting in increased adsorbability and with increasing molecular weight
Ethanol	46.1	0.020	
Propanol	60.1	0.038	
Butanol	74.1	0.107	
n-Amyl alcohol	88.2	0.155	
n-Hexanol	102.2	0.191	
Aldehydes			
Formaldehyde	30.0	0.018	Aldehydes, like alcohols, are highly polar compounds; polarity decreases as molecular weight increases, resulting in increased amenability to adsorption
Acetaldehyde	44.1	0.022	
Propionaldehyde	58.1	0.057	
Butyraldehyde	72.1	0.106	
Amines			
Di-*N*-propylamine	101.2	0.174	Amenability to adsorption again limited by polarity and solubility
Butylamine	73.1	0.103	
Di-*N*-butylamine	129.3	0.174	
Allylamine	57.1	0.063	
Ethylenediamine	60.1	0.021	
Diethylenetriamine	103.2	0.062	
Aromatics			
Benzene	78.1	0.080	Low polarity and low solubility of compounds makes them relatively easy to remove by adsorption onto carbon; removal is also enhanced by bonding between compound and carbon surface
Toluene	92.1	0.050	
Ethyl benzene	106.2	0.019	
Phenol	94	0.161	
Hydroquinone	110.1	0.167	
Aniline	93.1	0.150	
Glycols			
Ethylene glycol	62.1	0.013 6	These compounds have multiple sites for hydrogen bonding, which gives them hydrophilic properties and makes compounds difficult to remove by adsorption
Diethylene glycol	106.1	0.053	
Triethylene glycol	150.2	0.105	
Tetraethylene glycol	194.2	0.116	
Propylene glycol	76.1	0.024	

The factors that control the selection of carbon adsorption include particle size, surface area of the carbon, pore volume of the carbon, abrasion number, and iodine and molasses numbers. Most of these terms are obvious and self-explanatory. The molasses and iodine numbers are relative index numbers, derived from standard tests, that are used to measure the affinity of a specific activated carbon to high molecular weight and low molecular weight compounds, respectively.

Although most of the carbon used is granular media, powdered activated carbon is often used for adsorption when the wastewater to be treated has a toxic or biological inhibitory component and biological wastewater treatment

Table 6.9 Summary of carbon adsorption capacities.

Compound	Adsorption[a] capacity, mg/g	Compound	Adsorption[a] capacity, mg/g
bis(2-Ethylhexyl)phthalate	11 300	Phenanthrene	215
Butylbenzyl phthalate	1 520	Dimethylphenylcarbinol[b]	210
Heptachlor	1 220	4-Aminobiphenyl	200
Heptachlor epoxide	1 038	β-Naphthol[b]	200
Endosulfan sulfate	686	α-Endosulfan	194
		Acenaphthene	190
Endrin	666	4,4′ Methylene-bis-	
Fluoranthene	664	(2-chloroaniline)	190
Aldrin	651	Benzo(k)fluoranthene	181
PCB-1232	630	Acridine orange	180
β-Endosulfan	615	α-Naphthol	180
Dieldrin	606	4,6-Dinitro-*o*-cresol	169
Hexachlorobenzene	450	α-Naphthylamine	160
Anthracene	376	2,4-Dichlorophenol	157
4-Nitrobiphenyl	370	1,2,4-Trichlorobenzene	157
		2,4,6-Trichlorophenol	155
Fluorene	330		
DDT	322	β-Naphthylamine	150
2-Acetylaminofluorene	318	Pentachlorophenol	150
α-BHC	303	2,4-Dinitrotoluene	146
Anethole[b]	300	2,6-Dinitrotoluene	145
		4-Bromophenyl phenyl ether	144
3,3-Dichlorobenzidine	300		
2-Chloronaphthalene	280	*p*-Nitroaniline[b]	140
Phenylmercuric acetate	270	1,1-Diphenylhydrazine	135
Hexachlorobutadiene	258	Naphthalene	132
α-BHC (lindane)	256	1-Chloro-2-nitrobenzene	130
		1,2-Dichlorobenzene	129
p-Nonylphenol	250		
4-Dimethylaminoazobenzene	249	*p*-Chlorometacresol	124
Chlordane	245	1,4-Dichlorobenzene	121
PCB-1221	242	Benzothiazole[b]	120
DDE	232	Diphenylamine	120
		Guanine[b]	120
Acridine yellow[b]			
Benzidine dihydrochloride	220	Styrene	120
β-BHC	220	1,3-Dichlorobenzene	118
N-Butylphthalate	220	Acenaphthylene	115
N-Nitrosodiphenylamine	220	4-Chlorophenyl phenyl ether	111
		Diethyl phthalate	110
2-Nitrophenol	99	Bromoform	20
Dimethyl phthalate	97	Carbon tetrachloride	11
Hexachloroethane	97	bis(2-Chloroethoxy)	
Chlorobenzene	91	methane	11
p-Xylene	85	Uracil[b]	11
		Benzo(ghi)perylene	11

(Continued)

Table 6.9 Summary of carbon adsorption capacities (continued).

Compound	Adsorption[a] capacity, mg/g	Compound	Adsorption[a] capacity, mg/g
2,4-Dimethylphenol	78	1,1,2,2-Tetrachloroethane	11
4-Nitrophenol	76	1,2-Dichloropropene	8.2
Acetophenone	74	Dichlorobromomethane	7.9
1,2,3,4-Tetrahydro-naphthaene	74	Cyclohexanone[b]	6.2
		1,2-Dichloropropane	5.9
Adenine[b]	71	1,1,2-Trichloroethane	5.8
Dibenzo(a,h)anthracene	69	Trichlorofluoromethane	5.6
Nitrobenzene	68	5-Fluorouracil[b]	5.5
3,4-Benzofluoranthene	57	1,1-Dichloroethylene	4.9
1,2-Dibromo-3-chloro-		Dibromochloromethane	4.8
propane	53	2-Chloroethyl vinyl ether	3.9
Ethylbenzene	53	1,2-Dichloroethane	3.6
2-Chlorophenol	51	1,2-*trans*-Dichloroethene	3.1
Tetrachloroethene	51	Chloroform	2.6
o-Anisidine[b]	50	1,1,1-Trichloroethane	2.5
5 Bromouracil	44		
Benzo(a)pyrene	34		
2,4-Dinitrophenol	33	1,1-Dichloroethane	1.8
Isophorone	32	Acrylonitrile	1.4
Trichloroethene	28	Methylene chloride	1.3
Thymine[b]	27	Acrolein	1.2
		Cytosine[b]	1.1
Toluene	26		
5-Chlorouracil[b]	25	Benzene	1.0
N-nitrosodi-*n*-propylamine	24	Ethylenediaminetetraacetic acid	0.86
bis(2-Chloroisopropyl)		Benzoic acid	0.76
ether	24	Chloroethane	0.59
Phenol	21	*N*-dimethylnitrosamine	6.8×10^{-5}
Not adsorbed			
Acetone cyanohydrin		Adipic acid	
Butylamine		Choline chloride	
Cyclohexylamine		Diethylene glycol	
Ethanol		Hexamethylenediamine	
Hydroquinone		Morpholine	
Triethanolamine			

[a]Adsorption capacities are calculated for an equilibrium concentration of 1.0 mg/L at neutral pH.
[b]Compounds prepared in "mineralized" distilled water containing the following composition:

Ion	Conc., mg/L	Ion	Conc., mg/L
Na^+	92	PO_4	10
K^+	12.6	SO_4	100
Ca^{++}	100	Cl^-	177
Mg^{++}	25.3	Alkalinity	200

Table 6.10 Selected properties of activated carbon.

Property	Importance
Particle size distribution	Rate of adsorption increases as particle size decreases; head loss through packed column increases as particle size decreases; size distribution expressed as percentage passing various size sieves
Surface area	A measure of the area available for adsorption; the larger the surface area, the greater the adsorptive capacity; measured by determining the amount of nitrogen adsorbed by the carbon and reported as m^2/g
Pore volume	Measure of total macropore and micropore volume within the carbon particles; measured in cc/g
Iodine number	Refers to milligrams of iodine adsorbed during standard test; measures the volume present in pores from 10 to 28 Å in diameter; carbons with a high percentage of pore sizes in this range would be suitable for adsorbing low-molecular-weight substances
Molasses number	Refers to milligrams of molasses adsorbed during standard test and measures the volume in pores greater than 28 Å in diameter; carbons with a high percentage of this size pores would be suitable for adsorbing high-molecular-weight substances
Abrasion number	Measures ability of carbon to withstand handling and slurry transfer; this property is of limited value because measuring techniques are not reproducible
Bulk density	Useful in determining the volume occupied by a given weight of carbon

Table 6.11 Typical properties of two powdered carbons.

	Powdered carbon	
Property	**HydroDarco H**	**HydroDarco C**
Surface area, m^2/g	475 (min)	550 (min)
pH	10.5	10.5
Tamped density, g/mL	0.70	0.70
Molasses RF	40	95
Density, kg/m^3 (lb/cu ft)	704.9 (44)	704.9 (44)
Sieve analysis through 325 mesh, %	70	70
Water solubles, %	3.5	5.5

Table 6.12 Advantages and disadvantages of carbon adsorption systems.

Advantages	Disadvantages
A widely used technology with well-established performance levels	Somewhat higher complexity of operation than other removal technologies
Can be used for low concentration inlet streams	Larger space requirements than other technologies because of the need for contact time between carbon and material to be removed
Can achieve high removal efficiencies	Greatest absorption rates per weight of carbon are with vapor phase systems; water absorption lowers the carbon's adsorption capacity for all organic materials
Commercial regeneration is available as a contract system	Carbon replacement and/or regeneration is expensive; many carbons are currently selling at more than \$3.50 per kg (\$1.50 per pound); regeneration is only slightly less expensive than replacement, and the capacity of regenerated carbon is less than that of virgin carbon
Can efficiently handle fluctuations in wastewater flow rates and organic materials concentrations	Caution may be necessary when handling the spent carbon and the regenerate materials; depending on the organics being handled, the organics could be hazardous and/or explosive
Will produce a consistent effluent right up to breakthrough and exhaustion	Breakthrough can occur suddenly and will require close monitoring of effluent

Table 6.13 Granular activated carbon column design criteria.

Parameter	Value[a]
Vessel diameter	0.6–3.7 m (2–12 ft)
Area loading	1.3–6.8 L/m^2·s (2–10 gpm/ft^2)
Organic loading	0.04–0.14 kg (0.1–0.3 lb BOD or COD/lb-carbon)
Backwash	8.1–13.6 L/m^2·s (12–20 gpm/ft^2)
Air scour	8.1–13.6 1/m·s (3–5 ft^3/min/ft)
Bed depth	1.5–9.1 m (5–30 ft)
Contact time	600–3000 s (10–50 min)
Land area	Minimal
Spent carbon	1.4–4.5 kg/kg (3–10 lb/lb COD removed)

[a]BOD = biochemical oxygen demand and COD = chemical oxygen demand.

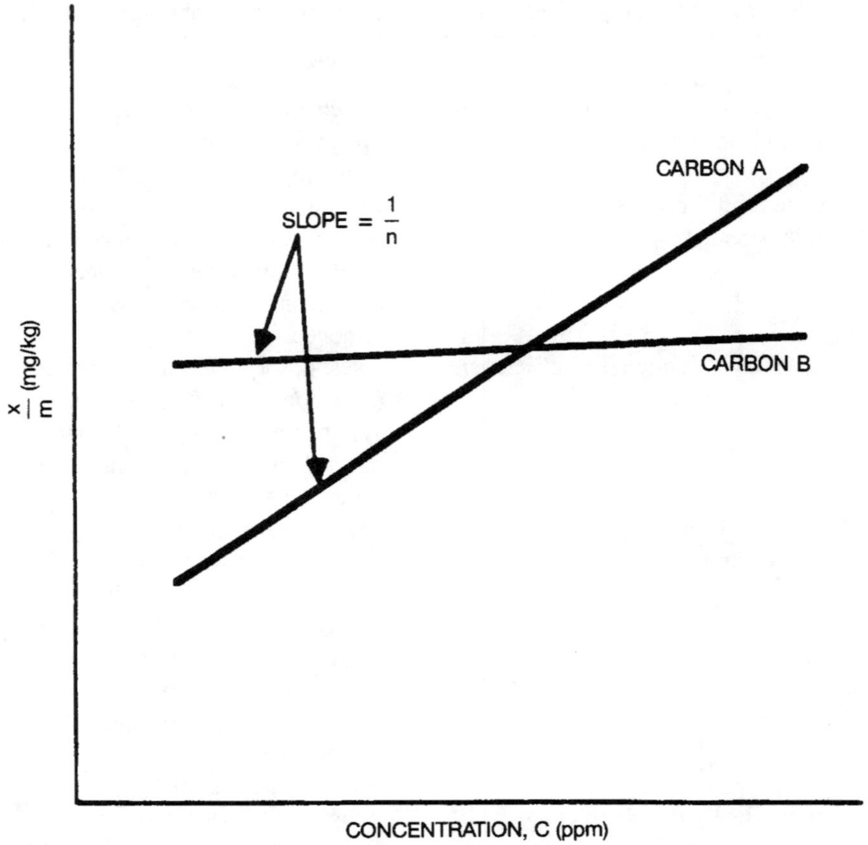

Figure 6.12 Isotherms for carbon adsorption.

systems are to be used. Carbon adsorption has often been used as a polishing element in water treatment (for taste and odor control) and is used for removing dilute quantities of benzene and gasoline components from drinking water.

When activated carbon is used to adsorb compounds from water, much more carbon is required per weight of contaminant removed than for air systems working on identical pollutants. For example, a carbon adsorption system adsorbing benzene from water will use between 625 and 1250 grams of carbon per 1 gram of benzene adsorbed, while an air or vapor phase system adsorbing 1 gram of benzene would require approximately 10 grams (Noonan and Curtis, 1990).

Physically, carbon adsorption equipment resembles filtration equipment. The principal difference is that the carbon systems are designed with larger access areas to permit removal of regeneration when it is necessary. Carbon adsorption beds can also be operated as fluidized systems, and filter beds cannot. However, fixed beds are more common in industry.

Multiple, fixed bed, carbon adsorption systems are typically used so that while one or more beds are adsorbing, at least one bed can be regenerating.

Regenerating a bed of activated carbon typically involves the thermal, solvent, or steam regeneration of carbon in the bed as discussed below.

CARBON ISOTHERMS AND PERFORMANCE. The basic equation for carbon adsorption was developed by Freundlich in the 1800s and has been in use ever since. The data are developed from pilot absorption tests and are plotted on a graph with the amount of waste constituent adsorbed divided by the weight of carbon on the vertical axis and with the concentration of waste remaining in solution on the other axis. The result is approximately a straight line with a slope of $1/n$. This all fits in the Freundlich equation as shown in Figure 6.12.

Experience with various types of carbon adsorption systems has shown that the waste contaminant in the effluent is essentially zero until the carbon begins to reach capacity (bed exhaustion). At this point the adsorbent begins to break through into the effluent. At some small value of waste absorbent in the effluent, perhaps as low as 1% or as high as 10% of the influent concentration, the carbon is said to be exhausted. The beginning of the breakthrough concentration and the pattern of breakthrough are shown in Figure 6.13. For development of the pilot data for carbon adsorption, the specific x axis in Figure 6.13 is pore volume of carbon, and, later after the system is up

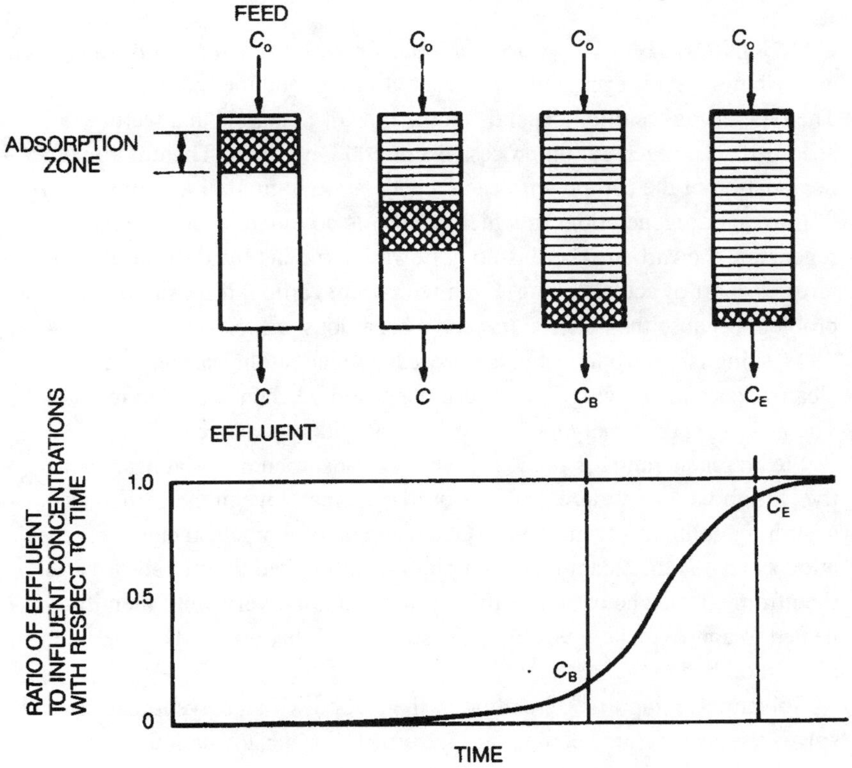

Figure 6.13 Plot of breakthrough curve.

and running and after one has obtained experience with the waste stream, the horizontal axis is often changed to either flow volume or length of filter run.

The final design of a carbon system is not unlike the design of a sand or filtration system, with a couple of important exceptions. Unlike the sand system in which head loss through the bed is used to gauge the end of the filter cycle, there is no good predictor of the carbon breakthrough. Granular media systems tend to be self correcting if flow patterns get uneven; carbon systems do not. Flow distribution patterns and residence time in the carbon bed are much more important in carbon systems than in sand systems. Also, unlike a sand system, the tankage must be designed for easy media removal.

In some carbon systems, which use relatively large volumes of carbon with low head loss systems, a waste roll off container has been piped and gasketed for use with carbon. Other commercially available large systems use a carbon charge at one time. These large systems are designed to fit the dimensions of a container truck that brings the carbon to the site. The carbon is typically transferred in slurry form from the tank to the truck and vice versa.

When spent, the carbon is dewatered and trucked to a carbon recycling station. Depending on the nature of the contaminant, the carbon may need to be in a closed container. The roll-off container system is principally designed for single-use operations when the carbon is to be used once and disposed without regeneration. Each of these systems have their own advantages.

CARBON REGENERATION. Carbon is regenerated in several ways, such as washing, steam regeneration, solvent cleaning, and thermal regeneration. The type of regeneration depends on the type of materials that were adsorbed. The regeneration process is not 100% efficient. There is some degradation of the carbon with each regeneration until after approximately four to eight regenerations at which point it is no longer economical to regenerate the carbon. There is no easy way to predict the desorption characteristics short of actual testing. Even with spent carbon, disposal may pose a problem because the material may be a hazardous waste.

Washing is desorption. It is the process of placing the carbon in a bed of clean water and allowing it to desorb the adsorbed chemicals into the liquid. The process is extremely inefficient and not widely practiced.

Steam regeneration is practiced when the adsorbed material has a relatively high vapor pressure, and it is used principally when the carbon is adsorbing pollutants from the air. One example of a practical steam regeneration system is the steam regeneration of a carbon bed used to absorb acetone from water. The carbon is first allowed to drain very well; then it is heated to approximately 90 °C by the steam, and that drives the acetone from the carbon bed.

Solvent cleaning is regeneration of the carbon with a solvent that dissolves the accumulated compounds adsorbed. For most installations, the regeneration involves a two-step process—first to remove the adsorbed material and then to remove the solvent from the carbon. There are a number of commercial systems that use methanol as a regenerant.

Thermal regeneration is a heat roasting of the carbon. It will restore the carbon to 80 to 95% of its original adsorption capacity. This process is not one that is undertaken on a small scale. The smaller thermal regeneration systems process between 5 and 10 tons of carbon per day. Larger systems are the rule. Air pollution control may be a problem with the regeneration. The dusts created by handling the carbon and desorbed chemicals must be treated to avoid nuisance odors and air pollutants. For smaller sites, regeneration services are typically considered because it is cheaper to treat and reuse the carbon than it is to throw it away.

DISSOLVED AIR FLOTATION. Dissolved air flotation (DAF) is used in a variety of applications including sludge thickening, grease and oil separation, and solids removal. In many instances it can be used to remove hazardous waste solids and oils from emulsion or solution when in combination with various types of chemical treatment. In concept the DAF unit is a large clarifier with two solids removal devices, one at the surface and one at the bottom. With the exception of the top removal device, the equipment looks very similar to a rectangular clarifier. In some instances, the bottom drag-out device is eliminated in lieu of periodic cleaning of the tank, but those instances are rare.

The DAF unit works on supersaturation of air into water under pressure, followed by its sudden release, causing a froth of bubbles that brings the suspended solids to the surface. There are a number of DAF units in operation, and almost all of them operate on the same set of principles. A portion of the main flow through the unit, between 10 and 30%, is diverted to a pressure tank where it is pressurized to between 137 and 415 kPa (20 and 60 psig). The pressure flow is saturated with air and then permitted to combine with the remainder of the flow just before the flotation chamber. The result is an almost explosive froth of bubbles that forms on all the suspended solids and at all the nucleation centers in the water and floats the solids to the surface.

A second type of DAF unit relies on beating the water into a froth with a large mixer, thus entraining the air and causing the particles to float. This type of system is primarily used in mineral separation, is often referred to as froth flotation, and is uncommon in hazardous waste treatment. In the froth flotation system, oils and protein solutions are added to the water, and these substances adhere to the solids, which also have an affinity for air. The violent agitation creates a froth that is removed from the water by skimming.

A third type of DAF unit, which was successfully manufactured in the U.S., was electroflotation. In this system a series of electrodes was spaced periodically across the bottom of the flotation tank and a dc current was applied at voltages between 4 and 20 volts. The resulting disassociation of the water formed extremely small bubbles of hydrogen and oxygen, and chlorine gas was also formed if there were any significant salts in the water. One configuration of the equipment even used the electrical generation of gas as the oxygen saturator in the pressure cell. The hydrogen is substantially less soluble that the oxygen, and when the pressure was released, it too was part of the flotation system. The flotation system worked best in areas where the conductivity of the water was greater than 3000 μmho/cm. With lower conductivity the voltages applied across the electrodes often exceeded 30 volts, and the power consumption of the equipment increased proportionally. The significant disadvantage of the electroflotation system was that the hydrogen and oxygen in the foam would be trapped and had some minor explosion potential until the foam had the opportunity to release the gas. Under conditions of good ventilation, this system was not dangerous to use.

One common factor in almost all DAF units is the chemical pretreatment system. The chemical pretreatment is responsible for at least half the success of the DAF unit. The chemical pretreatment system is typically either an emulsion breaking or a coagulation unit, or occasionally both. When there is oil in the waste, the oil emulsion is first broken using a combination of acids and metal salts and cationic polymers followed by a neutralization slow mixing/coagulation step, which is then followed by the flotation cell. Figures 6.14 through 6.16 show the design characteristics, data interpretation, and configuration flow diagram for a typical DAF system.

Dissolved air flotation can, depending on influent concentration, remove more than 99% of the suspended solids and more than 90% of the oils. The effluent quality from a typical DAF unit, when operating properly, may often contain suspended solids, less than 10 mg/L, and grease and oil, less than 5 mg/L.

Because the physical time required to float a particle to the surface of the liquid depends on the flotation tank geometry and a number of other factors, DAF units perform poorly under widely fluctuating hydraulic loads. The chemical regimen for floc formation, emulsion breaking, mixing, and flotation should always be kept as uniform as possible. Many plants that operate DAF units have one or more equalization basins capable of handling a single shift or a single day of wastewater flow, and the DAF unit is operated at as close to constant conditions as is possible.

The treatment dosage to be used in a DAF unit is determined by jar testing immediately before treatment. Some large commercial establishments even have DAF pilot cells that allow the operator to simulate the flotation and separation conditions so that he or she can select the correct chemical dosage. However, in most industrial applications, the treatment plant operator can select the correct treatment dosage by using jar test procedures, especially after he or she has had some experience operating the system.

Substantial changes in the chemical composition of the waste during holding and substantial variation in the chemical influent loadings can degrade the performance of the DAF system. Changes in chemical composition often occur for a variety of reasons, but one of the easiest to prevent is the change caused by bacterial action during holding or equalization. Disinfection of the waste can stop bacterial action before it starts, and at other times a simple stream of air bubbled through the holding tank will be sufficient to maintain the wastes in an aerobic condition.

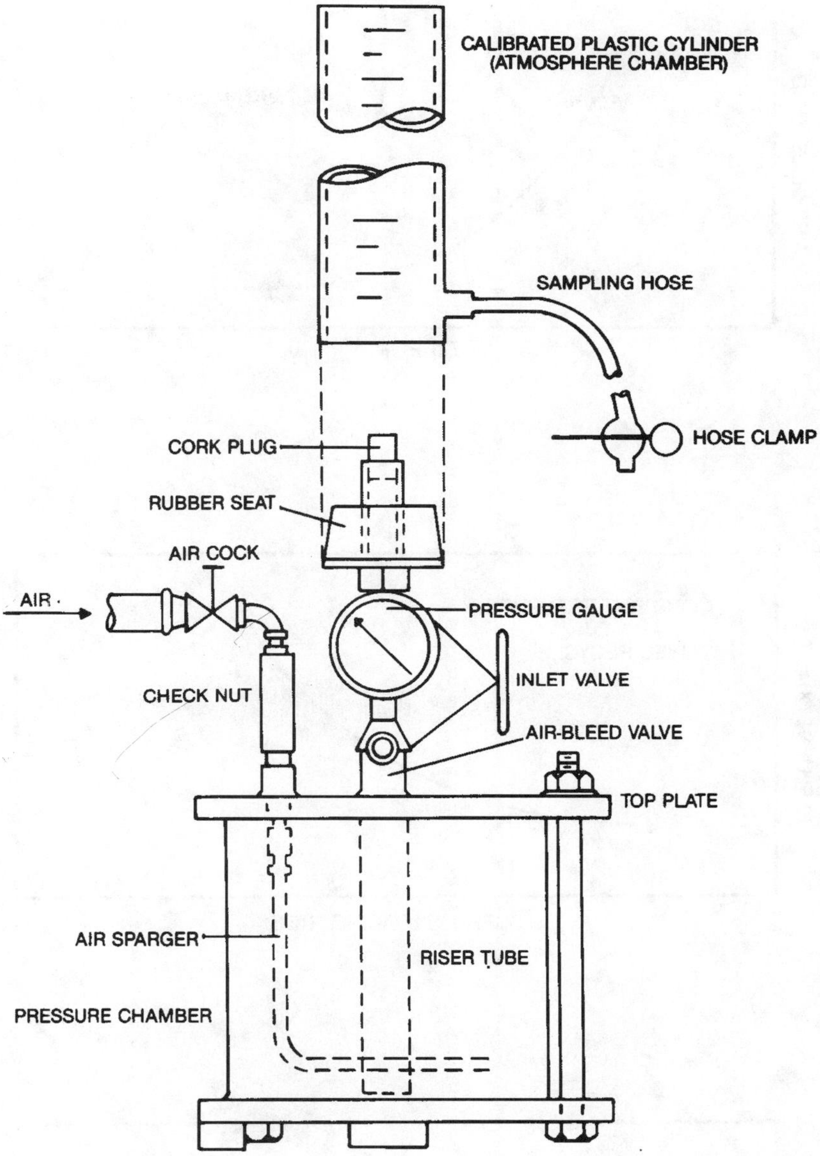

Figure 6.14 Laboratory flotation cell.

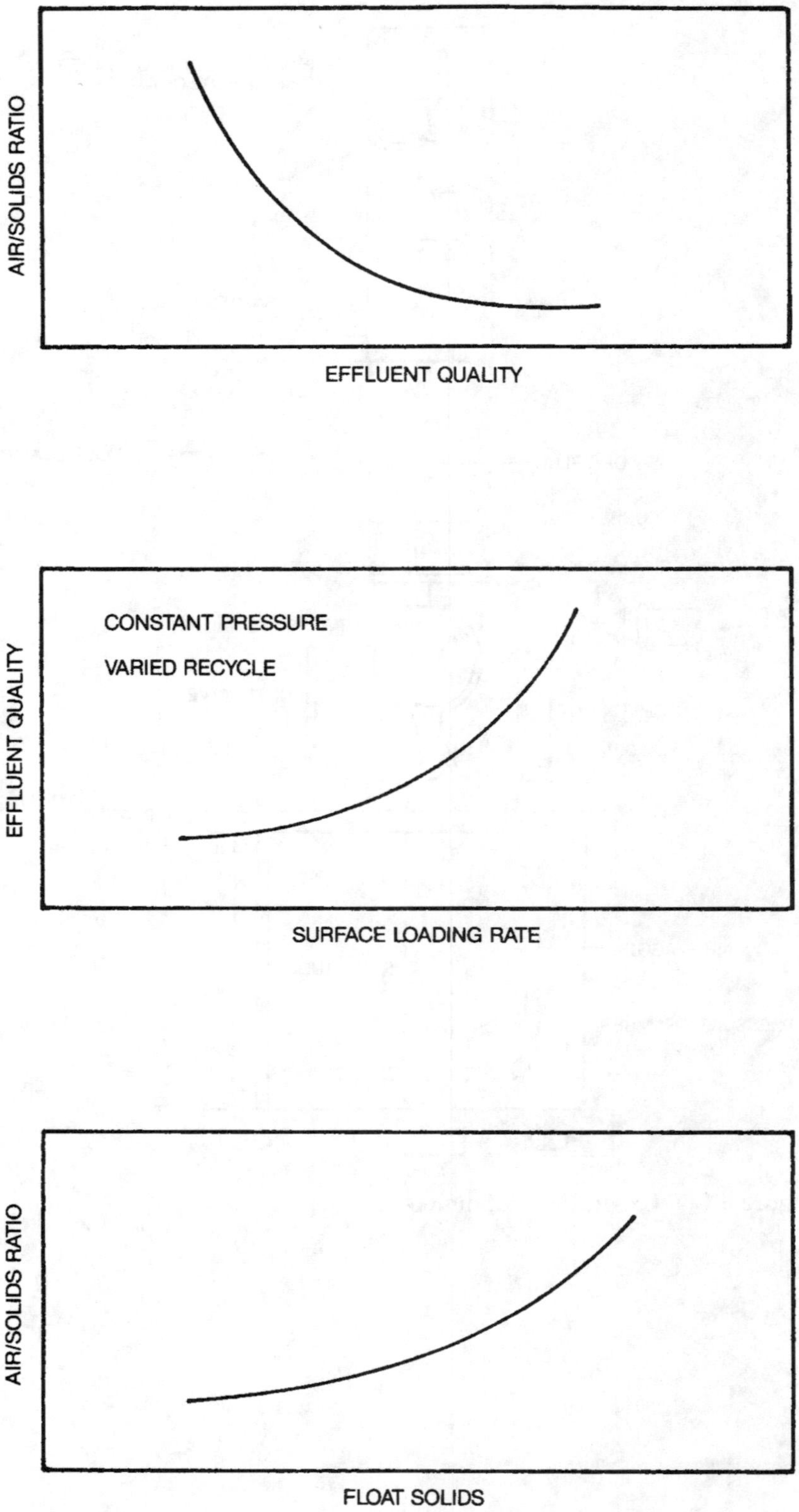

Figure 6.15 Treatability study data interpretation.

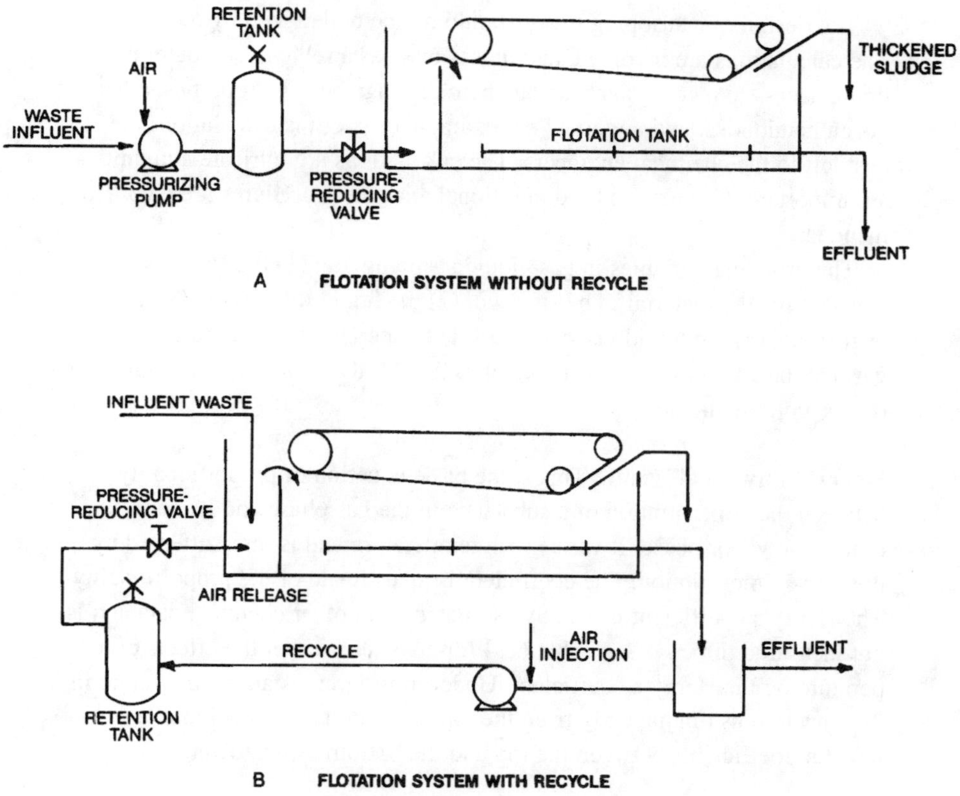

Figure 6.16 Flotation systems.

If the wastes have a large organic component, they may become anaerobic before or after despite the massive injection of air from the flotation system. If there are substantial quantities of volatile organic materials in the waste stream, the organics may be stripped from the waste, creating an air pollution, odor, or potential explosion hazard. If the wastes are allowed to become septic or anaerobic before they are treated by flotation, odor problems should be anticipated. Special precautions may be required to prevent the nuisances or hazards associated with this type of release.

AIR STRIPPING. Air stripping is yet another way of treating hazardous wastes. The process is simple and relatively inexpensive. However, changes in the regulatory framework have greatly affected the use of this technology for hazardous waste treatment applications.

Air stripping is a classical mass transfer application for chemical engineering. The process involves spraying the waste across the top of a packed tower so that it trickles downward through the packing while a current of air is blown up through the tower. The idea is to have the wastes start out at a high concentration at the top of the tower and arrive at the bottom with a much lower concentration of the strippable material. The gas stream, which starts out with none of the stripped material, is saturated when it leaves the tower.

The design of a stripping tower should not be undertaken by the novice. The calculations are involved, and unless one has a reliable computer program, access to a good database on chemical engineering properties, a half dozen handbooks, and a worked out example or two, the calculations are best left to the chemical engineers. The calculations are intricate and units are important. Calculation by dimensional analysis procedures are recommended.

The principal variables in gas–liquid stripping are: (1) the Henry's law constant for the material to be stripped, (2) the liquid diffusivity, (3) the gas diffusivity, (4) the liquid viscosity, (5) the temperatures of the liquid and the gas, (6) the gas flow rate (air to water ratio), (7) the tower or tray height, and (8) the column diameter.

Henry's Law Coefficient. This is the basic equation of proportionality between the concentration of a substance in the gas phase and its concentration in the vapor phase. It varies with temperature and is often affected by the salts in the solution. The coefficient is an estimate of the proportionality. The Henry's coefficient is usually estimated from other chemical data and is not measured directly. See Table 6.14 for an evaluation of the effects of temperature on the Henry's coefficient. Under most circumstances, the liquid in the tower cools dramatically from the top to the bottom, reducing mass transfer coefficients between the top and the bottom of the tower.

Table 6.14 Effects of temperature on system parameters.

Parameter	Value at 10 °C	Value at 30 °C	Change, %
Gas viscosity, 10^{-5} kg·m^{-1}·s^{-1}	1.76	1.86	+5.7
Gas density, kg·m^{-3}	1.248	1.165	−6.7
Gas-phase diffusivity (example: TCE), 10^{-6} m^2·s^{-1}	7.22	8.21	+13.7
Liquid viscosity, 10^{-3} kg·m^{-1}·s^{-1}	1.307	0.798	−38.9
Liquid density, kg·m^{-3}	999.7	995.7	−0.4
Liquid-phase diffusivity (example: TCE), 10^{-5} cm^2·s^{-1}	0.700	1.23	+75.7
Surface tension of water, kg·s^{-2}	0.074 2	0.071 2	−4.0
Henry's constant (example: TCE), dimensionless	0.173	0.440	+154

Liquid Diffusivity. This is the ability of the contaminant to diffuse from the liquid phase into the gas phase. It is empirically determined and varies with temperature, salt concentration, and interfering compounds. In practice, a single compound is almost never in solution. Many of the compounds frequently encountered have reactions between them or are mutually soluble. This lowers their transfer coefficients. Liquid diffusivity is also affected by the polarity of the compound. Polar compounds, such as acetone, are mutually soluble in water and are more difficult to remove by stripping.

Gas Diffusivity. This is the ability of the compound to diffuse into the gas. It is affected by temperature and by the properties of the substance in the vapor phase. The changes in gas diffusivity in a tower are small when compared with the changes of liquid diffusivity.

Liquid Viscosity. This is primarily affected by temperature and less by concentrations of salts and other solvents unless they are greater than 10%.

Gas Flow Rate. In practice this is an adjusted variable. The gas flow rate is set high enough to remove the material from the column. Increasing the gas flow rate can sometimes increase the removal across the column but at a penalty in energy and temperature that may be counterproductive to the design objectives.

Physical Parameters of the Column. Some of these parameters include liquid loading rate, column diameter, transfer height, and packing material. These are dependent variables that are set by the designer and are properties of the packing and temperature. The column diameter and transfer height (height of trays) to volume ratios are both set with regard to the liquid loading rate. A high liquid rate will often create tower flooding, a bulk movement of liquid down the tower that substantially reduces the mass transfer across the tower.

The temperatures of the entering streams are also important. Stripping works best when the temperatures are warmer because the diffusivity of the liquid is greater. As the gas stream contacts the liquid, it cools the liquid by evaporation, and in winter, a tower can freeze or plug with ice, causing tower failure.

The selection of packing material is a balancing of the cost of the packing against the surface area of the packing. Classical packing materials include shapes in configurations of saddles, rings, and balls. Some of the newer shapes are predominantly open shapes that have been computer designed to provide a balance between surface area and open area. The packing materials are measured in specifc areas per cubic measure of packing materials, such as square feet of transfer area per cubic foot of packing material.

In the past few years, a relatively new type of stripper has appeared on the scene, the low-profile stripper. This unit is a series of slotted trays through which the water flows in a torturous path on separate trays. The trays are stacked on top of each other, and the water drains from the upper tray to the lower trays. Air is blown upward through the slots in the bottom of the trays, and bubbles of air help strip the volatile material from the water. In many ways, the slotted tray is analogous to the plate in a distillation column because it serves the same function and shares many of the same design features as the low-profile stripper tray.

The low-profile strippers are more commonly used to remove highly volatile compounds from water, such as gasoline. Their low profile allows the unit to be built behind a conventional building or fence, avoiding the tall tower configuration that is typical with most conventional strippers and that many communities and commercial establishements find unsightly and objectionable.

Maintenance of these low-profile units is also somewhat easier because of the low tower height and ease of access. In all towers, the units have to be periodically flooded and acid etched to remove the deposits from deposits of minerals in the fluids, to kill bacterial growths, and to rejuvenate the packing.

The mathematical modeling on the low profile units is somewhat similar to that of the more conventional units, and the modeling is somewhat similar. Good reference materials on stripping include Green (1984), Noonan and Curtis (1990), Rich (1964), and Wright (1986). Of the programs used, there are two good programs that provide numbers and initial design parameters. The first is the Koch Tower Design program available from Koch–Glitsch Engineering in Wichita, Kansas. The program is available for the asking and is quite comprehensive for a number of chemicals if one has the properties of the materials. The second program is Airstrip, which first appeared in a DOS version, but is now available in Windows. Airstrip is available from Airstrip, Inc., 3209 Garner Road, Ames, Iowa, by license.

In using any software program, it is impearative that one look closely at the documentation and the numbers and work through an example, if only to satisfy oneself that the program is correctly computing the variables. The number crunching on a stripper design system is often extremely complex, and when one looks at conventional systems with transfer height and refluxing problems, the calculations are best left to those familiar with stripping technology.

CONSIDERATIONS FOR AIR STRIPPING. Stripping has recently become a specialty area in the environmental field because it is a good way to separate gasoline, some hazardous wastes, and some chlorinated solvents from groundwater. Many stripping towers were designed and a large number of the towers were incorrectly designed and failed to perform as designed. A well-designed tower can remove many organics and volatile compounds from water, and that was fine until the regulatory community finally figured out that transferring hazardous compounds from the water to the air was solving one problem while creating another. With the advent of methy *tert-*

butyl ether (MTBE) additive to the gasolines, stripping for removal of gasoline from groundwater has become less popular because MTBE is difficult to remove and typically requires a carbon adsorber after the stripper, as well as a recirculating flow through the stripper and the carbon.

In many communities, there is a limit on the amount of volatile organic compounds that can be discharged to the air during any single day from any single location. If the component stripped is a potential carcinogen or toxic substance, the limit may be substantially less than 1 pound per day, effectively eliminating the value of the stripper to the industry. An air permit may be required for any discharges to the air from the stripping tower. Permitting can be a definite concern, especially if the facility treating the waste does not have an NPDES permit. In that instance, in absence of an NPDES permit, a RCRA Part B permit may be required.

Stripping towers are typically tall slender structures, but depending on tower design, they can also be short tray systems. Tall slender towers are extremely top heavy and too tall to put in most buildings. In all cold climates and in most moderately warm climates, the tower water or air may have to be heated to compensate for the evaporative cooling and stripping losses taking place within the tower. The temperature decrease of the liquid across the tower is often 20 °C or more. Winter operations without heat can freeze the tower and cause the packing to collapse or can cause the tower to become top heavy and fail structurally.

Stripping towers need some thought and maintenance. The liquid being stripped needs to be evaluated for its chemical content. Iron, magnesium, manganese, and certain carbonate salts in the water can deposit or plate out on the tower packing, blinding the packing and reducing the tower efficiency. When the iron in the water is greater than approximately 0.5 mg/L, this may be a concern, and pretreatment should be used. If iron or other salts are in the water, consider a zeolite or manganese treated green sand pretreatment before the tower.

A stripping tower is also a home for bacterial growth. If it is being used for organic removal, the certainty of organic growth in the tower is almost 100%. Periodically, the tower needs to be shut down and flooded with a chlorine solution to kill the bacteria and/or acid cleaned to remove iron and mineral deposits from the water. The cleaning process may generate a substantial volume of hazardous wastes that will require treatment and disposal. Major tower maintenance cleaning is relatively infrequent, averaging between 6 months and a year, depending on the chemicals in the water and the biodegradability of the materials in the water.

Most stripping tower packings are dumped into the tower. The type of packing and the volume and cost are functions of the surface area to volume of the packing. Packing can be almost any porous media that can be cleaned. This includes spheres, saddles, and a host of shapes that are specially designed and engineered for specific purposes. The packings with higher surface:volume ratios are typically more expensive than those with lower ratios.

Packings are also expensive to the extent that a larger surface:volume ratio may permit construction of a smaller diameter tower that uses less packing materials, which may be significant factor.

The mist eliminator on the tower cap is important whether one uses a low-profile or conventonal tower design. Towers are turbulent flow devices. The air moving upward through the tower will entrain water and liquids in the tower. In many locations where strippers were designed for groundwater remediation, they were installed in and around gas station parking areas with no mist eliminators. The entrained liquid containing small amounts of gasoline and other solvents sprayed out of the top of the tower and attacked the finish on cars parked nearby. The mist eliminator provides a method of removing the entrained water from the air in the tower.

Mist eliminators come in several widely used types, but most configurations are stacked corrugated plates, wire mesh, or plastic mesh designs. The effectiveness of the mesh-style separators may be greater, but their higher maintenance requirements and ease of plugging must also be considered.

A stripper will generate a plume of air that is rich in stripped materials, generally hydrocarbons. As previously mentioned, the effluent from the stripper may be limited to discharge of as little as 1 pound per day of material, depending on the material. Where hydrocarbons are to be stripped, the usual limit is between 15 and 125 pounds per day of material released, depending on whether the site is within an air pollution nonattainment area. If a low permit limit has been established, additional treatment may be required on the exhaust gas stream. This additional treatment may be in the form of an afterburner or thermal destructor, or it may be carbon or other chemical processing.

The treatment requirements for stripper discharges to air are being more closely examined by regulatory agencies, and as they become more familiar with the applications for these devices, the necessity of installing additonal treatment units after the tower will increase, which will increase the cost of the treatment.

The stripper will generate a plume of air, generally saturated, but the plume may be warmer or colder than the surrounding air. The final temperature of the plume may be a consideration because the plume will often contain odiferous compounds that some in the vicinity will find offensive. Plume modeling is almost always imperative in stripper design, if only to avoid a nuisance lawsuit later and to ensure the regulatory community that one has accounted for the community and neighborhood in design considerations. The modeling should be done by someone familiar with that art and not left to the general novice. The modeling need not be elaborate, but it should account for the effect of nearby obstacles and buildings as well as the nasal threshold of a casual observer.

ION EXCHANGE. Ion exchange is the process of exchanging one set of cations or anions for another. Examples of the process include the removal of chromium (hazardous wastes) from solution; the removal of cadmium, lead, and iron from solution; water softening; and demineralization of water. A typical example is provided by the following equation:

$$PB^{+2} + Na_2Z \rightarrow 2Na^{+} + PbZ$$

Where Pb is the lead ion in solution, Z is the ion exchange resin and Na is sodium, which is initially attached to the ion exchange resin. The equation could just have easily been written using cadmium, zinc, copper, or any other metal cation. A similar reaction can be configured with an anion exchange for sulfate, cyanide, nitrates, or similar anions. The reactions are reversible. Ion exchange should be viewed as a concentration process, taking a material from one waste stream and putting it into another waste stream of smaller volume.

Ion exchange resins are either cationic or anionic, and either weak or strong. A cation exchange system is more familiar to most Midwesterners as a water softener. When an cation exchange system is followed by an anion exchange system, the result is demineralized water. Certain types of mixed bed anion and cation exchange systems are also used for total demineralization of water.

Ion exchange resins are either natural or synthetic. Many organics, including activated carbon and some soils, have ion exchange capacity, but they regenerate poorly if at all. One of the most popular natural cationic resins is manganese-treated green sand. It is used for water softening and has a high exchange capacity and modest cost. The largest majority of ion exchange resins are specially formulated plastic resins that have a specific exchange capacity measured in grains per cubic foot. This antiquated measurement of specific capacity is still widely used. One grain is the equivalent of approximately 15.43 mg. The typical resins in commercial use have exchange capacities in the tens of thousands of grains per cubic foot of resin.

The reactions on most resins are reversible. In the example above, flooding a cation resin with a strong sodium chloride solution reverses the reaction gradient and the lead cation is released to the solution. However, one then has to determine what to do with a brine solution that contains lead. Weak acids can also be used as regenerates, and in this specific case, it might be possible to use a weak cation system that is acid regenerated. The lead would be either in the form of lead chloride, lead sulfate, or lead acetate, depending on the regenerant acid. The lead can be precipitated by neutralizing the acid. Cation exchange resins can be configured to be regenerated with salts and acids, while anion exchange resins are often regenerated using caustic solutions and ammonium hydroxide.

Many hazardous wastes cannot be placed directly in contact with a resin because they will attack or foul the plastic resin base and destroy the resin capacity. Wastes containing oils, solids, sludges, and solvents may require pretreatment before ion exchange.

Resins work best in conditions when the pH of the incoming waste stream does not vary widely and when there is a solution that does not have competing ions. Most resins are not specific and will exchange on all cations or anions in the liquid right up to the exhaustion of the bed. If a degree of selectivity of removal of the ion is required, weak ion exchange resins can sometimes be used for that application. Weak cation exchange resins have been used successfully to remove ammonia from wastewater without pH adjustment and stripping.

Ion exchange systems are very amenable to instrumentation. Specific conductance on the effluent from the ion exchange column is one frequently used gauge of resin exhaustion. When the flows and waste streams through the resins are relatively constant, a timer or flowmeter is used to determine bed exhaustion. The calculation is made on the equivalent number of bed volumes of waste treated or the total flow volume treated.

THERMAL DRYING. Thermal drying, as used here, is not incineration. It is a volume reduction technique that is often used in dewatering. It is used to dewater certain types of sludges and highly concentrated solutions to form semisolids that are then sent for disposal. One of the largest differences between thermal dryers and incinerators is that most thermal dryers do not cause an oxidative reaction, except by coincidence. The temperatures are typically under contact chamber temperatures of approximately 200 °C (392 °F).

Dryers can come in any number of configurations including spray dryers, drum dryers, and screw dryers. Most, if not all, are steam driven and use electric power to turn the motor, which drives the shaft or rotates the drum. Drum dryers can achieve relatively high mass drying rates upward of 10 pounds per hour per square foot of dryer surface, but rates on the order of 8 pounds per square foot per hour are more common. Spray dryers have a very low efficiency because they have to evaporate the carrier liquid as well as dry the solid. Tray and auger type dryers have drying rates similar to those of drum dryers.

Dryers can achieve final moisture contents down to less than 1%. At less than approximately 5 to 10% moisture, most solids become dusty and difficult to handle. Lower moisture contents also reduce the overall thermal transfer efficiency of the dryer.

Determine the sludge drying characteristics for a particular waste stream by running a pilot-scale drying test. Small-scale laboratory tests cannot often overstate the drying performance of a particular dryer by 20% or more. Dryer testing should be conducted using scale up factors that do not exceed 1:20.

A drying system may have some air pollution emissions if the sludge contains a quantity of volatile materials. The elevated temperatures from the drying process will release the volatile materials and may create problems when none existed. In some instances, the application of heat may burn or roast the sludges creating odor, explosion, or hazardous material release problems that were not evident when the waste was wet. Pilot-plant testing at a moderate scale is again recommended.

CENTRIFUGATION. A centrifuge is a settling device that generates high centrifugal forces by whirling the waste material through a rapidly revolving disk or bowl. One of the easiest explanations for the concept centrifugation is provided by *Perry's Chemical Engineering Handbook,* Chapter 19 (Green, 1984).

Devices that generate centrifugal force can have moving parts or no moving parts. A pressure centrifugal collector (a cyclone) has no moving parts. Disk and nozzle centrifuges have internal moving parts that are carefully balanced. The amount of force generated depends on the type of centrifuge used.

Centrifuges are often used on sludges to decrease water content. They are cheaper to operate than dryers but have an end product with higher moisture content. Centrifuges are used in municipal systems to prethicken the sludge before it is fed into a belt filter press or filter press. With the right centrifuge and correct polymer dosage, a centrifuge can produce a high solids waste stream that can approach the performance of a filter press. Centrifuges are also widely used as concentrators and solids removal devices in the chemical industry and in remediation. They have been used successfully in reclamation of oil sludges from old lagoons and pits where they produce a reasonably concentrated liquid stream suitable for recovery. Figure 6.17 is a sketch of an Alfa Laval Sharples decanter centrifuge.

CYCLONES. Cyclones are a fixed type centrifugal separator in which the flow enters tangentially to the separator and flows along the wall. Cyclones are believed to have been named by the cyclonic shape of the vortex created by the working fluid as it swirls around the chamber. Figure 6.18 is a sketch of a liquid cyclone.

Cyclones are often used in degritting and coarse solids removals. Cyclones are less effective in the removal of materials that have an extremely low settling velocity and removal of colloidal materials. The difference between the cyclone and the centrifuge is that the cyclone has no moving parts and the vortex is created by the head loss across the unit. Also, the cyclones develop only a few gravities of acceleration, while the centrifuges can develop several hundred to several thousand gravities. Cyclones are an excellent grit and larger solids removal device. Head losses across the unit are proportional to the efficacy of the solids removal and range from a few pounds per square inch to more than 20 psi depending on flow rates and removals.

Both centrifuges and cyclones have two output streams, one high solids and one low solids. In the case of oils for separation, one would be an oil-rich stream, and the other would be an oil-poor stream. Each of these streams represent a potential waste product and may need separate handling. Where there are significant organic or oil components to the waste, both of these pieces of equipment can create an emulsion that may be more difficult to treat than the original waste stream. In the both cases, the emulsion will

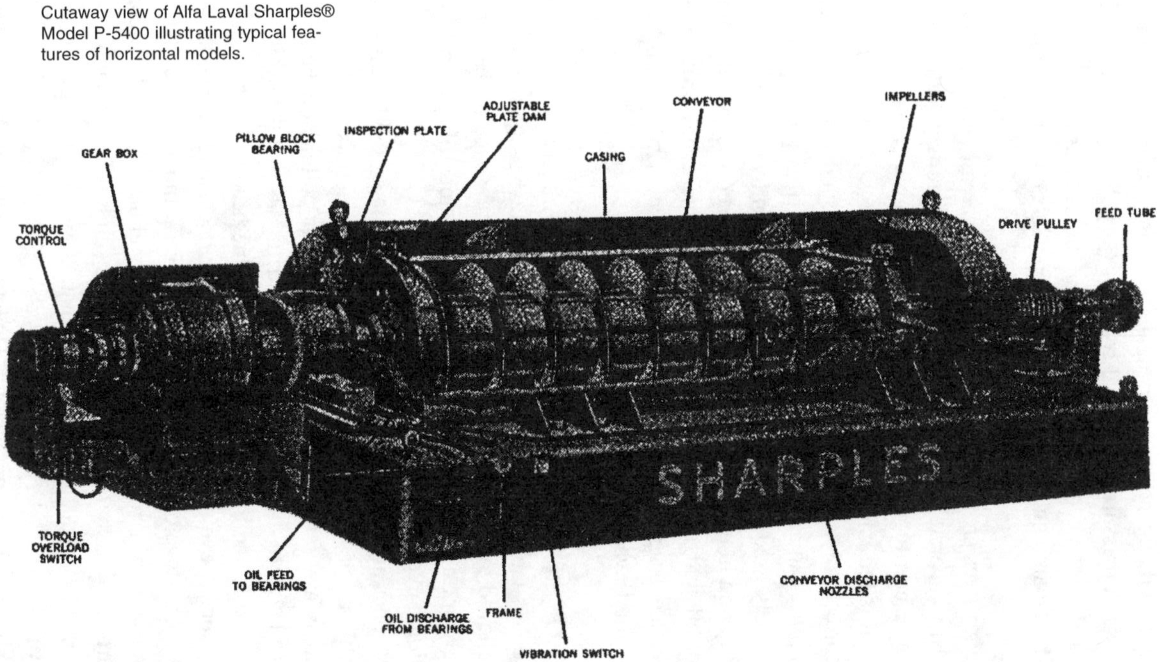

Figure 6.17 Cutaway picture of an Alfa Laval Sharples® decanter centrifuge. (Courtesy of Alfa Laval Separation, Inc.)

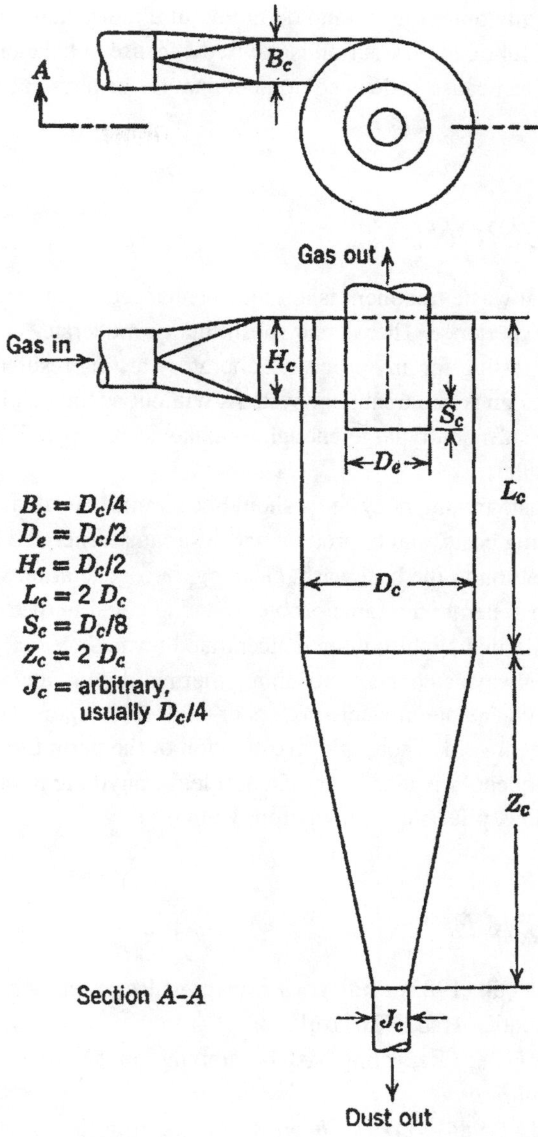

Figure 6.18 Sketch of a liquid cyclone.

be associated with the low volume waste, and the overall benefits of using the equipment may far outweigh any difficulties encountered.

Cyclones are inexpensive devices. Centrifuges, on the other hand, are expensive units. Centrifuges can easily cost more than a clarifier or filter press.

Cyclones are relatively easy to maintain. Centrifuges are extremely delicately balanced and fairly easily unbalanced. Factory service for maintenance is often necessary if high repair costs are to be avoided. Most centrifuges are typically quite rugged and do not require more than routine maintenance and lubrication. Caution is advised because unbalancing a high-speed centrifuge can cause serious equipment damage or physical injury to personnel.

CONCLUSION

Physical-chemical waste treatment is a science enhanced by the practitioner's art and experience. There is no substitute for moderate- to large-scale, pilot plant testing for any process. Laboratory testing results should not be used to design waste treatment systems without adequate pilot-scale testing at a scale sufficiently large enough to make accurate predictions of the full-scale results.

Hazardous waste treatment systems should be carefully evaluated for capital costs, operating costs, and byproduct waste streams generated by the process. Pay attention to the backwash, cleaning, or regeneration streams because they can generate the same problems encountered in the original waste stream, only in a slightly more concentrated form.

Permitting is always a concern. Stripping, thermal drying, and flotation processes all may generate unwanted odors or explosive vapors. Caution should be the watchword. Thorough investigation of the permitting requirements and consequences is mandatory. Do not leave anything to chance, test, measure, determine, pilot, and study before you build.

REFERENCES

Benefield, L.D., et al. (1982) *Process Chemistry of Water and Waste Water Treatment.* Prentice-Hall, New York.

Davies, R., et al. (1985) Recycling Metal Stamping Paint Wastes. *J. Water Pollut. Control Fed.*

Green, D.W. (Ed.) (1984) *Perry's Chemical Engineering Handbook.* McGraw-Hill, New York.

Metcalf and Eddy, Inc. (1979) *Wastewater Engineering: Treatment, Disposal, and Reuse.* McGraw-Hill, New York.

Noonan, D.C., and Curtis, J.T. (1990) *Groundwater Remediation and Petroleum.* Lewis Publishers, Chelsea, Mich.

Rich, L.G. (1964) *Unit Operations of Sanitary Engineering.* Van Nostrand Reinhold, New York.

U.S. Code of Federal Regulations (Year Not Provided) Title 40, 40 CFR, Part 261.24.

Wright, D. (1986) *Basic Programs for Chemical Engineers.* Van Nostrand Reinhold, New York.

Suggested Readings

Chen, W.F. (Ed.) (1994) *Civil Engineering Handbook.* CRC Press, Boca Raton, Fla.

Geyer, J.C., and Fair, G.M. (1954) *Water Supply and Waste Water Disposal.* Wiley and Sons, New York.

Chapter 7 Chemical Treatment

INTRODUCTION

Chemical hazardous waste treatment is a critical approach for solving many hazardous waste management problems. Using chemicals to adjust pH or enhance precipitation is an example of simple chemical treatment. Chemicals are often used to increase the efficiency of operation of other hazardous wastes unit operations, such as biological and physical treatment processes. Chemical treatment can be applied at the generator's site or at an off-site treatment, storage, and disposal facility (TSDF). Chemical treatment techniques are also used during the remediation of hazardous waste sites. Chemical treatment performed at the generator's site may require a permit to operate or require specialized worker training to minimize risk to human health and the environment.

This chapter will review several hazardous waste chemical treatment processes, including pH neutralization, precipitation, solidification, and oxidation. The chemical treatment processes to be discussed in this chapter and their potential applications are summarized in Table 7.1.

Table 7.1 Potential applications of chemical treatment.

Chemical process	Potential applications	Example source	Process example
pH adjustment	Wastewater	Steel manufacturing	Neutralization of waste pickle liquor
	Wastewater	Mining	Neutralization of acid mine drainage
	Wastewater	Electroplating	Neutralization of acid or alkaline cleaning baths or rinses
	Air	Aluminum extrusion	Die cleaning fume scrubber
		Sulfuric acid cleaning bath	Tank exhaust hood and caustic scrubber
	Soil/sludge	Superfund site	Remediation of lead-contaminated soils
	Ancillary processes		To enhance physical, biological, or other chemical treatment processes
Precipitation	Wastewater	Copper forming	Precipitation of copper from wastewater
	Wastewater	Auto manufacturing	Chromium and zinc precipitation
Carbon adsorption	Wastewater	Textile mill effluents	Adsorption of volatile organics
	Groundwater	Groundwater treatment	Adsorption of volatile organics
	Air	Drying ovens	Adsorption of volatile organics
Oxidation	Wastewater	Organic chemical production	Chemical oxidation of organics to carbon dioxide and water
	Wastewater	Electroplating	Chemical oxidation of cyanide
Reduction	Wastewater	Electroplating	Reduction of Cr(VI) to Cr(III)
	Wastewater	Paint and ink	Reduction of mercury
Solidification	Soils/sludges/sediments	Waste site remediation	Solidification of plating sludges
Solidification	Soils/sludges/sediments	Waste site remediation	Solidification of lead-contaminated soils with Portland cement

The type of waste treatment method chosen depends on the type of hazard present, the location of the treatment process to be used, and the method of disposal. Careful review of the regulatory aspects associated with each of these elements is necessary before the selection of an appropriate treatment alternative.

REGULATORY ASPECTS OF CHEMICAL TREATMENT

The federal Resource Conservation and Recovery Act (RCRA) was enacted in 1976 as an amendment to the Solid Waste Disposal Act of 1965. RCRA authorizes the U.S. EPA to regulate waste management activities, including the generation, storage, transfer, treatment, and disposal of hazardous wastes. The basis for determining if a waste is hazardous typically occurs in one of two ways—tests concluding that the waste exhibits one or more of the RCRA characteristics or lists classifying specific types of wastes as "hazardous." The goal of hazardous waste treatment by chemical techniques is to eliminate the characteristics that cause a waste to be classified as hazardous. These characteristics, defined in 40 CFR Parts 261.20 to 261.24, are ignitability, corrosivity, reactivity, and toxicity.

By definition, a waste is ignitable if it is a liquid that contains less than 24% alcohol (by volume) and has a flash point less than 140 °F (60 °C), if it is not a liquid but is capable of igniting under standard temperature and pressure, if it is an ignitable gas, or if it is an oxidizer. Ignitable hazardous wastes are often incinerated or blended with other fuels and combusted as a means of hazardous waste treatment. Venting a compressed gas into an absorbent solution is an example of a chemical treatment technique for an ignitable waste.

A waste is corrosive if it has a pH less than or equal to 2 or greater than or equal to 12.5 or if it is a liquid that corrodes steel (SAE 1020) at a rate greater than 6.35 mm per year at a test temperature of 130 °F (55 °C). Waste neutralization is the principal chemical treatment technique for eliminating corrosivity.

A waste exhibits the characteristic of reactivity if a representative sample of the waste has any of the following properties:

- It is normally unstable and readily undergoes violent change without detonating.
- It reacts violently with water.
- It forms potentially explosive mixtures with water.
- It generates toxic gases or fumes when mixed with water.
- It is a cyanide or sulfide bearing waste, which when exposed to pH values between 2 and 12.5 can generate a toxic gas or fume.

- It can explode if subjected to a strong initiating source or heated under confinement.
- It can explode under conditions of standard temperature and pressure.
- It is a regulated explosive.

Treating the waste in a controlled environment, such as by chemical oxidation or reduction, can eliminate reactivity concerns. The chemical oxidation of cyanide-containing wastes by the addition of ozone is an example of a chemical treatment process used to make a reactive waste nonhazardous.

A waste exhibits the characteristics of toxicity if, when using the toxicity characteristic leaching procedure (TCLP), the extract contains any of the contaminants listed in 40 CFR Part 260.24. The primary purpose of the TCLP test procedure is to determine if a waste can be disposed safely in a landfill. Solidification and stabilization are often used to meet the TCLP criteria before disposal of the waste. Solidification of lead-contaminated ash using Portland cement, for instance, yielded a TCLP lead concentration of less than 1.5 ppm (Boehmer et al., 1986).

If the treated waste cannot be recycled, reused, or incinerated, then the primary disposal option for chemically treated hazardous wastes is by land disposal. Land disposal is defined as placement in or on the land, such as in a landfill, surface impoundment, waste pile, or injection well (40 CFR Part 268). To reduce the likelihood that untreated hazardous wastes would be land disposed, the U.S. Congress included specific waste pretreatment requirements in the 1984 Hazardous and Solid Waste Amendments (HSWA). Specifically, before land disposal can occur, the toxic constituents in the waste must be sufficiently degraded or destroyed using the best demonstrated available technology (BDAT) to meet U.S. EPA's universal treatment standards (UTS). There is a UTS for each hazardous waste. The BDATs are listed in 40 CFR Parts 268.40 to 268.46. Because there are approximately 450 types of hazardous wastes, U.S. EPA has categorized the BDATs into three categories: treated constituent concentration in waste extractions (CCWE) based, technology based, and waste constituent concentration (CCW) based. Each classification is described in 40 CFR Parts 268.41, 268.42, and 268.43, respectively.

To determine the appropriate waste treatment option, it is imperative to perform treatability studies.

TREATABILITY STUDIES

Treatability studies are necessary to provide the design data and operating criteria required for each process in a treatment sequence. For example, these studies can determine the optimum chemical dosages and critical control points needed for pH adjustment of a corrosive waste, such as spent sodium hydroxide from an alkaline cleaning tank. The amount, type, and handling properties of raw materials, products, and byproducts, such as sludge from a precipitation process, can also be determined. The studies must be broad enough to evaluate a wide range of potential treatment options and their associated costs. This will provide a foundation for future studies of the most feasible, cost-effective options.

Characterization of the waste stream is also essential to protect workers and the environment. Formation of explosive or toxic mixtures is always a potential result of most chemical reactions. This is particularly true of waste streams that can no longer be counted on to react in the predictable manner of virgin chemicals of definite constituents and concentration. Care must be taken to ensure that representative samples are obtained and that a wide range of analyses are performed to provide an accurate characterization of the waste constituents.

Factors to be considered in designing a bench-scale treatability study are listed below:

- Review literature sources for the type of waste to be treated. The results and conclusions from other studies provide a useful starting point for site-specific studies.
- Determine the treatment objective(s). For example, if the purpose of a chemical treatment approach is to meet a TCLP of 5.0 ppm for lead, then evaluating options that cannot achieve this value are pointless. Also note that it is not typically cost effective to use options that achieve a much lower TCLP value.
- Identify the hazards of the proposed treatment process including explosiveness and toxic byproducts.
- Estimate the unit cost of the proposed treatment approach. Both capital cost and annual operation and maintenance costs should be estimated. Alternative chemicals and other waste products can be readily and economically evaluated at bench or pilot scale.

TREATMENT TECHNOLOGIES

Chemical treatment processes identified in this chapter include pH adjustment, chemical precipitation and coprecipitation, oxidation–reduction, carbon adsorption, and stabilization and solidification.

A wide range of chemicals are applied in these processes, ranging from common materials such as lime and Portland cement to less commonly applied chemicals such as hydrogen peroxide and ozone. Typically, more than one chemical will serve the intended purpose. Consequently, other factors will be involved in the selection of a chemical. These factors include

- Initial cost of chemical procurement,
- Safety of chemical handling and storage, and
- Quantity and characteristics of process byproducts.

Chemical treatment requires adequate mixing of the chemical into the waste stream for a sufficient time to enable completion of the chemical reactions. Mixing of the chemicals and the waste streams may take many forms, based on the phase of the treatment chemical and the waste stream. Aspirators may be appropriate for mixing a gaseous treatment chemical with a liquid waste, while a pug mill may be an appropriate device for mixing a stabilizing liquid or dry chemical with a granular waste. The required chemicals must be capable of being safely transported to and stored on the site. Any residual treatment chemicals must be capable of being safely disposed of at a reasonable cost.

After the chemical and waste have been adequately mixed together, it is likely that another vessel will be required to more economically provide the necessary reaction time.

Key design factors for chemical-waste processing equipment include

- Control and monitoring,
- Selection of materials for construction,
- Personnel safety, and
- Environmental protection.

Control and monitoring equipment, testing, and procedures must be appropriate for the reaction environment. Difficulty in monitoring the process may require that a treatment process be carried out on a batch basis rather than continuous flow. Also, the need for reprocessing must be taken into account. Analytical procedures appropriate for the reaction must be performed regularly to ensure that the treatment objectives are being met.

Many of the waste streams and chemicals will have damaging effects on common materials of construction such as concrete, carbon steel, and polyvinyl chloride. Careful selection of materials for construction of storage, reaction, and transport vessels, as well as piping, is necessary for the longevity of the units and for operator safety. This selection must be chemical specific. For example, certain types of stainless steel would be appropriate for ozonized gas but inappropriate for wastes with high chloride levels. Complicating factors in material selection include multicontaminant waste streams, temperature, and the effect of corrosion.

Personnel involved in treatment activities must be protected from the unique features of the chemical treatment process. Guidance can be taken from the chemical industry. However, many of the situations are unique to the hazardous waste management industry. Site remediation activities would likely make use of mobile treatment units erected on site on a temporary basis. In these situations, there may be a temptation to proceed without safety facilities or equipment that would only be used for a short time or that had not yet been delivered to the site. At TSDFs, receipt of unfamiliar waste streams may require additional worker training and modified monitoring programs. In all cases, comprehensive safety plans must be aggressively applied by trained professionals.

Chemical treatment of a waste has a potential to release pollutants to the environment. Adequate control of chemical spills, equipment leaks, reaction fumes, waste handling, and transportation must be included in the process design. Handling of waste material such as dust, vapors, and fumes should be provided for by hoods, recovery systems, or treatment units such as scrubbers. Personnel and environmental monitoring equipment is often needed to ensure that minimal contamination of the environment occurs during the treatment process.

pH ADJUSTMENT. The hydrogen ion concentration, as measured by pH, is a critical variable for successful treatment of most wastes. Adjustment of the pH of a waste stream is a process that is common to many other treatment processes sensitive to pH levels. The pH adjustment may be the only treatment required to eliminate the hazardous nature of the waste stream, such as neutralization of a corrosive wastewater. A summary of the types of hazardous waste treatment processes that are directly influenced by pH is given in Table 7.2.

The hydrogen ion concentration in an aqueous system typically is related to the extent that hydrogen ions dissociate in solution. Water will dissociate into hydrogen and hydroxyl ions as follows:

$$H_2O \leftrightarrow H^+ \; OH^-$$

Applying the law of mass action to this equation gives:

$$\frac{[H^+][OH^-]}{[H_2O]} = K$$

Where the brackets indicate concentration in moles per liter. Because the concentration of water in dilute systems is relatively constant, this concentration can be incorporated into the equilibrium constant, *K,* to give

$$[H^+][OH^-] = K_w$$

Where K_w is the ionization product of water, equal to 1×10^{-14} at 25 °C. The term pH is defined as:

$$pH = -\log[H^+]$$

Other species present in solution can dissociate in a similar manner causing complex equilibria relationships to influence the actual pH of the system. The design of a pH control system must be based on careful consideration of these "other" reactions. Often, treatability studies are needed to determine the appropriate chemical(s) and dosage(s) needed to adjust the pH of a complex waste stream. Design considerations for waste treatment by pH adjustment include the following:

Table 7.2 Examples of pH control for hazardous waste treatment.

Purpose	Strategy	Examples of hazardous waste
Neutralization	Adjust pH with acid or alkaline solution to a regulated limitation	Corrosive wastes ($2.0 \leq pH \leq 12.5$) acid pickling bath and rinse acid mine drainage sulfuric acid anodizing solution alkaline cleaning bath and rinse
Precipitation	Adjust pH with acid or alkaline solution to precipitate metals from solution as a metal–hydroxide, carbonate, chloride, or sulfide salt	Wastes containing metallic species metal processing wastes electroplating solutions tannery wastewater circuit board rinse waters spent anodizing baths photographic processing wastewaters
Co-precipitation–adsorption	Adjust pH with acid or alkaline solution to co-precipitate metals from solution and adjust surface charge of precipates to promote attraction	Wastes containing metallic species arsenic-containing wastes cadmium-containing wastes copper-containing wastes fluoride-containing wastes selenium-containing wastes
Pretreatment	Adjust pH to a desired set-point to enhance waste treatment by another technique	Reduction of hexavalent chromium oxidation of cyanide wastes air stripping of ammonia oxidation/ion exchange of selenium biological treatment of a low pH organic waste stream

- Phase of the waste (gas, liquid, or solid),
- Proposed treatment site (remediation site, TSDF, or generator location),
- Quantity of waste and byproducts produced, and
- Material status—finite quantity or continuing production.

Chemical treatment to adjust pH involves the application of acidic chemicals to lower the pH of caustic streams and caustic chemicals to raise the pH of acidic waste streams. A carefully designed and operated system is capable of achieving any desired pH level. However, the range of pH required should be realistically established in the interest of economic facility construction, practical operation, and maintenance. The first step in designing a pH control unit process is to titrate the waste with known aliquots of acid and base. A titration curve is then plotted from the milliequivalents (meq) of acid or base added versus pH. The optimum chemical dose and reaction time needed to achieve the target pH can then be determined. The design of required equipment, such as tank volume or detention time, can then be prepared.

Either virgin chemicals or another waste stream may be used for pH adjustment depending on the specific requirements of the reaction. Chemicals typically used to increase pH are pebble lime, hydrated lime, and sodium hydroxide. Sodium hydroxide can be purchased in drums at 50% concentration or as a bulk liquid at 20 to 25% concentration. If 50% caustic is to be used, care must be taken to avoid freezing conditions, which can occur near 50 °F. Chemicals used to lower pH include hydrochloric acid and sulfuric acid. Table 7.3 provides information on chemicals typically used for pH adjustment. Examples of waste products used in pH adjustment include carbide lime from the manufacture of acetylene gas and pickle liquor from steel processing.

Wastes requiring pH adjustment are most frequently in the liquid phase. Chemicals for pH adjustment may be in the gaseous, liquid, or solid (granular) phase. Carbon dioxide for pH reduction may be generated on-site by submerged combustion processes or transported to the site and stored as a liquified gas. Gases are dissolved in wastewater by applications of static mixers, venturi sections in pipelines, submerged diffusers, and submerged turbines. Solid chemicals, such as lime, are generally slurried before mixing with the waste stream.

Solid wastes requiring pH adjustment typically must be processed (homogenized) to a granular form for treatment. Processing may include a wide range of materials handling equipment such as grinders, shredders, and crushers designed to bring the material to a size and condition that it can be brought into contact with liquid or dry treatment chemicals. Often, leachate from the solids or sludge is the source of acidity or alkalinity. In this case, the solids must be blended with the leachate and pH adjusting chemicals. Hazardous waste gases can be treated with chemicals in liquid form or dissolved in a carrier liquid in scrubbers.

Table 7.3 Acid and alkali costs for neutralization.

Chemical	Specification	Cost,[a] cent/kg	Equivalent cost,[b] \$/kg equivalent
Sulfuric acid (H_2SO_4 [oil of vitriol])	98%	1.13[c]	0.59
Hydrochloric acid (HCl [muriatic acid])	20° Bé–32%	0.81[c]	0.96
Limestone (CO_3)	200 mesh, 93%	0.18	0.10
Quicklime (CaO)	Lump, 90%	1.11[d]	0.34
		0.79[c]	0.25
Hydrated lime ($Ca(OH)_2$)	Pulverized, 93%	1.04[d]	0.42
Sod ash (Na_2CO_3)	Light, 50% Na_2O	2.42[d]	1.41
		1.58[e]	0.92
Caustic soda (NaOH)	Dry	7.03[d]	2.81
	Dry	6.57[e]	2.63
	50% liquid	3.62[a]	1.45
Ammonia (NH_3)	Liquid	4.08[c]	0.69

[a]Cost is given in cent/kg of product as shipped, not including freight, except for 50% liquid caustic, whose cost is cent/kg NaOH.

[b]Cost per kg-equivalent is cent/kg × kg/kg equivalent; kg-mole/valence.

[c]Bulk cost in tank cars, tank trucks, or hopper cars.

[d]Cost in bags.

[e]Cost in drums.

CHEMICAL PRECIPITATION. Precipitation is applicable to the treatment of aqueous wastes and slurries containing heavy metals. The metals most commonly amenable to precipitation treatment include cadmium, copper, chromium, lead, mercury, nickel, and zinc. Precipitation is based on altering the chemical equilibria of the system by exceeding the solubility product for the species. The precipitate is then removed by a solids separation process, such as sedimentation, filtration, or centrifugation. Because metals are conservative, precipitation serves as a concentration process rather than a method of destruction. The product (for example, sludge) formed during the process may be reclaimed if the metals content is sufficient. Otherwise, sludge from the precipitation of a hazardous waste stream must be further treated, such as by solidification, or be disposed as a hazardous waste.

Both equilibrium and kinetic considerations are important aspects of precipitation treatment. Equilibrium relationships provide the theoretical basis for calculating the aqueous phase concentration of the metal at equilibrium and the mass of sludge formed as the result of treatment.

The theoretical concentration of heavy metal in solution after precipitation is described by the solubility product, in which a precipitate dissolves in water to form its corresponding ions:

$$A_aB_b(s) > aA^{b+} + bB^{a-}$$

The solubility product is defined as

$$K_{sp} = (A^{b+})^a(B^{a+})^b$$

Where ( ) denotes the activity of the species in solution. Common solubility products are listed in Table 7.4 (Snoeyink and Jenkins, 1980). In dilute solutions, concentrations can be used in place of activities.

Although equilibrium calculations describe treatment if the reaction proceeds to completion, in most cases kinetics govern the extent of reaction because insufficient time is available to achieve equilibrium. Precipitation kinetics involve three general steps: nucleation, crystal growth, and agglomeration and ripening of the solids. Precipitation sometimes does not occur in supersaturated solutions, not because of slow rates but because of the lack of nucleation sites. In such situations, precipitation can be induced by increasing the degree of supersaturation or the addition of a seed of fine particles to the solution.

The effectiveness of precipitation treatment is controlled by many factors specific to the type of process to be used and the waste stream to be treated. Factors such as ionic strength, the common ion effect, pH, complexation, and temperature are important. Ionic strength, for instance, lowers the activity of the metal in solution, thereby increasing its solubility. The common

Table 7.4 Solubility product constants.

Substance	Formula	K_{sp}
Aluminum hydroxide	$Al(OH)_3$	5×10^{-32}
Barium carbonate	$BaCO_3$	4.9×10^{-9}
Barium iodate	$Ba(IO_3)_2$	1.57×10^{-9}
Barium oxalate	BaC_2O_6	1.6×10^{-7}
Barium sulfate	$BaSO_4$	1.0×10^{-10}
Cadmium carbonate	$CdCO_3$	2.5×10^{-14}
Cadmium sulfide	CdS	1×10^{-28}
Calcium carbonate	$CaCO_3$	4.8×10^{-9}
Calcium oxalate	CaC_2O_4	1.9×10^{-9}
Calcium sulfate	$CaSO_4$	6.1×10^{-5}
Copper(II) hydroxide	$Cu(OH)_2$	1.5×10^{-19}
Copper(II) sulfide	CuS	8.5×10^{-45}
Copper(I) bromide	$CuBr$	5.9×10^{-9}
Copper(I) chloride	$CuCl$	3.2×10^{-7}
Copper(I) iodide	CuI	1.1×10^{-12}
Copper(I) thiocyanate	$CuSCN$	4×10^{-14}
Iron(III) hydroxide	$Fe(OH)_2$	1.5×10^{-36}
Lanthanum iodate	$La(IO_3)_3$	6×10^{-10}
Lead carbonate	$PbCO_3$	1.6×10^{-13}
Lead chloride	$PbCl_2$	1×10^{-4}
Lead chromate	$PbCrO_4$	1.8×10^{-14}
Lead hydroxide	$Pb(OH)_2$	2.5×10^{-16}
Lead oxalate	PbC_2O_4	3.0×10^{-11}
Lead sulfate	$PbSO_4$	1.9×10^{-6}
Lead sulfide	PbS	7×10^{-26}
Magnesium ammonium phosphate	$MgNH_4PO_4$	25×10^{-13}
Magnesium carbonate	$MgCO_3$	1×10^{-13}
Magnesium hydroxide	$Mg(OH)_2$	5.9×10^{-12}
Magnesium oxalate	MgC_2O_4	8.6×10^{-5}
Manganese(II) hydroxide	$Mn(OH)_2$	4×10^{-14}
Manganese(II) sulfide	MnS	1.4×10^{-15}
Silver arsenate	Ag_6AsO_4	1.0×10^{-22}
Silver bromide	$AgBr$	7.7×10^{-13}
Silver carbonate	Ag_2CO_3	8.2×10^{-12}
Silver chloride	$AgCl$	1.82×10^{-10}
Silver chromate	Ag_2CrO_4	1.1×10^{-12}
Silver cyanide	$AgCN$	2×10^{-12}
Silver iodate	$AgIO_3$	3.1×10^{-8}
Silver iodide	AgI	8.3×10^{-17}
Silver oxalate	$Ag_2C_2O_4$	1.1×10^{-11}
Silver sulfide	Ag_2S	1.6×10^{-49}
Silver thiocyanate	$AgSCN$	1.1×10^{-12}
Strontium oxalate	SrC_2O_4	5.6×10^{-8}
Strontium sulfate	$SrSO_4$	2.8×10^{-7}
Thallium(I) chloride	$TlCl$	2×10^{-4}
Thallium(I) sulfide	Tl_2S	1×10^{-22}
Zinc hydroxide	$Zn(OH)_2$	2×10^{-14}
Zinc oxalate	ZnC_2O_4	7.5×10^{-8}
Zinc sulfide	ZnS	4.5×10^{-24}

ion effect, that is, the presence of the same ion in the solution as the precipitant, increases the metal's solubility after precipitation treatment. Complexation may also inhibit precipitation treatment. The presence of some organic ligands can interfere with precipitation because these species can form soluble organometallic complexes. Humic substances and ethylenediaminetetraacetic acid (EDTA) are common organic compounds that increase the solubility of metals. Cyanide, an inorganic ligand, can also be highly effective in keeping metals in solution. Figure 7.1 shows that increasing the temperature from 25 to 75 °C increases the solubility of the copper(II) ion.

Because of the numerous variables present in a real system, the practical treatment of metal wastes typically relies on site-specific treatability tests rather than specific solubility product calculations. Because of the common ion effect, complexation, and kinetic limitations, excess precipitant is often needed to achieve the treatment objectives. Suspended and colloidal materials may also be removed by precipitation treatment. As a result, large quantities of sludge are produced. The sludge volume is typically greater than what is expected based on theoretical calculations.

Jar tests are the most common method for determining the optimum precipitant and chemical dosages. Design parameters that are determined by jar tests include optimum pH, mixing requirements, settling characteristics, and sludge production. Some materials, especially sulfide precipitates, settle as

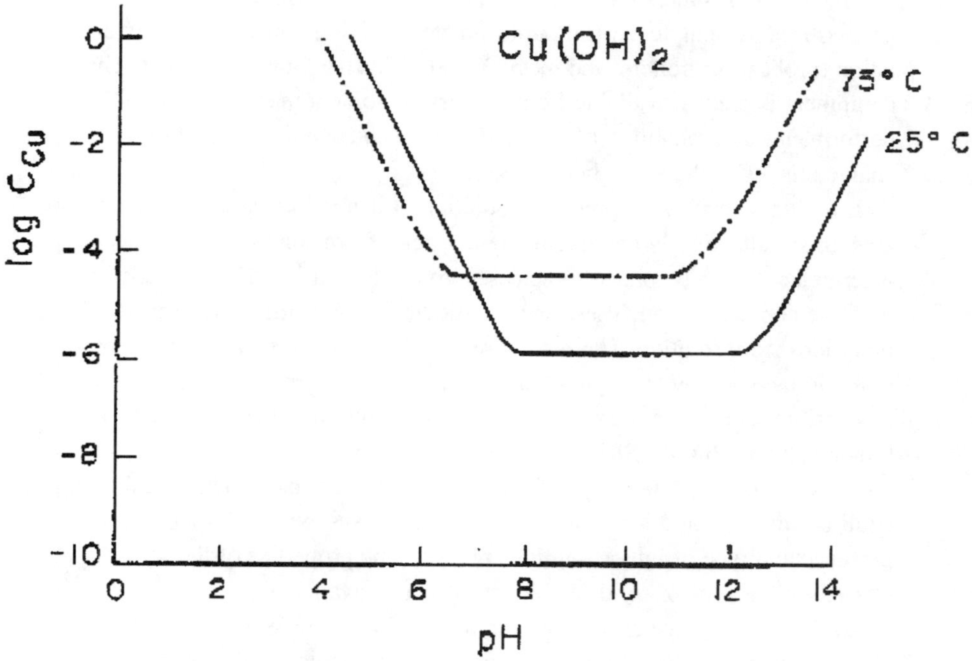

Figure 7.1 Effect of pH and temperature on copper hydroxide solubility (Cooney, 1991).

fine colloidal particles. Settling properties of these precipitates can be improved by the addition of flocculants such as alum or polyelectrolytes. Although a number of anions can be used to precipitate metals, hydroxide and sulfide are the most common. Both of these anions react with metals, producing low metal in concentrations after precipitation.

The hydroxide precipitation reaction uses an alkaline reagent to raise the pH of the wastewater, causing the solubility of the metal ions to decrease and thus precipitate from solution. Conversely, lowering the pH of a highly alkaline waste stream can achieve the same effect. This can be seen in Figure 7.1. The optimum precipitation of copper at 25 °C occurs between pH 8 and 12. The optimum pH at which metal–hydroxide complexes are least soluble is a function of pH, temperature, and concentrations of other ions present.

Some metal sulfide complexes are predicted to have lower solubilities than metal hydroxides and thus could achieve a greater dissolved metal ion removal from solution. Soluble sulfides, such as hydrogen sulfide and sodium sulfide, and insoluble sulfides, such as ferrous sulfide, have been used to precipitate heavy metals (Patterson, 1985). Because of the potential to form toxic concentrations of hydrogen sulfide during treatment, the use of a sulfide precipitation process requires precise control of chemical reagents and process pH. To avoid production of hydrogen sulfide gas, the operating pH should be elevated above pH 8.5 (Patterson, 1985). Table 7.5 presents examples of lead removal from hazardous liquid wastes using hydroxide and sulfide precipitation processes.

Precipitation is a relatively simple and well-established technology. The process requires only chemical pumps, meters, mixing tanks, pH instrumentation (probes, controllers, and meters), and solids handling equipment. This equipment is readily available from process equipment manufacturers. The performance and reliability of precipitation depends on the variability and characteristics of the waste being treated.

If multiple species are present in solution, it is possible, and often desirable, to simultaneously precipitate these species from solution. This process is termed coprecipitation. In some cases iron or aluminum salts are added to promote coprecipitation. These species provide a solid surface to remove other ions from solution. There are three mechanisms of removal: incorporation into the solid by adsorption on the surface of the growing particles, physical entrapment in the pore spaces, or substitution in the crystal lattice (Anderson and Rubin, 1981).

Effective coprecipitation is a function of many variables, including the equilibrium concentration of ions in solution, the suspension pH, and the presence and type of interacting ions present. The properties of the solid phase, such as type of solid formed, surface area available, solid age, concentration, and surface charge, are important. For example, Benjamin et al. (1982) showed that the effect of competing ions, such as lead, copper, and zinc, on the adsorption of cadmium onto aluminum oxide was to shift the cadmium adsorption edge to more alkaline pH values. Also, the removal of

Table 7.5 Examples of lead removal by various precipitation techniques.

Initial, ppm[a]	Wastewater source	Reaction sequence	Effluent, ppm	Reference
4.0	Storage battery manufacturing	pH = 6.5 with Na_2CO_3 clarification	0.088	Wallace and Singer (1981)
4.0	Storage battery manufacturing	pH = 6.5 with NaOH clarification	0.014	Wallace and Singer (1981)
5.0	Metal working	pH = xx with lime clarification	0.038	Robinson and Sum (1980)
5.0	Metal working	pH = xx with lime pH = yy with sulfide clarificatioin	0.027	Robinson and Sum (1980)
62.1	Foundry wastewater	pH = 10–10.5 with lime clarification and filtration	1.8	Ostantowski and Ruppersberger (1985)
1.16	Mixed electroplating wastewaters	Wastewater 1—pH > 11, add NaOCl to oxidize CN Wastewater 2—pH < 2.5, add $Na_2S_2O_5$ to reduce Cr(VI) to Cr(III) Combine wastewaters Add NaS + $FeSO_4$ at pH = 8 Sludge removal by clarification and filtration	<0.210	Pressman (1984)

[a]ppm × 1.000 = mg/L.

nickel was enhanced by the presence of cyanide because the cyan–nickel complex was more readily adsorbed. Merrill et al. (1986) found that increasing the iron oxide concentration was directly proportional to increasing arsenate removal.

Matsumoto et al. (1987) studied the removal of copper, nickel, and zinc from highly acidic hazardous wastewaters received by the CECOS International, Inc., Niagra Falls, New York, TSDF. Eight different jar tests were performed. Lime was first added to adjust the pH to 8; then various dosages of sodium sulfide were added. The contents were mixed for several minutes and filtered. Results from one study are presented in Table 7.6.

CHEMICAL OXIDATION AND REDUCTION. The objective of chemical oxidation or reduction is to detoxify a waste stream by adding a chemical to transform the hazardous constituents present either to nonhazardous species or to species that are more easily treated by another means. Examples of this latter purpose would be the oxidation of arsenite ion to arsenate followed by precipitation with sodium hydroxide (Patterson, 1985), the reduction of chromium(VI) to chromium(III) followed by hydroxide precipitation, and the oxidation of toxic organics before subsequent treatment by a biological treatment process.

Chemical oxidation–reduction (redox) of wastes is a well-established technology that is capable of detoxifying a number of contaminants in a hazardous waste stream. A redox reaction is a process in which there is an exchange of electrons from one substance to another. The reducing agent, or electron donor, exchanges electrons with an oxidizing agent, or electron acceptor. Theoretically, redox reactions are expressed as the sum of two half reactions, which balance the number of electrons gained or lost. The power of the oxidizing or reducing agent is determined from its electrode potential. An indication of how quickly the reaction will proceed, that is, spontaneously or slowly, can be determined from free energy considerations. The free energy of the reaction is related to the equilibrium constant and electrode potential by:

$$\Delta G^{o} = -nFE^{o} = -RT \ln K$$

Table 7.6 Co-precipitation of copper, nickel, and zinc from a TSDF wastewater.[a,b]

Contaminant	Initial, ppm[c]	Effluent, ppm
Copper	1 000	1.5
Nickel	603	4.7
Zinc	4.5	0.1

[a]Matsumoto, M.R., et al., 1987. (Used with permission.)

[b]Dosages of 60.7 g/L lime and 0.34 g/L Na_2S.

[c]ppm × 1.000 = mg/L.

Where

G^o = free energy of reaction, kcal/mole;
n = number of electrons exchanged during reaction per mole;
F = Faraday's Constant, 23.1 kcal/volt;
E^o = Standard electrode potential, determined from half reactions, volts;
R = Ideal Gas Constant, 1.98×10^{-3} kcal/mole-K; and
T = Absolute temperature, K.

If the value of G° is negative, then the reaction is spontaneous (Ford, 1992, and LaGrega et al., 1994).

Oxidation reactions can destroy volatile organic compounds, phenols, mercaptans, and cyanides. Reduction reactions have been used to treat chromium- and mercury-contaminated wastes (Patterson, 1985) and copper waste (LaGrega et al., 1994). Chemical oxidation of organic compounds may be attributed to the following reactions:

- Addition of oxygen,
- Withdrawal of hydrogen, and
- Withdrawal of electrons with or without withdrawal of protons.

The following oxidants are commonly used in reactions:

- Oxygen or air (O_2);
- Ozone (O_3);
- Hydrogen peroxide (H_2O_2);
- Potassium permanganate ($KMnO_4$);
- Chlorine (Cl2), hypochlorates, and chlorine dioxide (ClO_2); and
- Ultraviolet radiation (UV).

Reductants include sulfur dioxide, ferrous sulfate, sodium bisulfite, sodium metabisulfite, and sodium borohydride.

The choice of oxidant or reductant is based primarily on economics and materials handling. Caution must be used when applying a chlorine-containing oxidant to a waste stream containing organic contaminants because of the potential to form toxic chlorinated hydrocarbons during the reaction.

Chemical addition typically is conducted in completely mixed tanks or plug flow reactors. The oxidizing agent is injected into the system. Factors controlling the rate of reaction and the required reaction time include the concentrations of species present, the system temperature, the amount of impurities present, and pH.

Although redox processes are typically applied to hazardous liquid waste streams and contaminated groundwater, contaminated soils may also be treated. Oxidants, for example, can be added to soil in situ, or the soil can be excavated, slurried, and mixed with the oxidant in a reaction vessel. Ozone and hydrogen peroxide are the two most common oxidizing agents applied in hazardous waste treatment (LaGrega et al., 1994).

The nonhomogeneity of the waste streams frequently requires that bench-scale testing be performed before full-scale treatment is attempted. The presence of other contaminants in the waste stream must also be addressed in the interest of personnel safety and overall process economics. Control of a redox process is similar to pH adjustment in chemical management and mixing requirements. Reaction times are based on bench-scale testing. Process control is frequently possible in the closed loop mode by oxidation–reduction potential measurements or by a sensor measuring the primary contaminant under treatment.

One emerging technology uses supercritical water oxidation. Supercritical fluids are materials that, at elevated temperature and pressure, have properties between those of a liquid and a gas. In supercritical water oxidation, air and contaminated water are brought together above the critical point of water (374 °C and 218.4 atm) at which complete oxidation of organic contaminants to carbon dioxide occurs rapidly. This process is reported to be cost effective for aqueous wastes with organic concentrations in the range of 1 to 20%. In this range, the heat of combustion raises the supercritical fluid temperature from the critical temperature of water (374 °C) to 600 to 650 °C (LaGrega et al., 1994). Process design considerations include

- The ability of the waste to be pressurized,
- A potential to form char,
- Materials of construction,
- Concentration of suspended solids, and
- Potential to form salts.

Some typical oxidation and reduction reactions used during hazardous waste treatment are summarized in Table 7.7.

STABILIZATION AND SOLIDIFICATION. Hazardous waste disposal on land, even in "secure chemical landfills," may lead to serious groundwater pollution problems. Interest in the development of processes to render these wastes less dangerous before disposal has greatly increased. One process that has received a great deal of attention, primarily because of the Solid and Hazardous Waste Act amendments that ban disposal of liquids on land, is the stabilization and solidification of wastes. In this process, waste sludges, contaminated sediments, or soils are combined with various additives that chemically bind and physically solidify the hazardous materials, making them less susceptible to leaching (Bishop, 1988).

The terms "stabilization" and "solidification" are used here as defined by U.S. EPA (Malone et al., 1982). Both stabilization and solidification refer to treatment processes that are designed to accomplish one or more of the following results:

Table 7.7 Example oxidation–reduction treatment processes.

Process	Contaminant, waste	Initial, ppm	Final, ppm	Reference
Oxidation				
Ozonation	Cyanide	25	ND	Patterson (1985)
Ozone + U.V.	Cyanide	100–1000	0.6	Hassan et al. (1991)
Wet air ox.	Benzene	1–10	0.029	Randall et al. (1987)
Wet air ox.	Ethyl benzene	1–10	0.021	Randall et al. (1987)
Wet air ox.	Toluene	0.1–1	0.57	Randall et al. (1987)
Wet air ox.	Cyanide	100–1000	1.1	PEI Associates, Inc. (1990)
Wet air ox. + filtration	Cyanide	100–1000	0.13	PEI Associates, Inc. (1990)
Alkaline chlorination	Cyanide	28	2	Patterson (1985)
Alkaline chlorination	Cyanide	100–1000	150	Grosse et al. (1988)
SCOx	Carbon tetrachloride	0.1–1.0	<0.02	Staszak et al. (1987)
SCOx	2-Chlorophenol	>1000	<0.036	Staszak et al. (1987)
SCOx	PCBs	>1000	<0.082	Staszak et al. (1987)
SCOx	PCBs	>1000	<0.005 4	Staszak et al. (1987)
SCOx	Ethylene glycol	>1000	<0.075	Johnston et al. (1988)
Reduction				
Sodium sulfite	Chromium(VI)	10–100	<0.005	Robinson and Sum (1980)
Sulfur dioxide	Chromium(VI)	100	<0.05	Patterson (1985)
Ferrous sulfate	Chromium(VI)	1300	0.01	Patterson (1985)
Sodium borohydride	Mercury	10	0.22	Patterson (1985)

- Improve the handling and physical characteristics of the waste, as in the sorption of free liquids;
- Decrease the surface area of the waste mass across which transfer or loss of contaminants can occur; and
- Limit the solubility of any hazardous constituents of the waste, such as by pH adjustment or sorption phenomena.

Stabilization techniques are those that limit the solubility or mobility of the contaminants with or without change or improvement in the physical characteristics of the waste. Examples include the addition of lime or sulfide to a metal hydroxide waste to precipitate the metal ions and the addition of a sorbent to an organic waste. Stabilization typically involves adding materials that ensure that the hazardous constituents are maintained in their least mobile or toxic form.

Solidification implies that the beneficial results of treatment are obtained primarily through the production of a solid block of waste material that has high structural integrity—a product often referred to as a "monolith." The monolith can encompass the entire waste disposal site—called a "monofill"—or can be as small as the contents of a steel drum. The contaminants do not necessarily interact chemically with reagents but are mechanically locked within the solidified matrix. This is called "microencapsulation." Contaminant loss is limited largely by decreasing the surface area exposed to the environment, thereby minimizing moisture contact and leaching or isolating the contaminants from environmental influences by microencapsulating the waste particles. Wastes can also be "macroencapsulated," that is, bonded to or surrounded by an impervious covering (Wiles, 1989).

Processes. Many stabilization and solidification processes have been developed, including cement-based, lime-based, thermoplastic, organic polymer, encapsulation, glassification, and self-cementing techniques. These processes vary widely in their applicability to certain waste types, but most are best suited to inorganic wastes. The more typically used fixation processes use Portand cement, fly ash, silicates, or combinations of these to produce pozzolanic reactions resulting in heavy metal binding and conversion of liquids and sludges into solid waste forms.

Cement-based stabilization and solidification is a process in which waste sludges or contaminated soils are mixed with Portland cement. Water in the waste, or additional water added, causes hydration reactions that solidify the waste-cement matrix. Heavy metals in the waste typically react with cement constituents to form hydroxides, silicates, and other relatively insoluble products.

Small amounts of other additives, such as fly ash, clay, sodium silicate, or sulfides, may be added to improve the final product. Depending on the amount of cement and other additives used, the final product may vary from a granular, soil-like material to a solid monolith resembling hardened cement. The cement-stabilized mass typically has a high free lime content, buffering it against attack by acidic groundwater. This process is typically used for stabilization and solidification of plating wastes.

Lime-based processes involve pozzolanic reaction of lime with siliceous and aluminosilicate materials to form cementaceous substances. Commonly used pozzolans are fly ash, lime kiln dusts, and blast furnace slag, all of which contain significant amounts of silicates. The resulting reactions are similar to those for cement hydration, although they are generally slower. The final product is similar to that from cement-based processes.

Thermoplastic stabilization and solidification uses a thermoplastic material, such as asphalt or polyethylene, to bind the waste constituents into a solidified mass. Contaminants are physically entrapped (microencapsulated) within the mass, minimizing their potential to leach. These processes have been used for oil- or gasoline-contaminated soils and may have applications for some electroplating and refinery sludges. Organic polymers have been used in a few cases to complex the waste into a solidified mass. Urea formaldehyde is the most commonly used organic polymer for stabilization and solidification.

Glassification, in which the waste is mixed with molten glass and then allowed to solidify, has been used for radioactive wastes. In situ vitrification consists of inserting electrodes into soils, along with flaked graphite and glass, and applying an electric current. The soil is heated above its melting point and then cooled into a solid, vitrified mass incorporating inorganic contaminants. Organic contaminants are either destroyed by pyrolysis at the elevated temperatures or are vaporized and collected by a collection hood and scrubbers (Fitzpatrick et al., 1987).

Macroencapsulation is used to stabilize and solidify wastes in overpack drums. The wastes are typically agglomerated with lime, cement, or a polybutadiene binder and then encapsulated with a polyethylene resin overpack drum.

Waste Characterization and Treatability Studies. Wastes to be stabilized and solidified must be characterized before and after treatment. Pretreatment testing is performed to determine basic information on treatability of the waste material. Poststabilization and solidification characterization is needed to ensure that the process has been successful.

Both physical and chemical tests are used to evaluate waste stabilization and solidification processes. Physical tests used to evaluate waste stabilization and solidification processes include

- Index property tests, which provide data that are used to relate general physical characteristics of a material (for example, suspended solids) to process operational parameters (for example, pumpability);
- Density tests, which are used to determine weight-to-volume relationships of materials;

- Permeability tests, which measure the relative ease with which fluids (water) will pass through a material;
- Strength tests, which provide a means for judging the effectiveness of a stabilization and solidification process under mechanical stresses; and
- Durability tests, which determine how well a material withstands repeated wetting and drying or freezing and thawing cycles.

Specific tests associated with each of these are shown in Table 7.8 (U.S. EPA, 1986).

Chemical tests used with stabilization and solidification processes include those for pH, alkalinity, oxidation–reduction potential, total organic carbon, oil and grease, elemental analysis for metals, and analyses for specific organic compounds.

The most important characterization procedure is for contaminant leaching potential. Numerous leaching tests have been developed to test solid wastes, including specific tests for those wastes that have been stabilized and solidified. Table 7.9 lists some of the more common leaching tests (U.S. EPA, 1986). Each of these has its advantages and disadvantages; normally, more than one test is needed to determine the leaching properties of a specific waste.

The extraction procedure toxicity (EP Tox) test was mandated by U.S. EPA as the procedure to classify wastes as hazardous or nonhazardous. It does have limitations though and cannot be used to determine leaching rates or leaching of volatile organics. The TCLP has replaced the EP Tox test for defining hazardous and nonhazardous wastes as required by RCRA, 40 CFR Part 261.24. It has the advantage of evaluating leaching of volatile and semi-volatile organic compounds. The CalWET procedure is used in the state of California. It is similar to TCLP, except that it uses sodium citrate as the

Table 7.8 Physical testing methods for stabilization–solidification.

Test procedure	Reference	Purpose
Index property tests		
Particle size analysis	ASTM D422-63	To determine the particle size distribution of a material
Atterberg limits		
Liquid limit	ASTM D4318-84	To define the physical characteristics of a material as a function of its water content
Plastic limit	ASTM D4318-84	
Plasticity index	ASTM D4318-84	
Moisture content	ASTM D2216-80	To determine the percentage of free water in a material
Suspended solids	EPA method 209C	To determine the amount of solids that do not settle from a column of liquids

(Continued)

Table 7.8 Physical testing methods for stabilization–solidification (continued).

Test procedure	Reference	Purpose
Paint-filter test	EPA method 9095-SW846	To determine the presence of free liquids in a representative sample of bulk or noncontainerized waste
Density testing		
Bulk density drive-cylinder method	ASTM D2937	To determine the inplace density of soils or soil-like materials
Bulk density sand-cone method	ASTM D15556-82	To determine the inplace density of soils or soil-like materials
Bulk density nuclear methods	ASTM D2022-81	To determine the inplace density of soils or soil-like materials
Compaction testing		
Moisture density Relations of soil–cement mixtures	ASTM D558-82 ASTM D1557	To determine the relation between moisture content and density of a material
Permeability testing		
Falling head permeability	EPA method 9100-SW846	To measure the rate at which water will pass through a soil-like material
Strength testing		
Unconfined compressive strength of cohesive soils	ASTM D2166-85	To evaluate how cohesive, soil-like materials behave under mechanical stress
Unconfined compressive strength of cylindrical concrete specimens	ASTM D1633-84	To evaluate how cement-like materials behave under mechanical stress
Compressive strength of hydraulic cement mortars	ASTM C109-86 ASTM C109-86	To measure the compressive strength of hydraulic cement
Flexural strength	ASTM D1635-87	To evaluate a material's ability to withstand loads over a large area
Cone index	ASTM D3441-79	To evaluate a material's stability and bearing capacity
Durability testing		
Freeze–thaw durability	ASTM D560-82	To determine how materials behave or degrade after repeated freeze–thaw cycles
Wet–dry durability	ASTM D559-57	To determine how materials behave or degrade after repeated wet–dry cycles

Table 7.9 Leaching test methods for stabilized–solidified wastes.

Test method	Leaching medium	Number of extractions
EP Tox	0.04 M acetic acid (pH=5.0)	1
TCLP	Acetic acid	1
Cal WET	0.2 M sodium citrate (pH=5.0)	9 or more
MEP	Same as EP Tox, then with synthetic acid rain	9 or more
MWEP	Deionized water	4
Equilibrium leach test	Distilled water	4
Acid neutralization capacity	HNO_3 solutions of increasing strength	1
Sequential extraction test	0.04 M acetic acid	15
AMS-16.1	Deionized water	12
Dynamic leach test	Deionized water	Varies

leachant and therefore is a more stringent leach test for some metals. All three of these test procedures use a single extraction of the waste and consequently do not provide leaching rate data.

Many of the remaining leach tests listed in Table 7.9 are dynamic leach tests in that the leaching solution is periodically replaced with new solution. These tests simulate leaching of a monolithic waste form under nonequilibrium conditions and can often be used to predict flux or leaching rates from the waste. None of the leaching tests will produce results that are directly applicable to leaching behavior in the field. Nevertheless, results from several leach tests or from leach tests combined with physical tests can be used as indicators of field performance (Bishop, 1986, and PEI, 1990). Bench-scale testing is needed to determine proper waste:binder ratios and to evaluate the durability of the finished material.

Processing. The stabilization and solidification process can be used to process newly created wastes for disposal as well as for contaminated soils at hazardous waste sites. Mixing of the waste and binder can be done in a drum or plant process or in situ for contaminated soils, although it is typically easier to provide proper mixing in a reactor than to provide solidifying in a pit, pond, or lagoon (Wiles, 1989).

To date, very little specialized equipment has been developed for the application of stabilization and solidification processes. Equipment in use today has been adapted from either the building construction or chemical processing industries. Mixing units similar to those used for mixing concrete are commonly used. Others include pug mills, ribbon blenders, mixers, extruders, and screw conveyors. Many of the stabilization and solidification activities at hazardous waste sites have used mixing. The reagents are mixed directly into the soil or waste using backhoes or screw augers. This method has the benefits of not requiring removal of the waste from the site and of low cost, but mixing is often incomplete and inadequate.

Placement of the stabilized and solidified waste depends on its form. Wastes can be stabilized in place (in situ) or at another location (ex situ). Waste stabilized and solidified in situ can range in consistency from loose soil to concrete blocks. Soil-like materials should be placed in lifts using earth-moving equipment and then compacted. Monolithic materials must be placed before setting occurs.

Applications. Stabilization and solidification processes are already widely used to treat hazardous wastes. A 1986 TSDF survey conducted by the U.S. EPA (1990) found 107 stabilization and solidification units in operation in the U.S., of which 50 were cement-based and 37 were pozzolanic processes. Approximately 658 000 tons of wastes were being processed annually. These processes are also being used to treat contaminated soils and other wastes at Superfund sites. Approximately 25% of all Superfund sites were using some form of stabilization and solidification in 1988. The U.S. EPA SITE program has evaluated six stabilization and solidification processes to date for use at Superfund sites.

Stabilization and solidification technology has been used for more than 20 years to treat industrial and radioactive waste. Most recently the technology has received greater interest as a treatment approach for a variety of RCRA hazardous waste, contaminated soils, sludges, and municipal combustion residuals (ash). The results from several example solidification–stabilization processes are summarized in Table 7.10. Some specific case studies are presented below.

Stabilization was applied at a Superfund site in Morganville, New Jersey, by Solidtech, Inc. (U.S. EPA, 1990). Past activities at the site included chemical processing and oil reclamation. Heavy metals, such as arsenic, chromium, copper, lead, nickel, and zinc, and organic contaminants, including polychlorinated biphenyls (PCBs) and petroleum hydrocarbons, were present at the site in soil, a waste filter cake pile, and groundwater. The solid materials were treated by a several stabilization processes using Portland cement and proprietary additives. The unconfined strength of the treated samples ranged from 390 psi to 860 psi for the filter cake–oily sludge mixture. The permeability, wet–dry durability, and freeze–thaw durability results were also within acceptable guidelines. Table 7.11 shows some of the results of using this process to stabilize the waste sludge at this site.

A stabilization and solidification process developed by Hazcon, Inc. (U.S. EPA, 1989) was recently demonstrated at a former oil reprocessing plant in Douglassville, Pennsylvania. Soils at the site were contaminated with organic and inorganic contaminants, including oil and grease, volatile and semivolatile organics, PCBs, and heavy metals (particularly lead). After excavation, the contaminated soil was screened and fed to a mobile field blending unit consisting of soil and cement holding bins, a chloronan (proprietary nontoxic additives that encapsulate organic molecules, rendering them ineffective

Table 7.10 Sample lead stabilization–solidification processes.

Waste type	Process	Before treatment, TCLP, mg/L	After treatment, TCLP, mg/L	Reference
Ash	Cement 2:1	30	1.5	Boehmer and Larsen (1986)
Ash	Cement + Silicate	440	0.300	Boehmer and Larsen (1986)
Dry waste	5% Cement	45	1.2	Bricka and Cullinane (1990)
Dry waste	5% Kiln dust	45	0.79	Bricka and Cullinane (1990)
Dry waste	10% Lime + fly ash	45	0.095	Bricka and Cullinane (1990)
Liquid	140% Kiln dust	100	1.0	Bricka and Cullinane (1990)
Soil	Cement, 28-day cure	70	0.370	Weitzman and Hamel (1988)
Soil	Hazcon, 28-day cure	32	0.005	U.S. EPA (1989)
Soil	Chemfix	610	<0.050	U.S. EPA (Sept., 1900)
Sludge	Cement + additives	5.4	<0.050	U.S. EPA (Feb., 1990)
Sludge	20% Cement	1400	0.540	Bricka and Cullinane (1990)
Sludge	40% Kiln dust	1400	7.2	Bricka and Cullinane (1990)
Sludge	40% Lime + fly ash	1400	31	Bricka and Cullinane (1990)

Table 7.11 Sample results from solidification of waste sludge at Morganville, New Jersey, Superfund site.[a]

Contaminant	Untreated, mg/kg	Untreated TCLP, mg/L	Treated TCLP, mg/L
Tetrachloroethylene	1.6	0.008 9	<0.002 0
Toluene	8.4	0.055	<0.024
Trichloroethylene	3.3	0.027	<0.013
Xylenes	32	0.057	<0.014
PCB (Arochlor-1242)	16	<0.000 43	<0.000 11
PCB (Arochlor-1260)	27	<0.000 86	<0.000 21
Cadmium	1.0	0.009 3	<0.005 0
Copper	34	0.023	<0.020
Chromium	5.7	<0.030	<0.030
Lead	2500	5.4	<0.050
Nickel	3.0	0.027	<0.020
Zinc	150	1.3	<0.020

[a]U.S. EPA, 1990.

in interfering with the solidification process), and a blending process. Here the waste and pozzolanic material (cement, fly ash, and kiln dust) were added along with chloronan and water to produce a slurry. The slurry was then allowed to harden into a concrete-like mass before disposal. Results of the project are summarized below and in Tables 7.12 and 7.13.

- The volume of the solidified mass was more than double the undisturbed soil.
- Permeabilities after 28 days curing were 10^{-8} to 10^{-9} cm/s.
- Unconfined compressive strength ranged from approximately 220 to 1570 psi.
- Wet–dry and freeze–thaw weathering tests showed small weight losses (0.5 to 1.5% after 12 cycles).
- Volatile organics found in leachate appeared to be approximately the same in both the untreated and treated soils (less than 1 mg/L).
- Leachate from the solidified soils showed metal levels at or near detection limits, with lead the predominant metal lower by a factor of 500 (from between 20 and 50 mg/L in untreated soil to 0.1 mg/L in treated soil).

A number of companies are proposing stabilization and solidification processes that use proprietary reagents to enhance the stabilization of both organic and inorganic wastes. Other companies are investigating the use of organophilic clays along with cementaceous materials to bond organic wastes (Gibbons and Soundararajan, 1988). This technology looks very promising and may extend the use of stabilization and solidification beyond its current primary use as a treatment method for inorganic wastes.

Table 7.12 Chemical properties.

	Leachate concentrations, mg/L					
	Untreated soil			28-day cores		
Sample	VOC[a]	BNA[b]	Lead	VOC[a]	BNA[b]	Lead
1	0.92	ND[c]	1.5	0.38	ND[c]	0.007
2	0.02	1.02	31.8	0.06	1.45	0.005
3	1.03	2.81	17.9	0.72	2.79	0.400
4	5.10	0.01	27.7	0.37	0.10	0.050
5	1.10	0.01	22.4	0.84	0.11	0.011
6	0.06	0.01	52.6	0.11	0.73	0.051

[a]Volatile organic carbon.

[b]Base neutral/acid extractable.

[c]Not determined.

Payne et al. (1992) used a process with quicklime (CaO) to successfully stabilize oil-laden wastes (>15% oil) containing organic contaminants. The quicklime was treated with a fatty acid to change it from a hydrophilic compound to a hydrophobic one. The "pretreated" quicklime was then used to adsorb and bind the oils when mixed with the waste. Following this initial interaction, the CaO underwent an exothermic hydration step to produce calcium hydroxide. This solid, dispersed throughout the waste mixture, slowly reacted with natural CO_2 to form calcium carbonate. At this point, the contaminants have been irreversibly bound. For example, the TCLP for toluene was reduced from 1100 ug/L before treatment to 8 ug/L after 8 days of curing by this technique.

Costs. Costs for treatment and disposal using stabilization and solidification processes depend on waste-specific and site-specific conditions. Important

Table 7.13 Physical properties.

	Untreated soil			28-day cores		
Sample	Oil and grease, %	Bulk density, g/mL	Permeability, cm/s	bulk density, g/mL	Permeability, cm/s	UCS, psi[a]
1	1.0	1.23	5.7×10^{-1}	1.95	1.8×10^{-9}	1110
2	16.5	1.40	1.8×10^{-3}	1.61	4.0×10^{-9}	520
3	25.3	1.60	Impermeable	1.51	8.4×10^{-9}	220
4	4.3	1.68	2.0×10^{-2}	1.84	4.5×10^{-9}	940
5	4.5	1.73	7.7×10^{-2}	2.07	5.0×10^{-10}	1570
6	7.8	1.59	1.5×10^{-5}	1.70	2.2×10^{-9}	890

[a]psi × 6.895 = kPa.

Table 7.14 Cost of solidification–stabilization of contaminated soils.

Item	Estimated cost
Mobilization	\$100 000–200 000/site
Excavation	\$8–38 m^3
Processing	\$38–76 m^3
Disposal	\$76–191 m^3

factors are waste characteristics, type of process, disposal requirements, and other factors. Table 7.14 gives estimated ranges for stabilization and solidification processing costs for Superfund site work (U.S. EPA, 1989). In 1992, costs for employing stabilization using Portland cement at an NPL site ranged from \$84 to nearly \$150 per yard, depending on site conditions such as moisture content and accessability to the contaminated soil.

REFERENCES

Anderson, M., and Rubin, A. (1981) *Adsorption of Inorganics at Solid–Liquid Interfaces.* Ann Arbor Science Publishers, Ann Arbor, Mich.

Benjamin, M., et al. (1982) Removal of Toxic Metals from Power Generation Waste Streams by Adsorption and Coprecipitation. *J. Water Pollut. Control Fed.,* **54,** 1472.

Bishop, P., (1986) Prediction of Heavy Metal Leaching Rates from Stabilized/Solidified Hazardous Wastes. Paper presented at 18th Mid-Atlantic Ind. Waste Conf., Blacksburg, Va.

Bishop, P. (1988) Leaching of Inorganic Hazardous Constituents from Stabilized/Solidified Hazardous Wastes. *Haz. Waste Haz. Mater.,* **5,** 129.

Boehmer, A.M., and Larsen, M.M. (1986) Hazardous and Mixed Waste Solidification Development Conducted at the Idaho National Engineering Laboratory. Rep. No. EGG-WM-7225, EG&G Idaho, Inc.

Boehmer, A.M., et al. (1986) Drum-Scale Fly Ash Stabilization Development. Rep. No. EGG-WT-7393, EG&G Idaho, Inc.

Bricka, R.M., and Cullinane, M.J., Jr. (1990) Solidification/Stabilization as a Best Demonstrated Available Technology for Resource Conservation and Recovery Wastes. *Proc. 15th Annu. Res. Symp.,* EPA-600/9-90-006.

Cooney, E.J. (1991) The Effect of Ferric Oxide Age on Adsorption and Coprecipitation of Copper from Aqueous Solution. Unpublished Ph.D. thesis, Dep. Environ. Eng., Illinois Institue of Technology, Chicago, Ill.

Fitzpatrick, V., et al. (1987) In-Situ Vitrification—An Innovative Thermal Treatment Technology. Groundwater Treatment Conf., New York.

Ford, D.L. (Ed.) (1992) *Toxicity Reduction: Evaluation and Control.* Water Quality Management Library, Vol. 3, Technomic Publishing Company, Inc., Lancaster, Pa.

Gibbons, J., and Soundararajan, R., (1988) The Nature of Chemical Bonding Between Modified Clay Minerals and Organic Waste Materials. *Am. Lab.,* **20,** 38.

Grosse, D.W., et al. (1988) Evaluation of Technologies for Treating Metal/Cyanide Bearing Hazardous Wastes (F007). EPA-600/D-88-195, U.S. EPA.

Hassan, S.Q., et al. (1991) Ozone-Ultraviolet Light Treatment of Iron Cyanide Complexes. *Proc. 17th Annu. RREL Haz. Waste Res. Symp.* EPA-600/9-91-002, U.S. EPA.

Johnston, J.B., et al. (1988) Destruction of Pharmaceutical and Biopharmaceutical Wastes by the MODAR Supercritical Water Oxidation Process. In *Biotechnology.* Vol. 6.

LaGrega, M.D., et al. (1994) *Hazardous Waste Management.* McGraw-Hill, New York.

Malone, P., et al. (1982) Guide to the Disposal of Chemically Stabilized and Solidified Waste. SW-872, U.S. EPA, Washington, D.C.

Matsumoto, M.R., et al. (1987) The Removal of Metal Contaminants from Aqueous Hazardous Wastes Prior to Biological Treatment. AIChE 1989 Spring Natl. Meeting, Houston, Tex.

Merrill, D., et al. (1986) Field Evaluation of Arsenic and Selenium Removal by Iron Coprecipitation. *J. Water Pollut. Control Fed.,* **58,** 18.

Ostantowski, R.A., and Ruppersberger, J. (1985) Upgrading Foundry Wastewater Treatment. *Proc. 39th Ind. Waste Conf.,* Purdue Univ., W. Lafayette, Ind.

Patterson, J.W. (1985) *Industrial Wastewater Treatment Technology.* 2nd Ed., Butterworth, Boston, Mass.

Payne, J.R., et al. (1992) DCR Treatment of Oily Wastes and Oil-Contaminated Soils. *Proc. R&D 92—Natl. Res. Dev. Conf. Control Haz. Mater.,* San Francisco, Calif.

PEI Associates, Inc. (1990) Characterization and Treatment of Wastes from Metal Finishing Operations. EPA-600/9-90-055, U.S. EPA.

Pressman, M. (1984) Pilot Plant Demonstration of Sulfide Precipitation Process for Metal Finishing Wastewater Treatment. Rep. 2408, U.S. Army, Belvoir Research and Development Center, Fort Belvoir, Va.

Randall, T.L., et al. (1987) Solvent Waste Cleanup by a Combination of Wet Air Oxidation and Biophysical Treatment. Paper presented at the 60th Annu. Conf. Water Pollut Control Fed., Philadelphia, Pa.

Robinson, A.K., and Sum, J.C. (1980) Sulfide Precipitation of Heavy Metals. U.S. EPA-600/2-80-139, U.S. EPA.

Snoeyink, V.L., and Jenkins, D. (1980) *Water Chemistry.* Wiley and Sons, New York.

Staszak, C.N., et al. (1987) The Pilot Scale Demonstration of the MODAR Oxidation Process for the Destruction of Hazardous Organic Waste Materials. *Environ. Prog.,* **6,** 1, 39.

U.S. Code of Federal Regulations (1996) Title 40, 40 CFR Parts 260 to 268.

U.S. Environmental Protection Agency (1979) *Survey of Solidification/ Stabilization Technology for Hazardous Industrial Waste.* EPA-600/2-79-056, U.S. EPA, Washington, D.C.

U.S. Environmental Protection Agency (1986) *Test Methods for Evaluating Solid Waste.* 3rd Ed., SW-846, U.S. EPA, Washington, D.C.

U.S. Environmental Protection Agency (1989) Immobilization Technology Seminar—Speaker Slide Copies and Supporting Information. CERI 89-222, U.S. EPA, Cincinnati, Ohio.

U.S. Environmental Protection Agency (1990) *Technology Evaluation Report: Chemfix Technologies, Inc. Solidification/Stabilization Process, Clackamas, Oregon.* EPA-540/5-89-011a.

Wallace, J.R., and Singer, P.C. (1981) Precipitation of Lead from Storage Battery Manufacturing Wastewater. *Proc. 35th Ind. Waste Conf.,* Purdue Univ., W. Lafayette, Ind.

Weitzman, L., and Hamel, L.E. (1988) Evaluation of Solidification/ Stabilization as a BDAT for Superfund Soils. Acurex Corp. Project No. 8304, EPA-RREL, Cincinnati, Ohio.

Wiles, C. (1989) Solidification and Stabilization Technology. In *Standard Handbook of Hazardous Waste Treatment and Disposal.* H. Freeman (Ed.), McGraw-Hill, New York.

Chapter 8
Thermal Processes for Hazardous Waste Treatment

INTRODUCTION

The disposal of waste materials by thermal processing has been practiced for centuries. Farmers have routinely burned agricultural wastes and animal byproducts as a way to produce energy and as a housekeeping practice. The coming of the machine age and the industrial revolution brought about the creation of manufacturing wastes, many of which could be disposed of by thermal processing. Government regulation of these waste disposal practices did not occur until the early 1960s. The initial regulations were primarily concerned with the generation of air emissions from the combustion processes. As the public became more aware of the environment and sought to clean up the air, land, and water, governmental regulations increased. The passing of the Resource Conservation and Recovery Act (RCRA) in 1975 dramatically changed the requirements for the treatment of wastes using thermal processing.

Although recognized as an effective means of waste disposal, thermal treatment has faced strong public opposition. The mere mention of the "I" word (incineration) spawns protests by environmental activists and generates citizen coalitions. Thermal treatment of wastes had caused such a firestorm of concern, that on May 18, 1993, Carrol Browner, head of the U.S. EPA under President Clinton issued a moratorium on the issuance of hazardous waste incineration permits. Although not absolute, the moratorium put a low priority on evaluating hazardous waste incineration permit applications. This included both those units designed and operated solely for the destruction of hazardous waste, as well as boilers and industrial furnaces that burned hazardous waste as a fuel source.

U.S. EPA issued a final strategy for hazardous waste minimization and combustion in 1995. It contains eight strategic goals.

(1) Public outreach and state coordination. The purpose of the goal is to maintain the broad and open dialogue among all the stakeholders in the hazardous waste disposal process. It gives top priority to the federal government working with the states as co-regulators.

(2) Waste minimization. The purpose of the goal is to encourage and reinforce the strong preference for waste source reduction rather than waste treatment and disposal.

(3) Role of combustion and alternative technologies. The purpose of the goal is to define the appropriate role for combustion in the hazardous waste disposal process.

(4) Emission controls and standards. The purpose of the goal is to develop more efficient air pollution controls for waste combustion processes.

(5) Enforcement and compliance. The purpose of the goal is to encourage the continuing compliance with hazardous waste disposal regulations and the strict enforcement against violations of the regulations.

(6) Public involvement in permitting. The purpose of the goal is to maintain the involvement of the public as stakeholders in the hazardous waste disposal process.

(7) Facility permitting priority. The purpose of the goal is to give higher permitting priority to facilities for which a final permit decision would result in the greatest environmental benefit or greatest reduction in health risk.

(8) Risk assessment. The purpose of the goal is to encourage the use of scientific understanding and evaluation on combustion issues to ensure that permits are issued that protect against unacceptable risks to health and environment.

POTENTIAL APPLICATIONS

A hazardous waste is any material that meets the definition of a hazardous waste as defined in 40 CFR Part 261. Determining whether a waste is hazardous can be an extremely complicated process, and extra care should be taken to ensure it is done correctly.

According to U.S. EPA, approximately 306 million tons of hazardous waste are generated every year. Approximately 1% of that is currently destroyed by incineration. The types of incineration and facilities that treat waste vary. Table 8.1 shows a listing of system type and permit status.

WATER. Technically, aqueous waste streams containing hazardous organic contaminants can be successfully treated with thermal processes. Typically, economics will be the governing factor as to whether this approach is used. Some key factors that need to be examined and evaluated to determine if thermal processing is the best approach include organic content and solids content.

An aqueous waste is defined as a liquid stream containing at least 40% (wt) water. In general, the concentration of organics in industrial aqueous wastes can vary from a few percent to as much as 60%. At concentrations less than 20%, self-sustained, high-temperature combustion without an assist fuel is impossible. This corresponds to a higher heating value of the waste stream of approximately 29 000 BTU/gal (3500 BTU/lb). Concentrations less than this will require assist fuel to maintain temperature.

Aqueous waste streams that contain inorganic dissolved solids will generate fine particulate (typically less than 1 micron in size) when processed at high temperatures. These particulates, which are often metal salts or metal oxides, must be collected before discharging the combustion gases into the atmosphere.

Table 8.1 Incineration permit status.

Facility type	Permitted	Interim status	Proposed	Total
Commercial incinerator	21	7	5	33
Noncommercial incinerator	120	14	21	155
Commercial BIF[a]	0	38	1	39
Noncommercial BIF[a]	0	98	2	100
Remedial	0	0	2	2
Chemical munitions incinerator	2	0	7	9

[a]Boiler or industrial furnace.

AIR. Determining whether a gas is classified as a hazardous waste can be a confusing and convoluted process. The distinction must be made to determine whether the control of the gaseous substance is regulated under the hazardous waste regulations or the Clean Air Act. The Clean Air Act Amendments of 1990 have focused on reducing the emissions of toxic air pollutants, also known as hazardous air pollutants (HAPs). A total of 188 HAPs are targeted for control by the Clean Air Act. The regulations typically specify a maximum achievable control technology (MACT) standard for each of the compounds according to the industrial category that emits them. Many of the MACT standards use thermal processing as their basis for control standards. Thermal processes have been successful in controlling HAP emissions from various industrial processes subject to the MACT standards of the Clean Air Act Amendments. These include the following:

- Asphalt manufacturing and processing,
- Petroleum and natural gas production and refining,
- Gasoline and organic liquids distribution operations,
- Surface crating operations,
- Manufacture of paints and coatings,
- Agricultural chemicals production,
- Fibers production,
- Pharmaceutical production,
- Polymers and resins production,
- Synthetic organic chemical manufacturing,
- Pulp and paper production and wood products industry,
- Rubber manufacturing,
- Explosives production,
- Photographic chemicals production, and
- Semiconductor manufacturing.

SOILS AND SLUDGES. Treatment of contaminated soils has been successfully done with both high-temperature and low-temperature thermal processing. On-site treatment is typically more economical than transporting the material to a central treatment facility.

SPECIFIC CHEMICAL COMPOUNDS. There are many specific hazardous waste chemicals generated as a result of manufacturing processes. In many instances, the most effective way to dispose of these substances is with thermal oxidation. Some of these include the following:

- Fluids from organic acid manufacturing,
- Byproducts from pesticide manufacturing,
- Chlorosilanes from pharmaceuticals manufacturing,
- Gases and liquids generated from the manufacturing of vinyl chloride monomer, and
- Organometallic compounds used in the glass industry.

Treatment of such wastes at the point of manufacture with thermal oxidation is an economic and effective means of disposal.

REGULATORY REQUIREMENTS

The treatment of hazardous waste is controlled by a number of federal regulations. These include the following:

- The Resource Conservation and Recovery Act of 1976 (RCRA). The Resource Conservation and Recovery Act provides regulations that control the storage, transport, treatment, and disposal of hazardous waste. The act is written to provide "cradle to grave" tracking of hazardous wastes.
- The Toxic Substance Control Act of 1979 (TSCA). Section 6(e) of this act prohibits the further manufacture of polychlorinated biphenyls (PCBs), established limits on PCBs used in commerce, and established regulations for their disposal.
- The Comprehensive Environmental Response Compensation and Liability Act of 1980 (CERCLA). The Comprehensive Environmental Response Compensation and Liability Act provides for the cleanup of uncontrolled waste sites resulting from poor disposal practices of the past. This act provides for the establishment of the Superfund to assist with the remedial actions.
- The Superfund Amendments and Reauthorization Act of 1986 (SARA). The Superfund Amendments and Reauthorization Act expands the provisions and funding of CERCLA.
- The Hazardous and Solid Waste Amendments of 1984 (HSWA). The Hazardous and Solid Waste Amendments significantly amend and reauthorize RCRA. It establishes a strict timeline for the restriction of untreated waste from land disposal. It also establishes pretreatment levels achievable by best demonstrated available technology (BDAT).
- The Clean Air Act Amendments of 1990 (CAAA). At the time of this writing, proposed MACT standards for hazardous waste combustors are being negotiated. These standards will be codified as a new Subpart EEE to 40 CFR Part 63. Although the new, more stringent MACT standards would supersede existing RCRA standards, the current opinion is that hazardous waste combustors will require both RCRA and CAAA permits.

The 1993 U.S. EPA moratorium on issuing hazardous waste incineration permits has made it almost impossible to build a new hazardous waste incineration system. The moratorium is not absolute, however, because 46 out of the 50 states have RCRA permitting authority.

Incineration standards per 40 CFR Part 264.343 require the following:

- A 99.99% destruction and removal efficiency for each principal organic hazardous constituent (POHC),

- Hydrochloric acid (HCl) emissions no greater than the larger of either 1.8 kg/h or 1% of HCl in the stack gas, and
- Particulate matter emissions no greater than 180 mg/dscm (0.08 grains per dry standard cubic feet) corrected to 7% oxygen. It is extremely difficult to obtain a permit for waste characterization.

WASTE CHARACTERIZATION

Hazardous wastes can be divided into three categories: solids, liquids, and gases. Sludges and slurries will typically be classified as liquid if they can be pumped and atomized, solids if they cannot. Each of these categories has a number of candidates, which include the following:

Solids

- Off-spec product,
- Contaminated combustible solids (wood, plastic, cloth), and
- Contaminated noncombustible solids (dirt, rock).

Liquid

- Off-spec or expired product,
- Chemical and pharmaceutical manufacturing byproducts, and
- Contaminated water.

Gas

- Compressed gases and
- Hazardous waste off-gases.

Hazardous Waste

- Industry generated (current and ongoing).

PHYSICAL STATE PARAMETERS.

Solids

- Material handling properties (size uniformity),
- Ash content, and
- Heating value.

Sludge

- Pumpability,
- Atomizable,
- Ash content,
- Heating value, and
- Corrosivity.

Liquid

- Viscosity,
- Atomizability,
- Solids content (suspended and dissolved),
- Heating value, and
- Corrosivity.

Gas

- Pressure,
- Oxygen content, and
- Heating value.

COMBUSTION THEORY. Combustion is the rapid combination of oxygen with a fuel, resulting in the release of heat. A fuel can be defined as any substance that, when heated in the presence of oxygen, will upon reaching its auto-ignition point oxidize and release heat. The process is a chemical reaction and can be written and balanced. For toluene (C_7H_8), the reaction equation would be as follows:

$$C_7H_8 + (9)O_2 + (9)\,3.76\,N_2 \xrightarrow{\text{HEAT}} 7CO_2 + 4H_2O + (9)\,3.76\,N_2$$

One of the requirements for the combustion of any fuel is that the fuel be in a gaseous state. This makes it necessary to change the state of solid and liquid fuels before they can be burned.

For solid fuels and combustible wastes, this process is fairly complex and is perhaps best illustrated by following the path of a piece of carbonaceous material into a hot combustion chamber. As a fuel particle is blown into an incinerator combustion chamber, the hot gases transfer energy to the particle according to the following equation:

$$Q = \frac{4\pi(T_s - T_g)}{(1/r_s - 1/r_g)} \qquad (1)$$

Where

Q = energy transferred,
π = gas thermal conductivity,
T_s = temperature of the particle,
T_g = temperature of the gas surrounding the particle,
r_s = radius of the particle, and
r_g = external radius of the boundary layer.

After being heated above a certain temperature, the particle enters the thermal decomposition stage. During this stage, volatile matter is released and char residue begins to form. The volatile matter is mixed with the surrounding air through molecular and turbulent diffusion. If the stoichiometry is correct and the temperature is high enough, the gases ignite. It can be shown from equation (1) that the heat flux to the particle is proportional to the square of the particle's radius. Therefore, the smaller the radius the less heat required to achieve the decomposition temperature and the quicker the combustion reaction proceeds.

Like the solid particle, when a liquid fuel (waste) droplet enters a hot combustion chamber, it is heated and begins to evaporate. The rate of evaporation is defined by the following equation:

$$m = 4\pi r^2 \beta (C_s - C_g) \tag{2}$$

Where

m = vaporization rate,
r = radius of the droplet,
β = mass transfer coefficient,
C_s = concentration of saturated vapors at the droplet surface, and
C_g = concentration of saturated vapors in the surrounding gas.

The vapors mix with the surrounding air through molecular and turbulent diffusion. If the stoichiometry is correct and the temperature is high enough, the gas will ignite.

As can be seen from equations (1) and (2), the size of the particle or the droplet is a critical parameter, which dictates the rate and efficiency of the combustion process. When possible, solid fuels should be ground or pulverized and injected in suspension to the combustion chamber. Liquid fuels (wastes) should be finely atomized into the combustion chamber or burner.

For waste material used as a fuel, as with conventional fuels, increasing the intensity of the combustion process decreases the amount of time required for the reaction to take place. It will also increase the completeness of the reaction and reduce the required reactor volume. The well-known "three Ts of combustion" (time, temperature, and turbulence) are three variables that help define and control the combustion process.

TIME. The more time allowed for a reaction to take place, the more complete the reaction will be. In combustion reactions, however, this is a law of diminishing returns. The majority of the combustion reaction (90%) occurs in less than 100 m/s. Squeezing out the last 10% of the reaction can take 5 times as long (see Figure 8.1).

TEMPERATURE. When an organic material is combusted, heat is produced. This heat raises the temperature of the products of the reaction. If the reaction is started and completed in a perfectly insulated vessel, the heat generated would only be transferred to the products of the combustion. The resulting temperature of the products would then be referred to as the adiabatic or theoretical flame temperature. In combustion kinetics, the rate constant depends on temperature alone. Therefore, the higher the temperature, the faster the reaction. It is not always practical, however, to operate at the

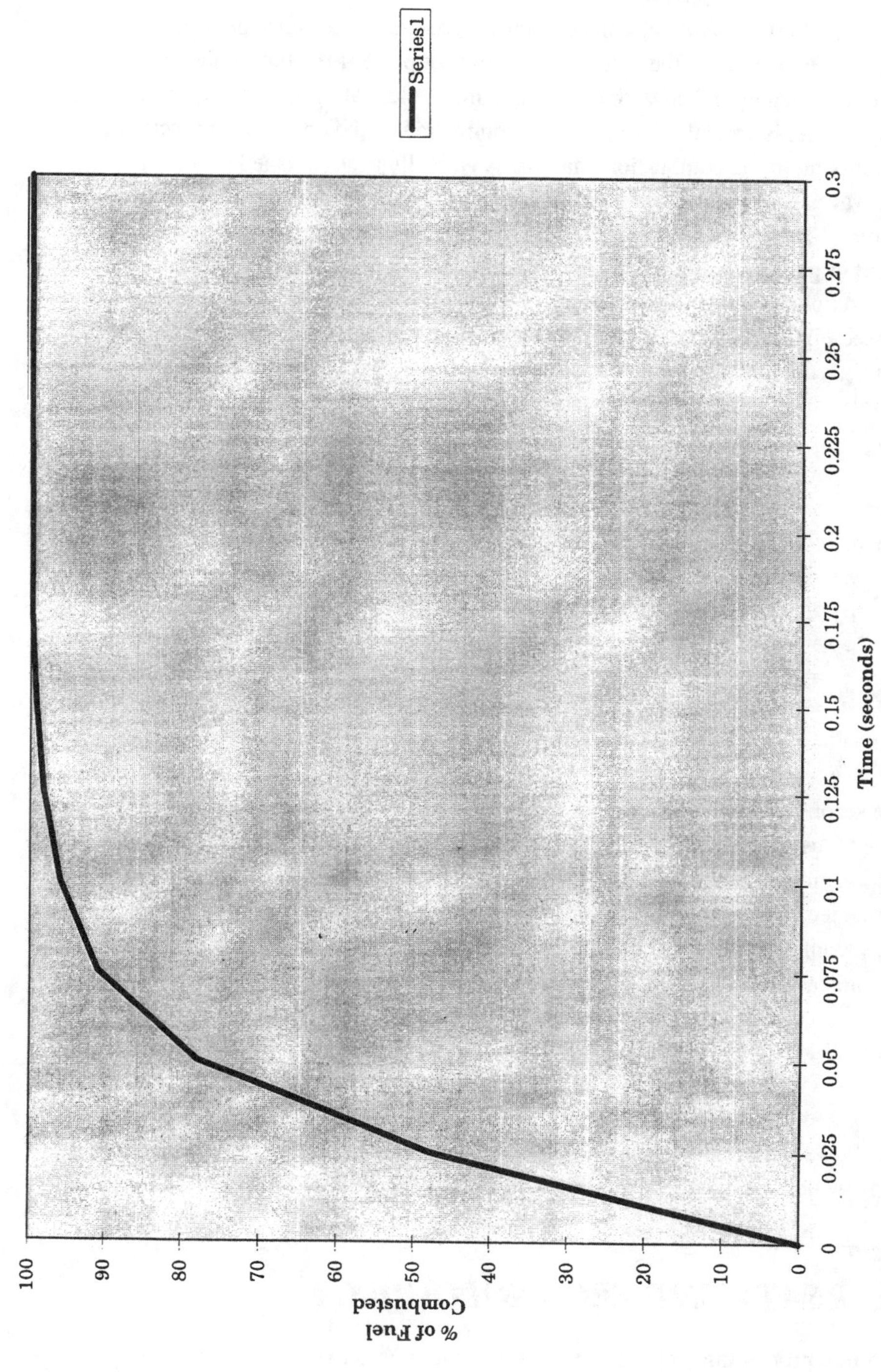

Figure 8.1 Duration of the combustion process for methane.

highest possible temperature. Economic considerations, material constraints, and emission generation must all be considered in the design.

TURBULENCE. Although the combustion reaction rate constant depends solely on temperature, the reaction will not take place if the fuel molecules are not intimately mixed with the oxygen molecules. Mixing, like time and temperature, is crucial to optimize the completion of the combustion reaction. Gas mixing in the combustion chamber is typically accomplished in one of two ways. The first method is introduction of turbulent-free gas jets to the chamber. These jets can be created by the auxiliary burner, the injection of liquid or gaseous waste streams, and the injection of combustion air. Mixing is accomplished by a free boundary layer that develops outside the core of the jet (see Figure 8.2). In this boundary layer region, momentum and mass are transferred perpendicular to the direction of flow. The amount of mixing afforded from a free jet can be estimated by the following equation:

$$m_e = [0.32(\rho_a/\rho_o)^{0.5}(x_c/d_o)m_o] - m_o$$

Where

- m_e = entrained mass,
- ρ_a = average gas stream density,
- ρ_o = jet density,
- x_c = distance from nozzle entry,
- d_o = nozzle diameter, and
- m_o = mass of the jet at the nozzle exit.

The second method is to impart a swirl to the combustion gases. This can be done by using an external vane assembly that imparts a swirl to the burner combustion air before combustion (see Figures 8.3a and 8.3b) or by firing burners or free jets tangentially into the combustion chamber. The degree of mixing in a gas stream using this method is directly proportional to the swirl number, which is a dimensionless number defined as the ratio of the angular momentum (*Gp*) of the gas stream to the linear momentum (*Gz*) of the gas stream.

$$S = Gp/Gz \tag{3}$$

Gas flows with high swirl numbers ($S \geq 0.6$) create intense recirculation patterns, which are excellent mixing devices (see Figure 8.4).

TREATMENT TECHNOLOGIES

Thermal processing technologies can be divided into two basic types, low temperature and high temperature. Low-temperature systems operate at 1000 °F (538 °C) or less. High-temperature systems operate above 1000 °F and can go as high as 3800 °F (2093 °C).

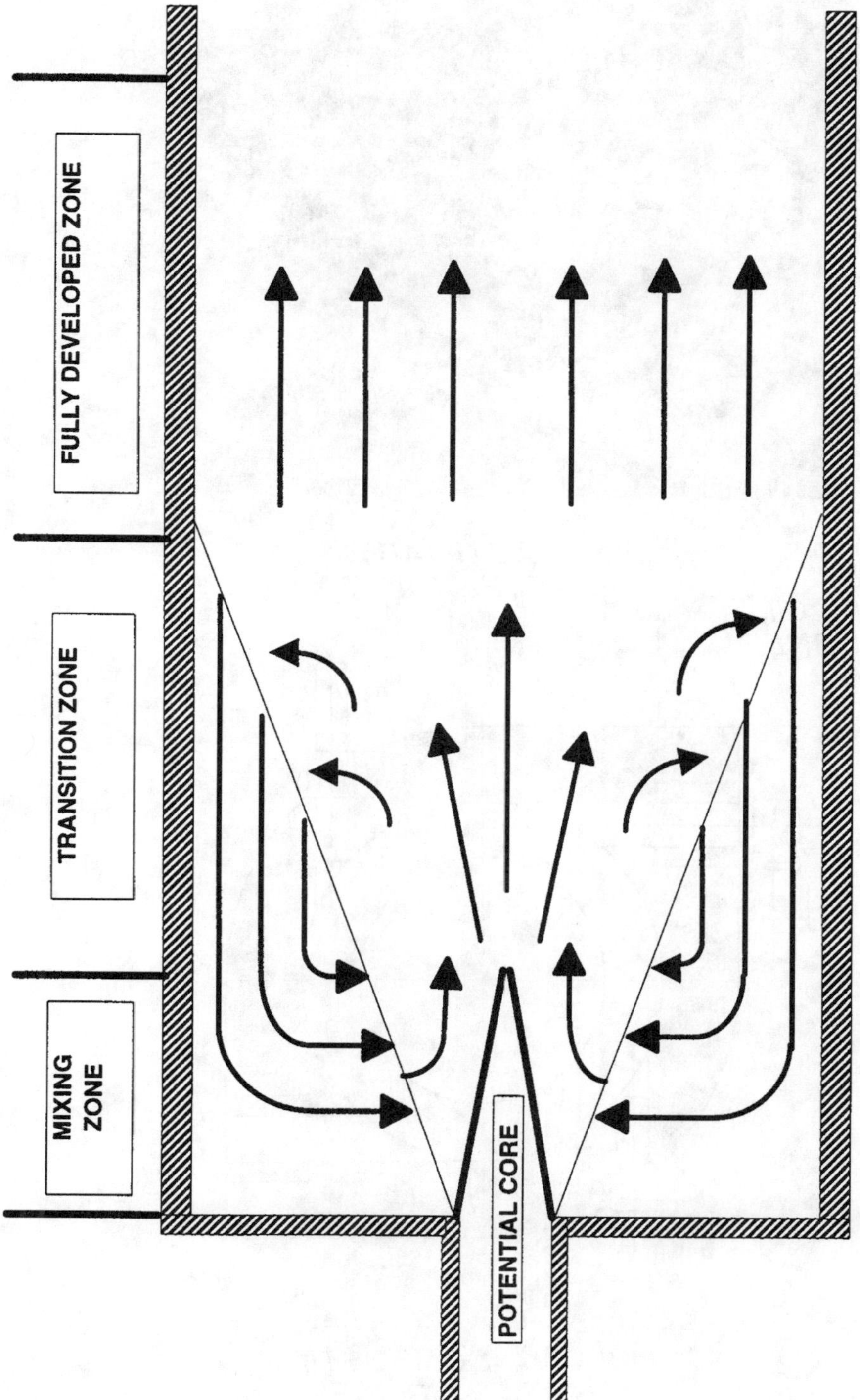

Figure 8.2 Combustion chamber mixing with free jets (Beer, 1972).

(a)

(b)

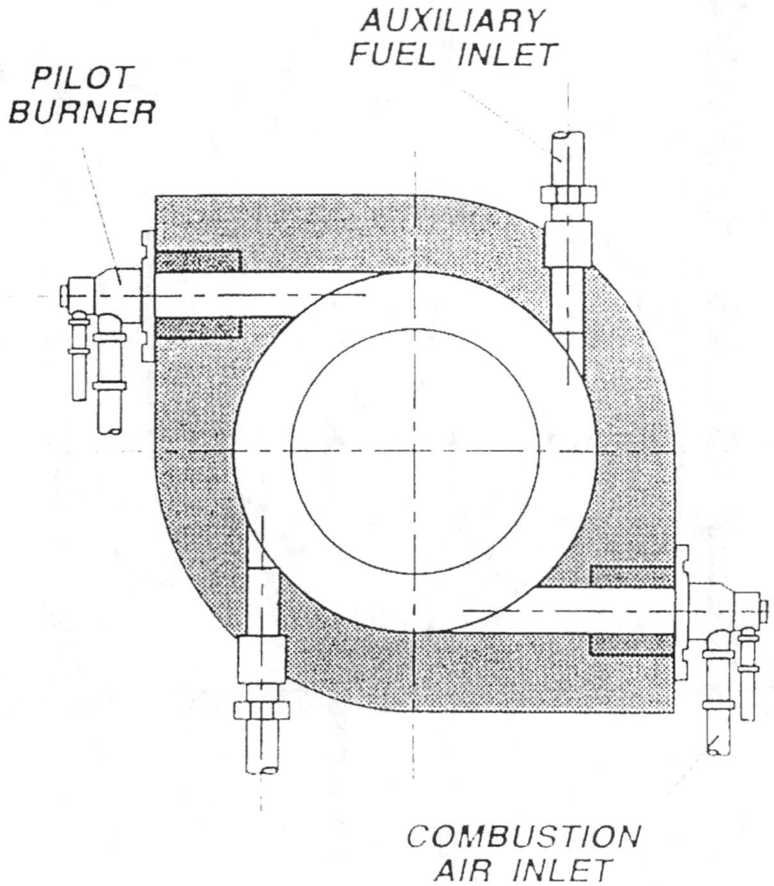

Figure 8.3 **(a) Burner cutaway showing combustion air swirler (Coen Company. Courtesy of Coen Company.) and (b) cross section of a thermal oxidizer combustion chamber using tangentially fired burners to generate swirl.**

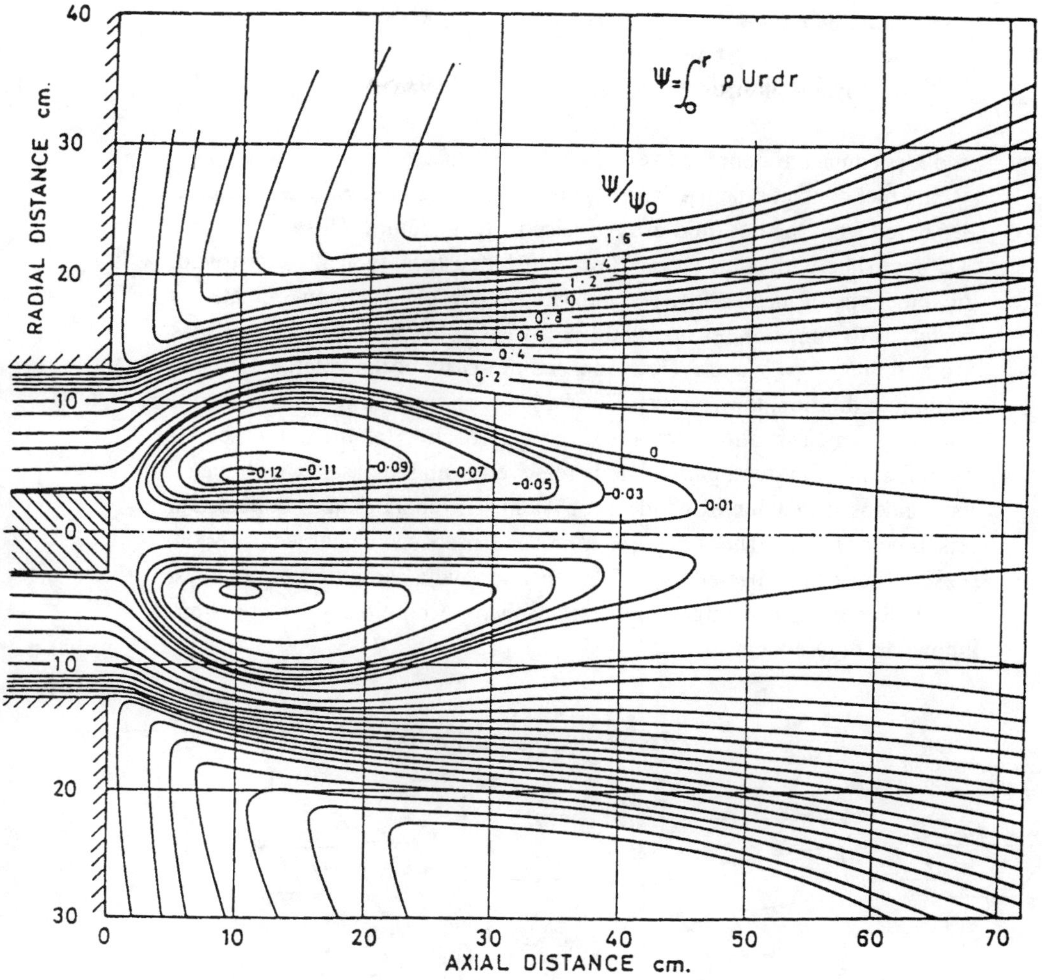

Figure 8.4 Streamlines of recirculation eddy in swirling jet (Beer, 1972).

LOW-TEMPERATURE SYSTEMS. Low-temperature systems are primarily used to treat noncombustible materials (such as soils) that have been contaminated with volatile waste materials. Low-temperature systems consist of two basic types, direct contact systems and indirect contact systems. Direct contact systems are used to treat materials contaminated with low boiling point wastes. These wastes may or may not be halogenated. As the name implies, direct contact systems operate by directly heating the contaminated material in a rotating drum using a counter-current flow of hot gas at a temperature of approximately 300 °F (149 °C). The rotation of the drum mixes the contaminated material, constantly exposing a fresh surface to the flow of the hot gas. The system off-gases composed of air, water vapor, products of combustion, volatiles, and entrained particulate are treated in several steps to remove and collect the contaminants. These steps include the following:

- Particulate removal,
- Acid gas neutralization, and
- Removal or destruction of organic contaminants.

A direct contact system is illustrated in Figure 8.5.

Indirect contact systems are designed to keep the contaminated materials separated from the hot fluid used to raise its temperature. This separation is accomplished with some type of heat transfer surface, typically constructed of some type of metal alloy. The indirect systems can heat contaminated materials to temperatures up to 800 °F (427 °C). Because the waste off-gases from indirect type systems are composed primarily of the volatized waste with little dilution, the volume of waste gases is much lower than that of the direct system. This allows for a recovery approach to treating the gases. As with the direct system, particulate is removed from the gases, which then can be condensed in a duplex condenser system. The condensate is collected and disposed or reused. The vent gases from condensers can be burned at 1800 °F (982 °C) or can be passed through a separator and a carbon adsorption system before being discharged to the atmosphere. An indirect contact system is shown in Figure 8.6.

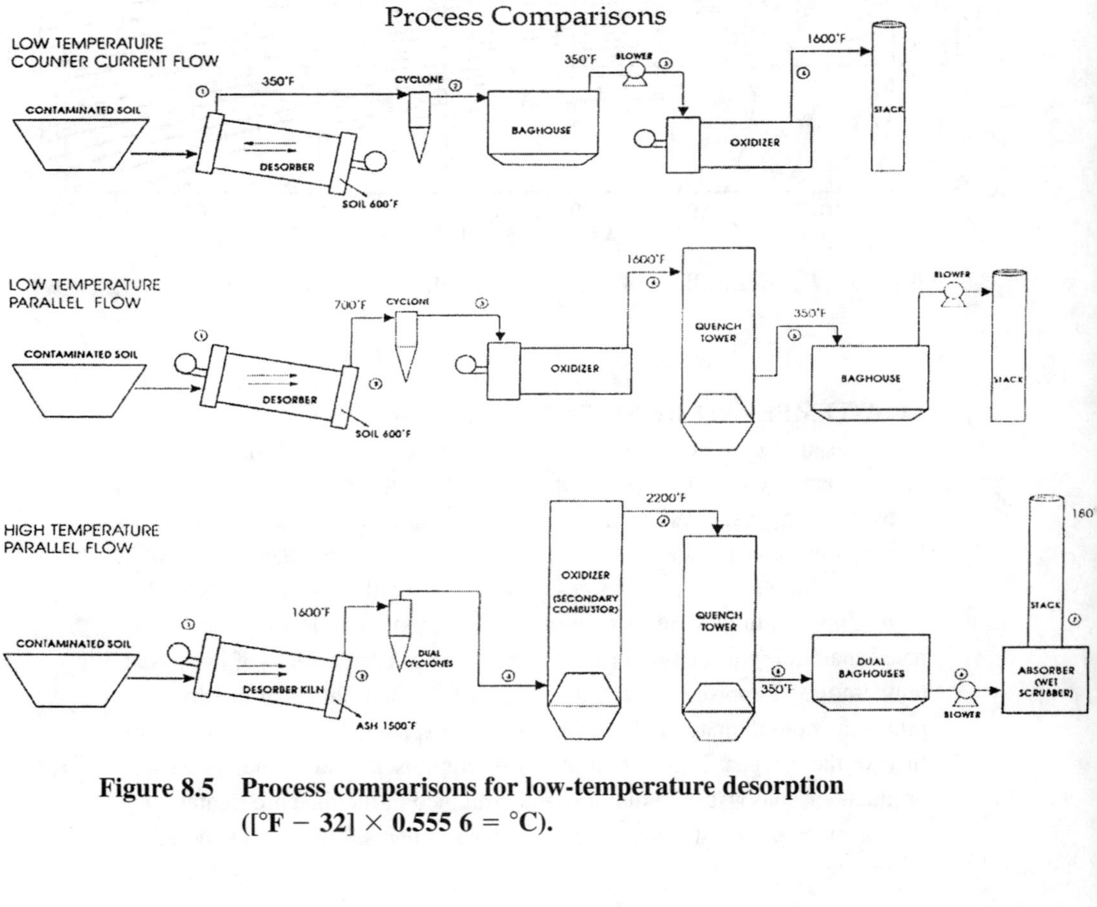

Figure 8.5 Process comparisons for low-temperature desorption ([°F − 32] × 0.555 6 = °C).

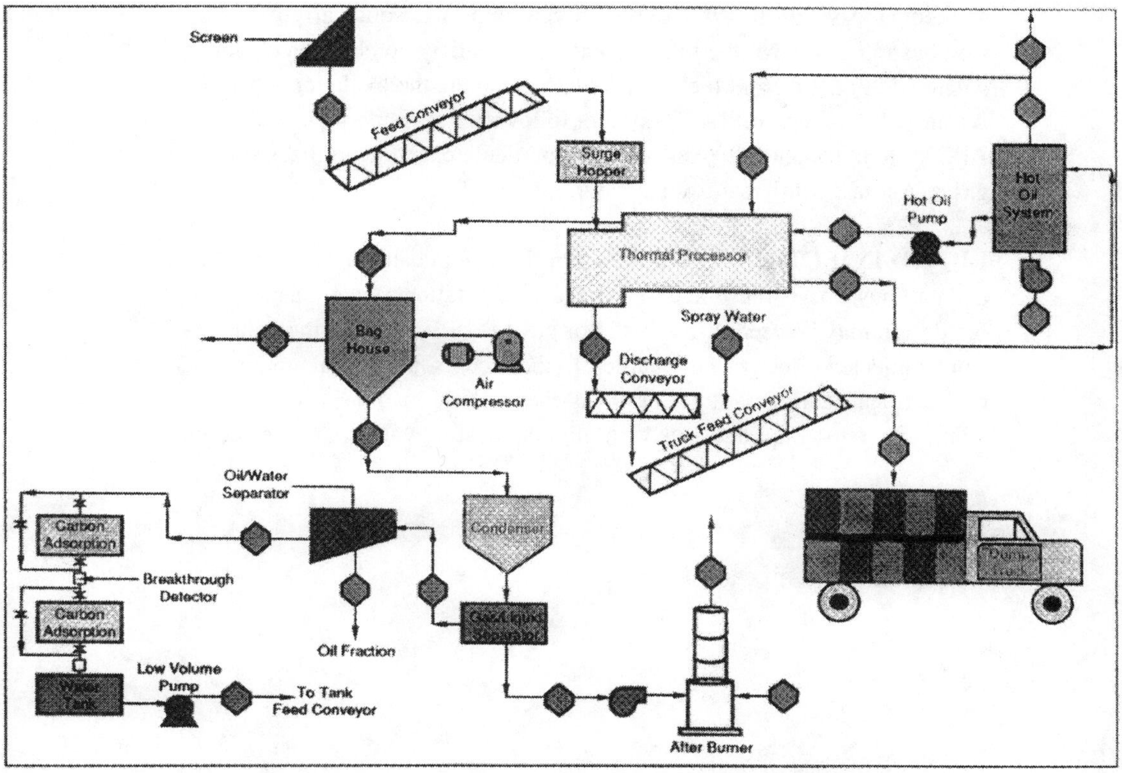

Figure 8.6 Low-temperature thermal treatment system process flow diagram.

Low-temperature thermal treatment is limited by the size of the material being processed as well as its physical state. Solid materials with a moisture content of 20% or less have the best results.

HIGH-TEMPERATURE SYSTEMS. High-temperature thermal processing systems burn waste contaminants by converting them to their appropriate oxides. There are several system designs, each having its own unique advantages and disadvantages.

Probably the most familiar high-temperature system is the rotary kiln. The primary processing chamber consists of a cylindrical, refractory lined vessel, inclined to the horizontal plane along the central axis. The cylinder has two riding rings that rest on trunnion roller assemblies and a drive mechanism that affects its rotation. Waste material is fed into the higher end of the kiln. Liquid wastes are injected through an auxiliary burner or discreet injectors depending on their heating value and fluid properties. Solids are screw fed or injected with a hydraulic ram depending on their physical characteristics. As the kiln rotates and the solids burn, fresh material is exposed, and the waste mass is moved toward the lower end of the kiln. The kiln is fired with conventional burners using natural gas or oil as fuel. Gas velocities in the kiln are kept at less than 10 ft/sec (3.048 m/s) to minimize solid carryover. Combustion gases exit the kiln and enter a secondary combustion chamber, where additional air and fuel are added to complete the oxidation reaction. An air pollution control (APC) system follows the secondary chamber. The APC system quenches the gas, removes particulate, and neutralizes acid gases. A rotary kiln system is shown in Figure 8.7.

LIQUID INJECTION SYSTEMS. For those applications when only liquid and gaseous wastes must be destroyed, a single stationary combustion chamber design may be used. If the liquid or gas has sufficient heating value, it can be injected directly to the burner. In these cases, a minimal amount of conventional fuel (natural gas or oil) is used to base load the system to prevent flame outs. Liquid wastes with minimal heating values are injected into

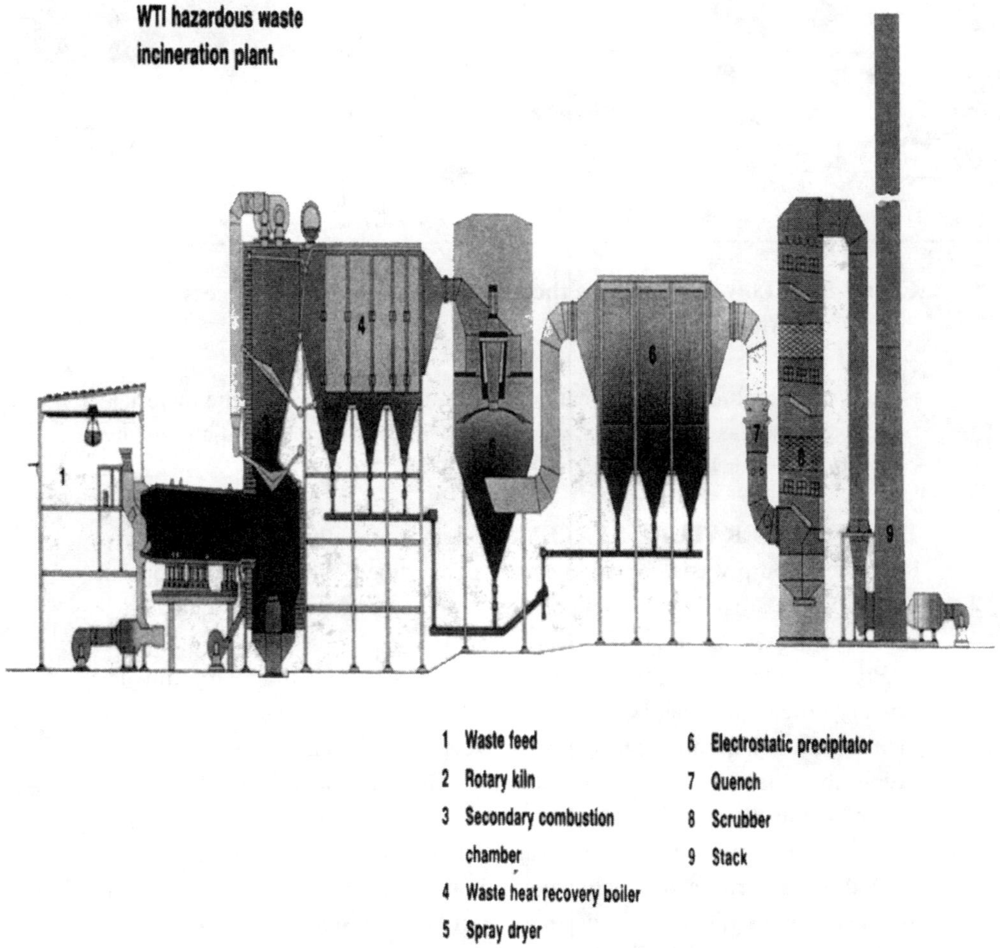

Figure 8.7 WTI hazardous waste incineration plant. (© Von Roll America, Inc. All rights reserved. Reprinted with permission.)

the chamber downstream of the main burner. In many instances, liquid wastes will contain dissolved or suspended solids or organic compounds that will generate inorganic particulate when oxidized. For these applications, the burner is down-fired into a vertical combustion chamber. This configuration minimizes fouling of the burner and injector systems and helps increase system reliability.

FLUIDIZED BED SYSTEMS. Fluidized bed systems can be used for a variety of wastes but are primarily used for sludges and wastes with low heating values (less than 5000 BTU/lb). A fluidized bed thermal processor is characterized by a bed of inert, granular material such as sand or alumina pellets. The material is contained in a refractory lined vessel and rests on a gas distribution plate. Combustion air is fed through the distribution plate and fluidizes the bed material. The bed typically expands from 1.5 to 2.5 times the static height of the bed material. The region in the combustion chamber above the fluidized bed is known as the free board. This space acts as a disentrainment area for the bed material and other solid particulate generated by the waste. Auxiliary fuel and waste are introduced to the bed. The turbulent mixing of the bed provides instantaneous heat transfer to the waste material resulting in thorough oxidation of the organics. The bed typically operates at temperatures between 1450 and 1600 °F (788 and 871 °C). Flue gases generated by the combustion process enter the freeboard region above the bed where additional combustion air may be added to complete the combustion process. A conventional fluidized bed is designed to keep the bed material from becoming entrained in the gas flow. It is common practice to enlarge the freeboard volume to reduce the velocity of the flue gases to minimize solids carryover. A circulating fluidized bed operates on the same principles as a stationary fluidized bed except that it is designed to carry the bed material from the main chamber. A percentage of the bed material and other particulate generated for the waste combustion is continuously entrained in the flue gas, carried from the combustion chamber, and collected in a cyclone. The material is then recycled back into the main bed. Circulating fluidized beds are suitable for wastes with higher heating values.

MULTIPLE-HEARTH INCINERATORS. Multiple-hearth incinerators have historically been used for the treatment of sludge (see Figure 8.8). Their use for hazardous waste treatment has required the use of an afterburner to ensure the complete destruction of toxic gases released and generated from the combustion process in the multiple-hearth incinerator.

The basic multiple-hearth incinerator consists of a refractory lined cylindrical chamber that stands vertically upright. A rotating air cooled shaft is positioned along the central axis of the cylinder. A series of solid flat hearths in the shapes of doughnuts and discs are positioned at various elevations up the cylinder. The doughnut shaped hearths are attached to the cylinder wall, and the disc-shaped hearths are attached to the shaft. As the shaft rotates, the

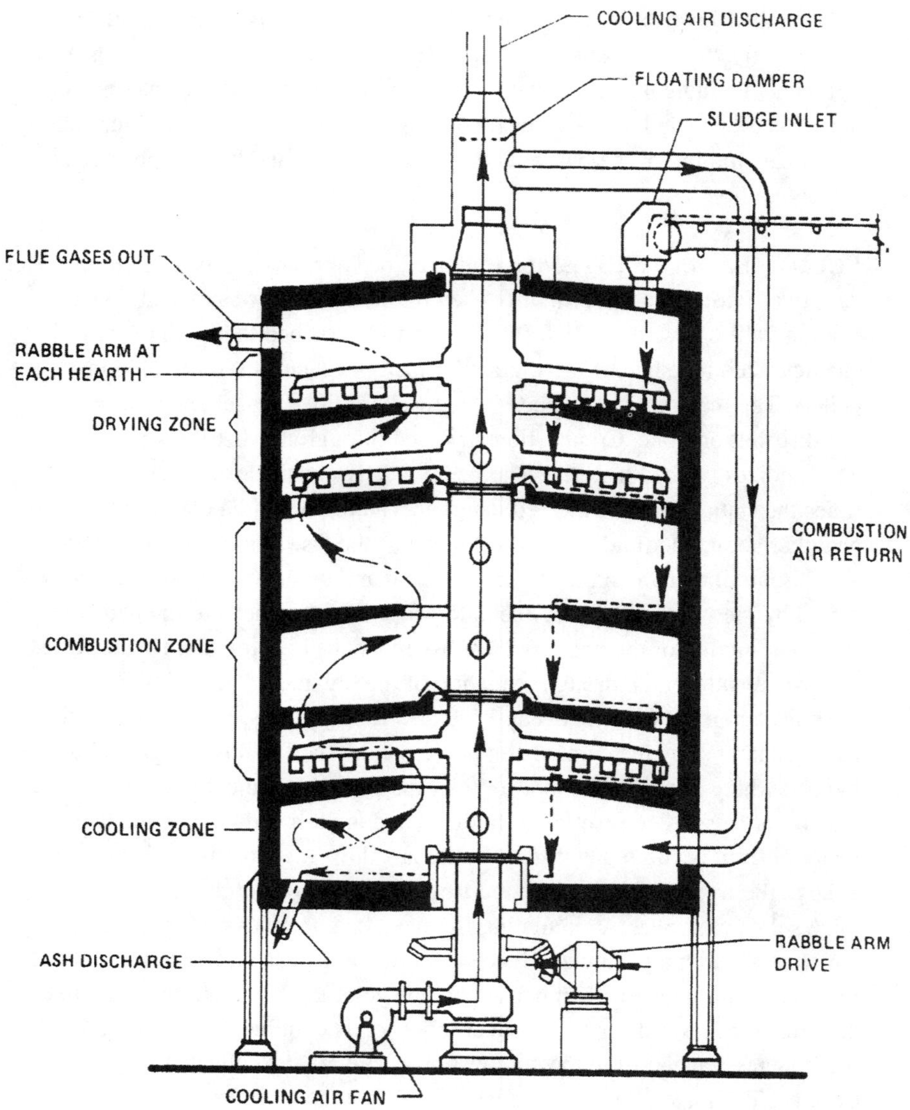

Figure 8.8 Multiple-hearth sludge incinerator.

hearths on the shaft also rotate. Stationary ramble arms are positioned over the hearths to move the waste along the hearths. As the shaft rotates, waste travels the length of the cylinder by cascading from one hearth to the next. Burners fired on conventional fuels are positioned at various elevations above the hearths to provide the energy to combust the waste material. Combustion air is added in the bottom of the chamber.

A multiple-hearth incinerator has three operating zones. The first zone is the upper hearths on which the waste feed is heated and dried. This zone typically operates between 700 and 1050 °F (371 and 566 °C). The middle zone is where the solids are combusted, and it typically operates at between 1500 and 1860 °F (816 and 1016 °C). If liquid waste and gas streams are treated in the unit, they would be injected in this zone. The third zone, at the bottom of the chamber, is where the ash is cooled by the incoming combustion air. This zone operates in temperature ranges of 400 to 700 °F (204 to 371 °C). Exhaust gases from the multiple-hearth incinerator enter a secondary combustion chamber where additional energy and combustion air are added to complete the reaction. Gases from the secondary combustion chamber are then treated to remove particulate and neutralize acid gases.

ADJUNCT SYSTEMS

AIR POLLUTION CONTROL SYSTEMS. The primary products of the thermal processing of hazardous waste are carbon dioxide and water vapor. However, because of the myriad of compounds contained in waste products, other substances are produced that must be treated or removed from the combustion products. The most typical of these substances include acid gases such as HCl, solid particulate such as metal oxides and inorganic salts, oxides of sulfur and nitrogen, and gaseous pollutants that are solids at ambient temperatures such as mercury and lead.

In most thermal processing systems, the combustion gases are conditioned before the pollutants are removed. This conditioning involves cooling the gases from combustion chamber temperatures (1400 to 2500 °F [760 to 1371 °C]) to more manageable temperatures (ambient to 500 °F [260 °C]). This quenching can be accomplished through indirect heat exchange devices such as boilers or through direct heat exchange such as injecting and evaporating water in the gas stream. The type of quench used is a function of the application. Some of the considerations for determining the best quenching method are considering if the energy can be recovered, considering whether the injection of water or scrubbing liquid can enhance pollutant collection, and considering whether the flue gas contains chlorine (which can form dioxin through improper quenching).

ACID GAS NEUTRALIZATION. The most familiar method of acid gas scrubbing and neutralization is wet absorption. Wet absorbers use the solubility properties of acid gases to remove them from the gas stream. This can be accomplished with plain water or with a scrubbing liquid containing a base such as sodium or potassium hydroxide. The neutralization reaction can be written as follows:

$$HCl + NaOH \rightarrow NaCl + H_2O$$

Most wet absorption systems use mechanical devices to provide intimate contact between the contaminated gas stream and a scrubbing liquid. These devices include dump packing, structural packing, bubble caps, and sieve trays. See Figure 8.9 for examples of wet absorption systems.

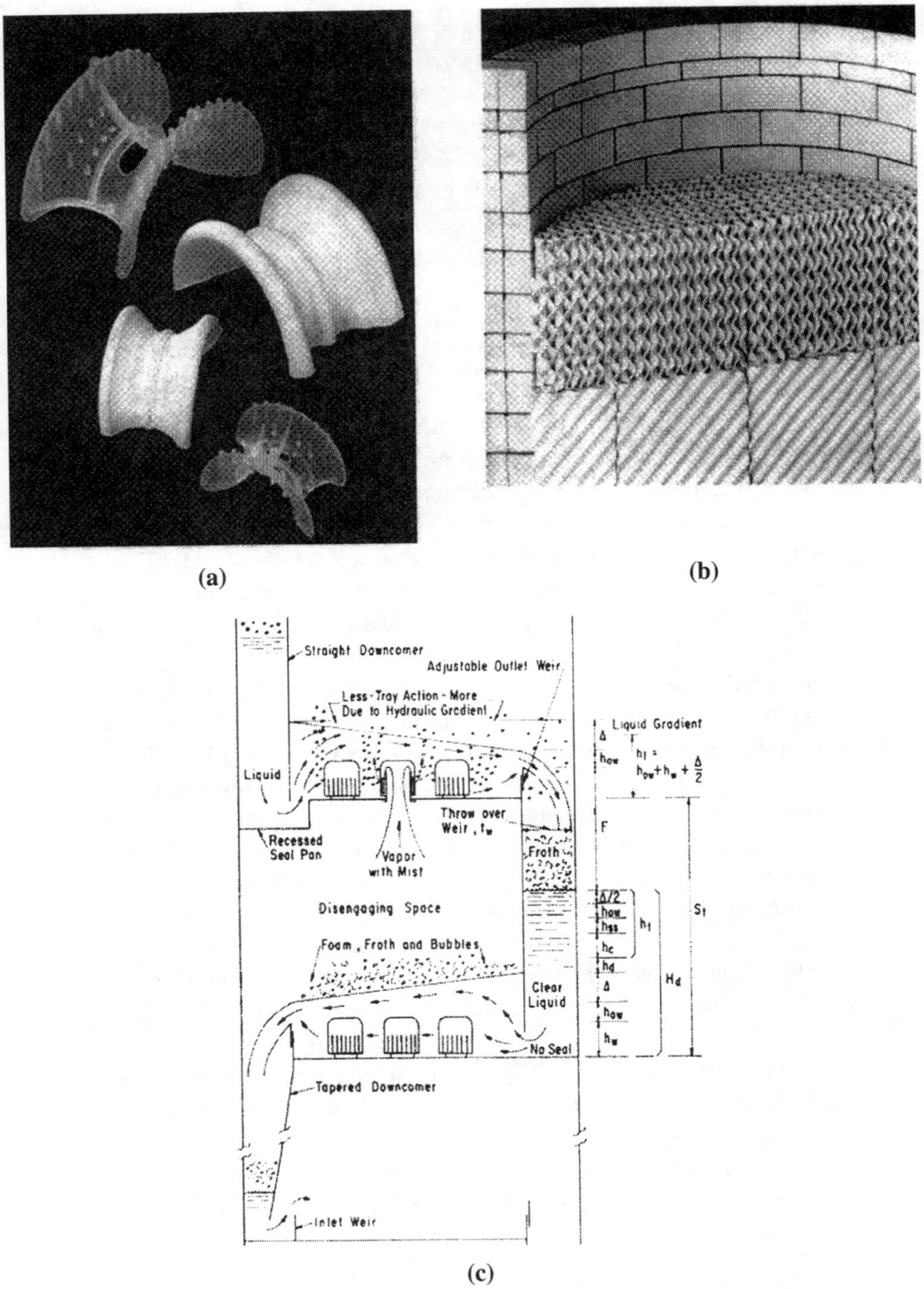

Figure 8.9 (a) Dump packing (Courtesy of Koch-Glitsch, Inc., Knight Division), (b) structured packing (Courtesy of Koch-Glitsch, Inc., Knight Division), and (c) bubble cap tray schematic—dynamic operation. (Ludwig, 1997. Reproduced by permission. From *Applied Process Design for Chemical and Petrochemical Plants, Volume 2, Third Edition.* Copyright © 1964, 1979 by Gulf Publishing Company, Houston, Texas, 800-231-6275. Copyright © 1997 by Ernest E. Ludwig. All rights reserved.)

Another typical method used to neutralize acid gases is dry absorption. Dry absorbers are divided into two basic types—wet and dry injection systems. Wet injection systems are typically used to quench the flue gas as well as neutralize the acid gases. They work by injecting lime slurries to the flue gas stream. The water evaporates, quenching the gas and drying the lime into fine particles. These particles come in intimate contact with the acid gases in the flue gas and begin the neutralization process. Dry injection systems inject powdered lime or sodium carbonate into the flue gas stream to begin the neutralization process. Both wet and dry injection systems typically use a bag filter collector downstream of the injected material. These baghouse filters perform two functions. First, they collect the unreacted lime, salts, and any other particulate that may be contained in the gas stream. Second, the filter cake that forms on the baghouse acts as a site where the neutralization reaction is completed.

PARTICULATE COLLECTION. The three primary mechanisms used to collect particulate matter are inertia, diffusion, and interception. These are illustrated in Figure 8.10. Inertia impaction relies on the particulate colliding with a collecting body. This mechanism works well for particulate with diameters greater than 10 microns. Diffusion uses the Brownian movement phenomenon of small particles to collect them. This mechanism works well for particles with diameters of 3 microns or smaller. Interception, like inertia impaction, relies on the particle coming in contact with the target collector.

BAGHOUSES. Baghouse collectors use all three mechanisms to collect particulate. The basic design consists of a matrix of fabric bags suspended from a tubesheet in a gastight housing. Dirty gas enters the housing and flows through the porous bags. The particulate in the gas collects on the bags through the mechanisms previously described. As more and more particulate is collected, a cake forms. This particulate cake enhances the performance of the bag acting itself as a filter and collection site. As the cake thickness increases, the gas pressure drops across it and the bag. This pressure drop is monitored and used as an indication of when to clean the bags. The particulate cake is removed from the bags in one of three ways. Shaking is probably the oldest method of cleaning. The shaker design has the ends of the bags attached to mechanical devices that physically shake the particulate from the bag. This design is illustrated in Figure 8.11. The reverse flow method of cleaning uses parallel bag compartments to keep the system continually in operation. After the pressure drop in one compartment has reached the desired level, the dirty gas flow is diverted to the parallel compartment, and air is blown through the dirty compartment in the opposite direction of the normal flow. Pulse cleaning is probably the most widely used method. For this method, the bags are supported by wire cages. Compressed air nozzles are positioned over the clean discharge holes in the bag tube sheet. When cleaning is required, a blast of compressed air is blown into the bag (typi-

Mechanism	Model	Separation number	Description
Inertial impaction	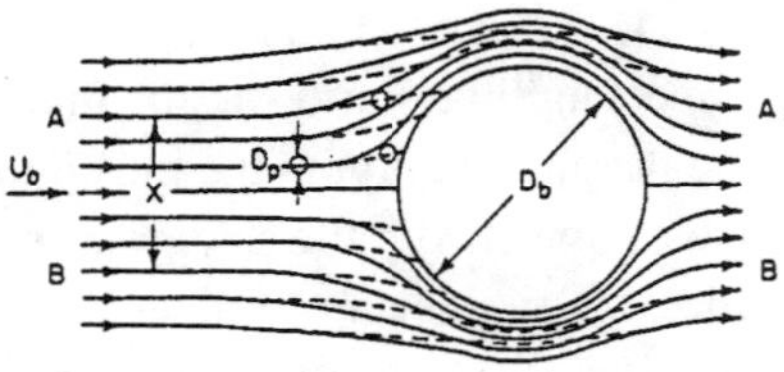	$N_{si} = \frac{K_m \rho_p D_p^2 U_o}{18 \mu_G D_b}$	In the case of inertia, a particle carried along by the gas stream, on approaching a collecting body (liquid droplet or fiber), tends to follow the stream but may strike the obstruction because of its inertia. Solid lines represent the fluid streamlines around a body of diameter D_b and the dotted lines represent the paths of particles which initially followed the fluid streamlines. For a flow around a spherical or cylindrical collector the quantity $\frac{X}{D_b}$, where X is the distance between limiting streamlines A and B, represents the fraction of particles initially present in a volume swept by the body which will be removed by inertial impaction.
Brownian diffusion	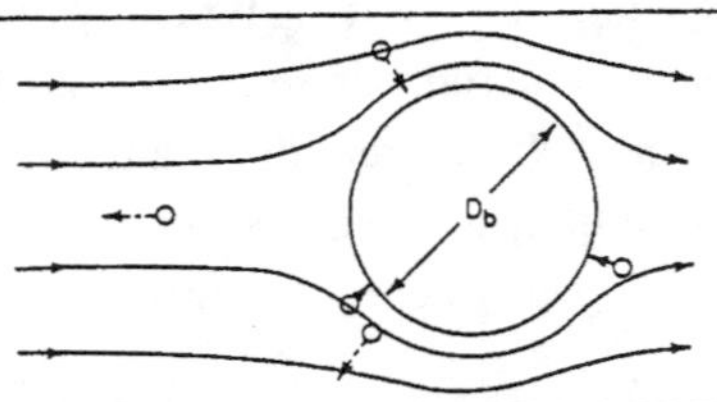	$N_{sd} = \frac{D_v}{U_o D_b}$	Smaller particles, particularly those below about 0.3 microns in diameter, exhibit considerable Brownian movement and do not move uniformly along the gas streamline. These particles diffuse from the gas to the surface of the collecting body and are collected.
Flow-line interception	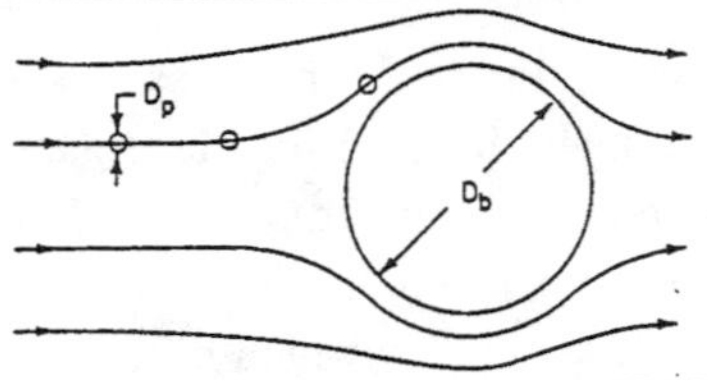	$N_{sf} = \frac{D_p}{D_b}$	Particles also may be collected by direct interception. The particle may follow a gas streamline and be collected without inertial impaction or Brownian diffusion if the streamline is fairly close to the collecting body. If a particle 1.0 micron in diameter follows a streamline which passes within 0.5 microns of the body, the particle will touch the body and will be collected.

—— Fluid streamline
---- Particle path

Figure 8.10 Particulate capture mechanisms (micron × 1.000 = μm).

cally through a venturi), resulting in a pressure wave that travels down the bag and "snaps" the cake from the bag. The dislodged cake falls into a cone hopper at the bottom of the housing and is removed. This type of baghouse is shown in Figure 8.12.

VENTURI SCRUBBERS. Venturi scrubbers have been successfully used to collect particulate. They primarily use inertial impaction and interception to collect the particulate. As the particulate laden gas enters the converging section of the venturi nozzle, its velocity increases. Simultaneously, water is introduced to the inlet section typically through an overflow weir assembly, which allows a sheet of liquid to flow down the wall of the converging nozzle section. As the gas and liquid meet in the throat of the venturi, the low pressure created by the increased velocity shears the water sheet into thousands of water droplets. These droplets impact the particulate in the gas stream and encapsulate the particulate in water droplets. These droplets are then removed in a centrifugal separator. A typical venturi scrubber system in shown in Figure 8.13.

ELECTROSTATIC PRECIPITATORS. Electrostatic precipitators (ESPs) use the charged characteristics of particles to collect them. Gases containing

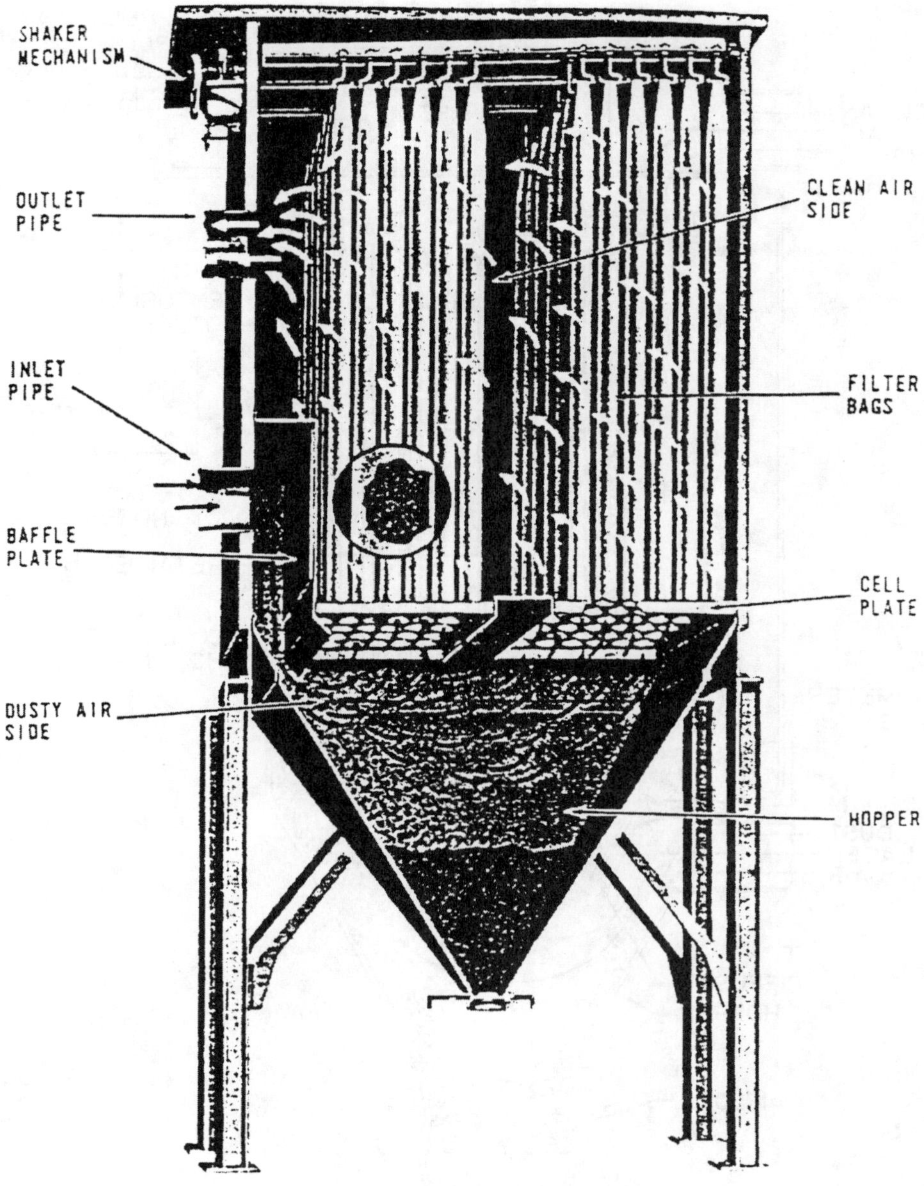

Figure 8.11 Typical simple baghouse with mechanical shaking.

the particulate are exposed to an electrical field. The charged particles are then collected on a collection surface with the opposite charge. The particles are removed and collected. In dry systems, the collectors are typically cleaned by mechanically "rapping" the collecting surface, allowing the particulate to fall and be removed from the system. Wet ESPs use a constant flowing film of water, which bathes the collector and carries away the particulate as it is collected.

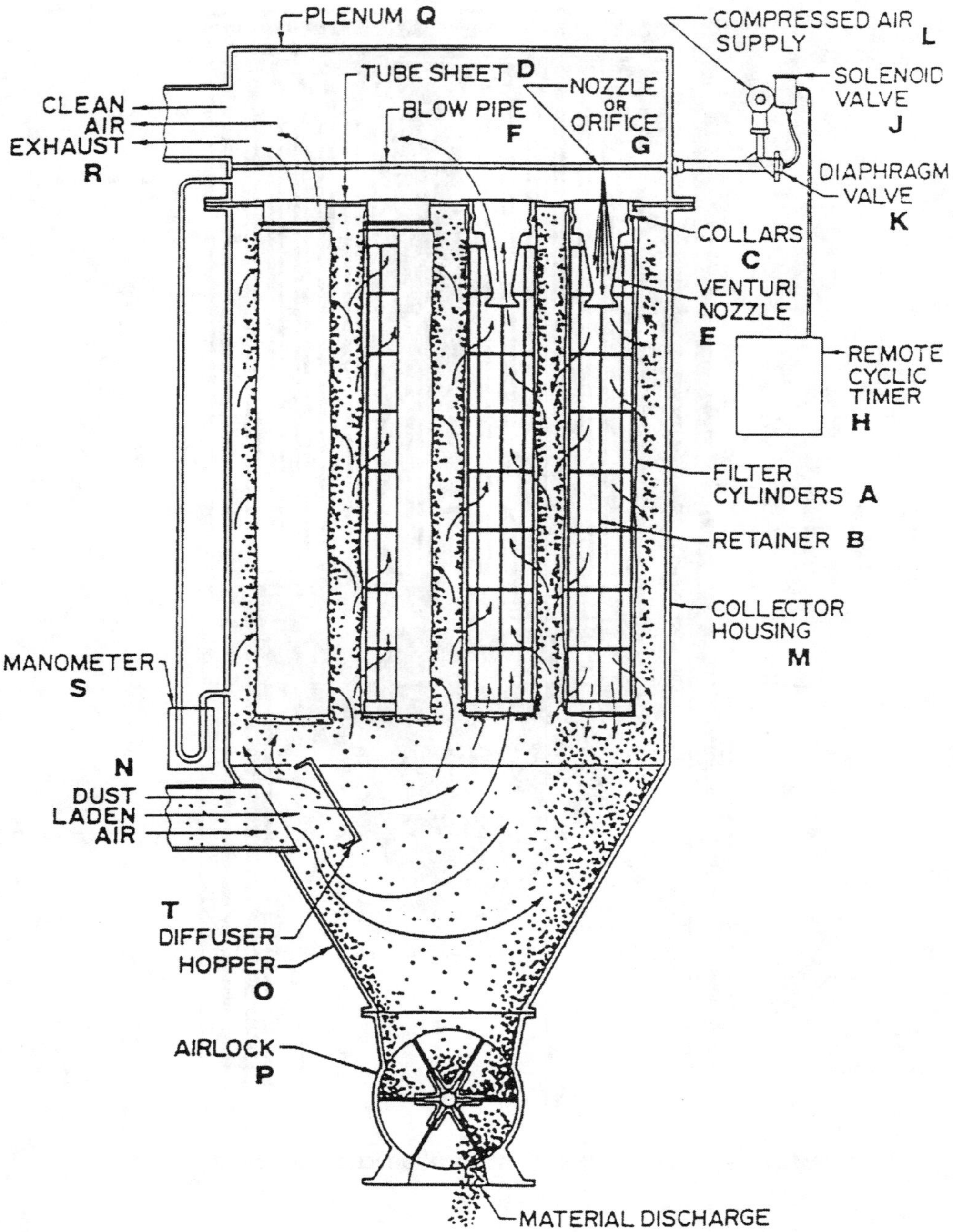

Figure 8.12 Typical pulse jet baghouse.

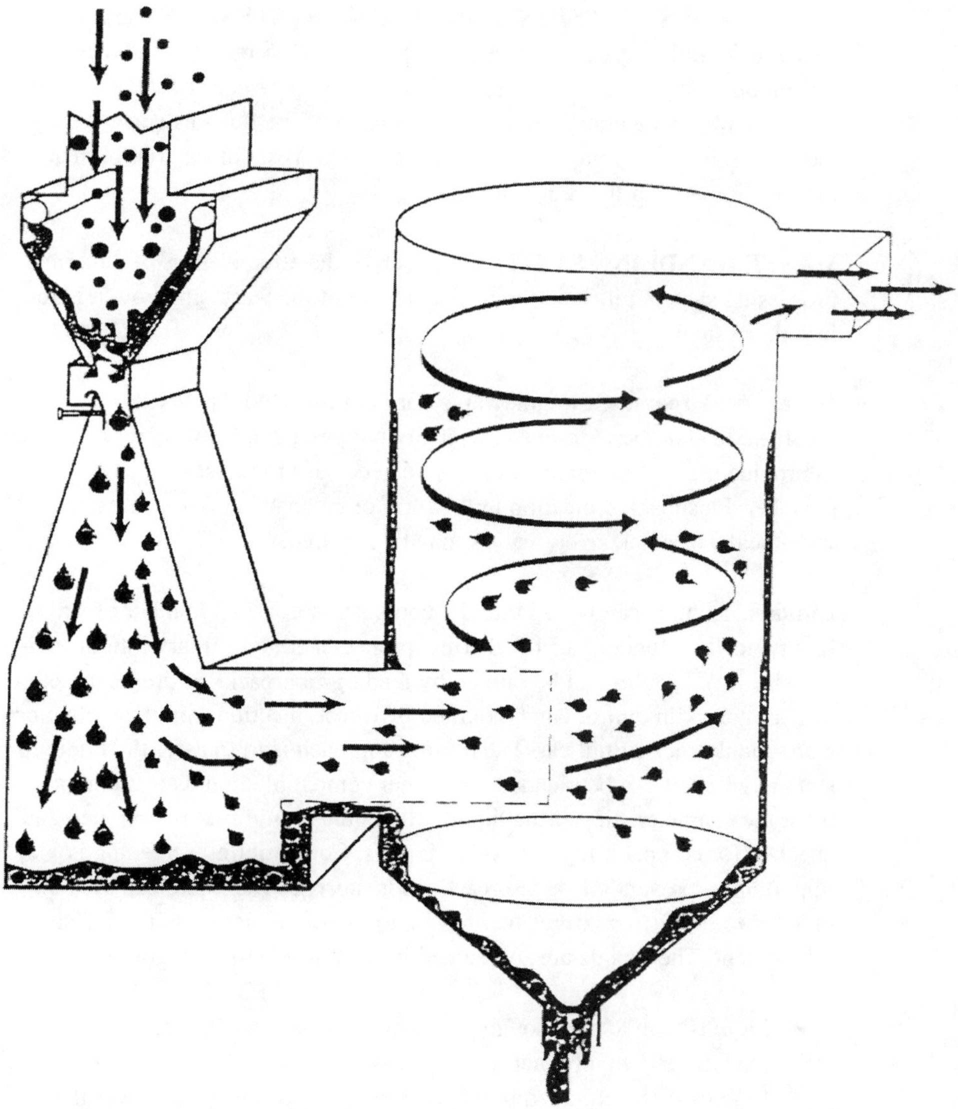

Figure 8.13 Typical venturi scrubber and cyclonic separator.

CONDENSATION. Condensation is a mechanism used to collect both fine solid particulate and gaseous compounds, which are solids or liquids at ambient conditions (for example, mercury). Some scrubbers have been designed to subcool the flue gas below the adiabatic saturation endpoint. This cooling not only enhances the removal of condensable solids, it can also be used to capture submicron solid particulate. As the gas is subcooled, condensation of a water droplet will preferentially occur around a nucleus. Submicron particles in the flue gas provide these condensation sites. The result is that the particulate becomes encapsulated in the water droplet and can then be easily collected.

CONTINUOUS EMISSION MONITORING SYSTEMS. Hazardous waste incineration systems are required by 40 CFR to monitor and record certain operating parameters. Maintaining certain parameters such as incineration chamber temperature and oxygen content of the flue gas verifies that the system is performing according to the standards set in the trial burn and the requirements of the regulations.

WASTE HANDLING SYSTEMS. Much of the effectiveness of a thermal processing system relies on the physical form of the waste and how it is fed into the system.

Gases and Vapors. Gases and vapors are typically fed directly into the combustion chamber. Storage is typically not practical. Blowers may be used to provide the motive force and pressure necessary to inject the gas to the chamber. Flashback protection in the form of mechanical flame arrestors, water seals, or quick safety valves may be required.

Liquids. Liquids can be fed into the combustion unit in a number of ways. It is typically a function of the thermal processor design (rotary kiln or multiple hearth). Liquids can be burned by feeding fiberpacks or drums directly into a rotary kiln unit or can be burned by direct injection from a pipeline or from a tank truck or rail car. The normal approach is to transfer the liquid to storage where it can be blended with other compatible liquids that complement the characteristics of the liquid. The liquid would first be sampled and checked for compatibility with other liquids. Compatibility screening is important to prevent unwanted reactions in storage. The liquid can then be blended to produce mixtures that have a low enough viscosity in BTU and ash content. The liquids are segregated into the following categories:

- Liquids with a low enough viscosity and high enough heating value for feed directly to a burner,
- Liquids that can be fed to a burner but must be heated to lower the viscosity,
- Liquids too viscous for feed to a burner so they are placed in storage that will allow feeding directly into a lance, and
- Water solutions that must be atomized into the incinerator.

Sludges. Sludges can be received, sampled, and blended in much the same manner as liquids. The main difference is that sludge will typically be fed through a lance into the incinerator instead of being atomized in a burner. The sludges will frequently require agitation in storage and will require circulation loops in any piping to keep the solids from settling out and plugging the system.

Solids. Solids can arrive in fiberpacks from within the plant, in fiberpacks inside overpack drums if shipment from outside the plant, or in dump trucks.

The feeding of solids is one of the most critical variables in the design of a thermal treatment system. The feeding of solids that have been segregated to smooth out variations can have a major affect on the capacity of the system.

Drums or fiberpacks are normally sampled before being fed to the incinerator. Any liquids would be decanted to minimize the potential for sudden heat releases when the drum is fed into the incinerator. The drums can also be segregated to smooth out waste feed. For example, a group of drums containing halides can be segregated and fed over a period of time to smooth out the generation of acid gases and thus stay within emission limits. If drums cannot be fed directly into the system, it may be necessary to transfer material from the drum into fiberpacks, shred the drum or fiberpack, or dump the material out of the drum or fiberpack.

ASH HANDLING SYSTEMS. In a properly designed thermal treatment system, the ash leaving the combustion system should not contain organic material. Inorganic components in the waste are not destroyed, although they may be oxidized into other compounds.

The inorganic materials exit the thermal treatment system either as bottom ash from the combustion chamber in scrubber water or as fly ash from the air pollution equipment. Unless demonstrated otherwise, these ashes should be considered as hazardous waste and handled as such.

The ash from the combustion chamber and the fly ash must both be discharged without letting significant quantities of air into the treatment system. For fly ash and treatment systems where all ash particles are small, removal can be accomplished using rotary air locks. When large solids are present, such as drums or the rings from fiberpacks, large discharge systems such as airlocks or water seals are required (the ash can drop directly into a vat of water).

Ash is normally either air cooled or quenched with water as it leaves the system. Air-cooled ash is then conveyed to drums or dumpsters for transfer to a landfill for disposal. Any ash that still contains organic material can be recycled back to the thermal treatment system. Some systems routinely recycle ash to make sure combustion is complete. If necessary, the ash can be chemically treated or stabilized before it goes to the landfill.

REFERENCES

Beer, J.M. (1972) *Combustion Aerodynamics.* Applied Science Publisher, Ltd.

Ludwig, E.E. (1997) *Applied Process Design for Chemical and Petrochemical Plants.* Gulf Publishing.

SUGGESTED READINGS

Beer, J.M., and Chigier, N.A. (1972) *Construction Aero Dynamics.* Halstad Press, Wiley and Sons, New York.

Glassman, I. (1977) *Combustion.* Academic Press, New York.

Stambuleann, A. (1976) *Flame Combustion Process in Industry.* ABACUS Press.

Chapter 9
Process Integration

INTRODUCTION

The selection of treatment options depends on the nature of the hazardous material, the required effluent concentration levels, and economic considerations. Hazardous materials can exist as solids, liquids, or vapors. The available treatment alternatives for hazardous materials include biological, physical, chemical, and thermal treatment. Each of these has been discussed previously in this manual. The integration of this information into treatment train options is the objective of this chapter.

Hazardous materials can be found in natural or manmade environments. Typical examples include groundwater, surface water, soils, construction materials, new and waste materials storage, wastewater streams, and the workplace and ambient air. Stringent environmental laws passed in recent years for the protection of human health and the environment have brought many formerly "safe" materials into the hazardous materials classification.

Traditional industrial wastewater treatment addresses the disposal of hazardous materials in liquid form from the relatively simplistic viewpoint of gross contaminant reduction. Many of the same treatment methods can be used, but greater personal protection and lower effluent concentrations are required.

Traditional solid wastes collection, processing, and disposal have not been totally applicable to hazardous materials. The requirements for solidification before disposal, manifesting, disposal, and cataloging at Class I dump sites have been a far more complex process than picking up solid wastes and transporting them to a sanitary landfill. Part of the reason for the stricter laws has been the environmental problems currently associated with former landfill sites.

Air pollution control technology has also been affected by the requirements of hazardous materials control. Detection levels have been lowered, the types of materials monitored has increased, and a general awareness of the dangers of air pollution in the workplace and community has resulted in improvements in control and monitoring technologies.

Economic considerations governing hazardous waste treatment must be applied on a total accounting basis so that all relevant capital, operating, and maintenance cost factors are considered. In addition to concern for product, process, and facility life, there may also be uncertainties as to treatment level and liabilities associated with permit violations. Designing for zero discharge of pollutants, if chosen by management, avoids the issues of future imposition of higher stringencies, the costs associated with permitting and monitoring, and other factors influencing costs. Thus, all treatment options should be considered when evaluating costs.

Hazardous waste processing frequently involves retrofitting into an existing plant structure and is highly site specific. Costs may exhibit bias based on layout because of long piping runs for utilities. Typically, a lack of accurate "as built" drawings adds to cost because of unknown interferences.

Industries have given a great deal of attention to avoiding the hazardous waste problem in a number of ways. These include product substitution of a nonhazardous manufactured product, process modification, or reagent change, resulting in the generation of smaller quantities or less toxic byproducts. Another is "bottling up," which results in zero discharge and includes attention to fugitive emissions, leakage and reuse of recycle, or recycle of materials elsewhere in the plant. It is a continuing process requiring sensitivity when dealing with the problem of managing the wastes that must be generated. The subject of waste reduction is discussed in Chapter 3.

An examination of economics is typically worthwhile when considering the question of waste minimization. Also, issues of expansion in existing treatment facilities can be addressed, as can the issue of pretreatment in a plant generating a unique waste load that could affect the larger system.

Consideration of any particular process on paper alone is particularly difficult because of the low concentrations of hazardous materials that typically must be managed. General references are available to suggest possible processes. These include the *Standard Handbook of Hazardous Waste Treatment and Disposal* (Freeman, 1989) and compendia from U.S. EPA and

WEF. More detailed consideration of alternatives may be possible using the published data in WEF's Literature Review (WEF, 1997). There is, however, no substitute for bench- or pilot-scale testing using actual waste streams, slip streams, or composite samples of the wastes to be managed. These data may be used to design a full-scale treatment process with the aid of process simulation software. These cover such topics as solids handling, reactors, physical properties, sizing, costing, and general unit operations. Treatability testing is necessary to demonstrate the feasibility, the control parameters, and the treatment costs for any potentially applicable treatment train. Site-specific costs for construction and operation of the treatment train are also important when selecting the combination of treatment technologies that achieves the desired results for the least cost.

WASTE CHARACTERIZATION

Characterizing a waste to determine the appropriate cost-effective treatment requires both a description of the waste composition in terms of its chemical nature, its liquid and solid matrix, and a definition of hazardous versus nonhazardous streams and a description of the possible segregation or combining of waste streams.

The selection of possible processes is largely determined by the chemical nature of the waste and the matrix within which it occurs. Regardless of the matrix, a mass balance is required to initiate development of an adequate process train to handle a waste stream.

Depending on the hazardous material of concern, the mass balance outlets may vary greatly. For example, the mass balance for copper from an electroplating operation focuses on wastewater, sludge, scrubbers, etch wastes, and product incorporation. A mass balance for Freon 113 focuses on fugitive emissions and still bottoms quantities.

A well-constructed material balance requires attention to flow and chemical variability and allows assessment of amenability of the streams to be segregated or combined for treatment. Variability is an important factor in the design of a process unit. Segregation has a less obvious economic effect. The concept is to treat substreams before combining them to affect processing on a smaller volume of waste and to facilitate or avoid interfering with downstream processing. A convenient example is plating waste treatment when hexavalent chrome is reduced to trivalent, and cyanide-bearing streams are oxidized upstream of the point at which these streams are combined with other metal-bearing streams, which then flow into a precipitation process.

There may be questions about the toxicity of a particular waste stream if constituents are not listed wastes or incorporate more than one waste listed as hazardous. Opportunities for delisting under appropriate, detailed U.S. EPA procedures should not be ignored, particularly at the discharge end of the waste treatment process.

TREATMENT PROCESS SELECTION

The standards to which wastes must be treated for release are closely related to considerations for ultimate disposal or reuse of material. If recovery, reuse, or delisting options are to be considered, they should be carefully evaluated. Options for zero discharge or nonhazardous wastewater discharge to a publicly owned treatment works or a receiving body should likewise be thoroughly understood. Disposal options for solids and sludges provide similar alternatives, such as composting, landfill cover and hazardous versus nonhazardous landfills, and options for on-site, permitted disposal via a properly licensed facility and contractor.

The Resource Conservation and Recovery Act (RCRA) applies to all considerations that have cradle-to-grave management of hazardous waste from generator to disposal. Use of environmental counsel is suggested in developing an overall plan. Most states are now authorized to administer RCRA. Standards higher than the federal regulations can now be required by the states.

The process for treating hazardous materials must be evaluated in light of multimedia interactions so that the hazardous components of the wastes are concentrated and isolated for treatment and transferred from one matrix to another. Items to consider during unit process selection are

- How to manage mixed waste streams with the potential of conflicting treatment goals,
- Equipment fouling,
- Generation of a toxic byproduct or intermediate, and
- Low residual concentration requirements.

An example considers volatile organic compounds (VOCs) and metals in a dilute waste stream. To remove volatiles with a stripping process, metals may have to be removed first to avoid fouling the stripping operation. Precipitation of the metals, however, would coprecipitate or sorb some of the organics in the flow of the metal precipitates. In this example, metals are concentrated by precipitation, but the presence of organics may affect how the solids are subsequently treated. The VOCs are stripped and transferred into a different media, an air stream or water vapor stream, and then concentrated by reabsorption onto carbon or by cooling into a condensate. The condensate or the carbon, now organic rich, requires further treatment. If the VOCs were accompanied by semivolatiles requiring further treatment, that treatment could also produce other waste streams, perhaps a solid phase, that could be combined with the metal precipitates of the earlier treatment step.

The challenge of simultaneously considering multimedia interactions, plus trying to avoid unnecessary generation of recycle or waste byproduct streams, is complicated by the very low concentrations of toxic compounds in waste streams. Frequently, concentrations in the part per billion (ug/L) range are intolerable for compounds of concern; parts per trillion (ng/L) is not unknown as a target treatment level. As analytical techniques are improved, detection levels are lowered, and health effects experience is gained, one can only expect greater concern for very low allowable limits of hazardous materials.

In selecting processes to remove contaminants of concern, one must look at process performance for a majority of a class or family of compounds that are present or specific removal for particular compounds of concern. The complexity and reliability of the treatment step must be considered next. The ability of the process to accept a wide range of stream variability and other off-design operating expectations need to be addressed. Ease of testing and demonstration are part of the selection process. Ease of integration with existing or planned facilities is important so that space requirements, piping runs, utilities, and waste stream utilization are optimized. All of these aspects can be related to economics.

When there is a regulatory interface, documentation, planning detail, and compatibility with existing permits must also be taken into account. If one is able to demonstrate a proven technology application in a situation, it will be easier to develop the desired working relationship with regulatory authorities.

LIQUIDS TREATMENT PROCESS. In considering a general scheme of treatment train selection, it is perhaps helpful to consider an overall processing strategy. The first step is to phase-separate and generate oil-free aqueous and water-free oil streams. Solids may settle out in the process and gaseous streams may be generated. All will be relatively free of the other phases or matrices; the aqueous stream will be diluted, and the organic will be concentrated and may consist of sinkers and floaters with respect to the aqueous phase. All will be relatively homogeneous permitting ease of further treatment.

The next step is to remove metals. Metals lead to poisoned biological treatment and interfere with physical treatment because of scaling. Removal of volatile metals will remove constraints on possible thermal processes later in the sequence. Removal of organic-complexed metals from the oil stream will likewise not inhibit use of thermal processes.

Next, the various treatment combinations with organics that may be present must be considered. Segregation into categories such as the following may be helpful:

- Volatile and semivolatile,
- Biorefraction polynuclear aromatic hydrocarbons (PNAs), dioxins, and pesticides, and
- Chlorine-containing compounds.

Finally, polishing treatment may be required to attain the requisite levels of dissolved anions and cations. Proven technology exists in membrane, thermal, and ion exchange processing.

DILUTE ORGANICS. The organics portion of a waste stream can be treated by a number of oxidation processes, including biological, wet air oxidation, chemical and advanced, ultraviolet (UV) and ozone, or peroxide. They can also be concentrated by stripping into overheads and by absorption onto carbon, resin, or zeolite and subsequently treated as part of the matrix to which they are removed. Contaminant-specific processes have been developed for problems such as polychlorinated biphenyls (PCBs). When PCBs are present, the major process will be PCB-specific, and the others will provide pretreatment to enable that process to work.

Activated carbon has been used extensively for removing dilute organics. It can be applied with relatively little engineering because it can be "changed out" and replenished easily with fresh material on initial breakthrough. If carbon is the selected major process step, then pretreatment becomes focused on extending the life of the carbon between change outs. For example, one might air strip VOCs to avoid loading the absorption sites with those organics.

Table 9.1 shows some of Calgon's operating experience with carbon. In this table, benzene is readily removable by carbon; benzene is also readily removable by air or steam stripping. Phenol is shown as readily sorbed; it is also easily biologically oxidized. One has to choose judiciously among competing processes so that all the compounds of concern are removed most economically.

CONCENTRATED ORGANICS. The fate of toxic organics is logically thermal destruction. As with conventional combustion processes, atomization parameters, heating value, and flame temperature are important parameters in hazardous liquid incineration. With hazardous wastes, elevated flame temperatures and long residence times are the rule; pyrolytic or two-stage combustion is typical. When chlorine-containing organic waste is present, elaborate gas cleaning systems are used. The presence of volatile metals in the organic can further complicate combustion to the point at which potential incineration contractors cannot accept anything but low levels of contained metals according to their air permits. When considering a multiplicity of organic wastes, one must look at the miscibility or homogeneity of the mixture. Also, the flash points must be considered because the inherent safety depends on the most volatile component in the mixture.

SOLIDS. Industrial solid waste streams that contain hazardous wastes cover such a wide range that generalization is difficult. Examples include electric air furnace (EAF) dust, machine shop floor sweepings, and refinery oil and water separator sludges. The ultimate fate of these materials is unclear because of U.S. EPA's land ban. Maximum attention will be given to resource recovery. The EAF dust might be resmelted to remove zinc, lead, and cadmium so that the residual nonhazardous focus will be on volume and toxicity reduction. The ultimate fate of these wastes will depend on economic factors. Processing by incineration in combination with ash fixation will become more widely available as the land ban is implemented.

Table 9.1 Carbon influent and effluent.

Compound	Occurrences	Influent concentration range	Carbon effluent concentration achieved
Carbon tetrachloride	4	130 μg/L–10 mg/L	<1 μg/L
Chloroform	5	20 μg/L–3.4 mg/L	<1 μg/L
Dichlorodiphenyl dichloroethane	1	1 μg/L	<0.05 μg/L
Dichlorodiphenyl dichloroethylene	1	1 μg/L	<0.05 μg/L
Dichlorodiphenyl trichloroethane	1	6 μg/L	<0.05 μg/L
cis-1,2-Dichloroethylene	8	5 μg/L–4 mg/L	<1 μg/L
Dichloropentadiene	1	450 μg/L	<10 μg/L
Disopropyl ether	2	20–30 μg/L	<1 μg/L
Tartiary methyl-butylether	1	33 μg/L	<5.0 μg/L
Diisopropyl methyl phosphonate	1	1250 μg/L	<50 μg/L
1,3-Dichloropropene	1	10 μg/L	<1 μg/L
Dichlorethyl ether	1	1.1 μg/L	<1 μg/L
Dichloroisopropylether	1	0.8 μg/L	<1 μg/L
Benzene	2	0.4–11 mg/L	<1 μg/L
Acetone	1	10–100 μg/L	<1 μg/L
Ethyl acrylate	1	200 μg/L	<10 μg/L
Trichlorotrifloroethane	1	6 mg/L	<1 μg/L
Methylene chloride	2	1–21 mg/L	<100 μg/L
Phenol	2	63 mg/L	<1 μg/L
Orthochlorophanel	1	100 mg/L	<1 μg/L
Tetrachlorethylene	10	3 μg/L–70 mg/L	<1 μg/L
Trichlorethylene	15	5 μg/16 mg/L	<1 μg/L
1,1,1-Trichloroethane	6	60 μg/L–25 mg/L	<1 μg/L
Vinylidiane chloride	2	3 μg/L–4 mg/L	<1 μg/L
Toluene	1	3–7 mg/L	<10 μg/L
Xylene	1	0.2–10 mg/L	<101 μg/L

INTEGRATED TREATMENT PROCESS

An integrated treatment process is a concept. It eludes precise definition but may be said simply to fit the situation. Its characterization, however, is useful, and that may include several factors such as the following. First, treatment process elements should be applied to segregated streams to effect treatment. Thus, streams to be treated are combined or separated so that the largest stream flows receive commensurate treatment. Conversely, contaminants that are comparably difficult to remove should not be commingled with larger streams and may first be concentrated or separated so that minimal quantities receive more costly treatment occasioned by more rigorous processing techniques. Second, treatment should be combined if possible to effect multiple removals by a single process element. Finally, contaminants that would hinder a downstream process element should be removed first. Of course, the least expensive process should be considered to effect a given removal. One has to evaluate the myriad options.

Conceptually, integrated treatment process elements are presented in Figure 9.1. An aqueous influent is assumed to contain organics, metals, and settleable solids all at low concentrations. This general case easily leads one to consider floating and sinking organics (DNAPLS and LNAPLS), dissolved and suspended metals, as well as emulsions as suggested by Figure 9.1.

The initial treatment step is typically to remove solids because their presence hinders downstream treatment. Floaters and sinkers can frequently be

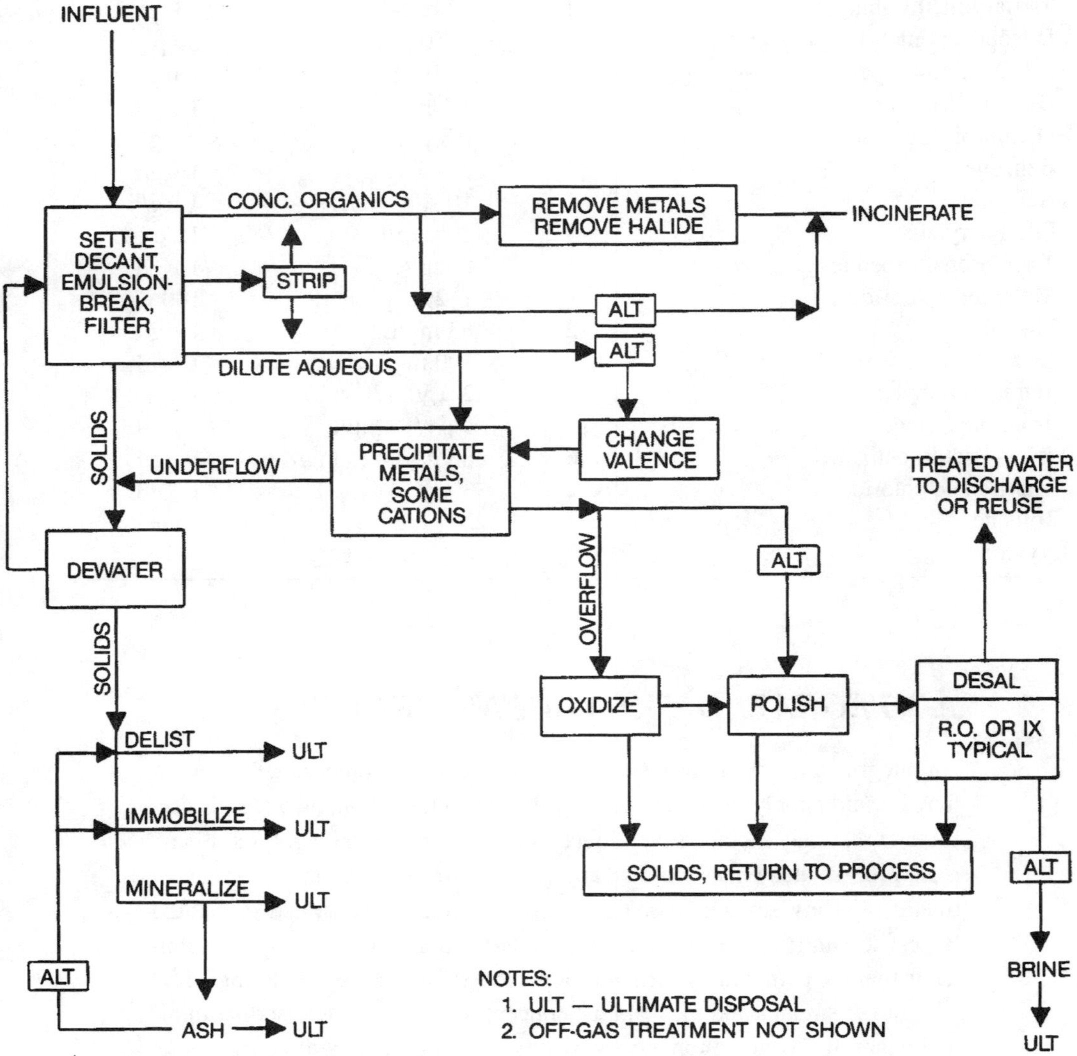

Figure 9.1 Hazardous waste treatment process flow diagram.

separated in this treatment step, particularly if gravity separation is used. Separated solids may contain adsorbed organics or metals that are chemically or otherwise combined with solids. The API separator illustrates equipment that accomplishes the multiple-phase separation. It can produce overflow and underflow nonaqueous phase organics, an aqueous layer, and sludge. These streams all require consideration for reuse or further treatment.

The aqueous phase in this example may contain dissolved organics, heavy metals, and water quality parameters requiring treatment as shown in Figure 9.1. Options for treating organics include air stripping, adsorption, or oxidation (which can be biological or chemical). Treatment off-site could be an option if there is capacity in an existing municipal wastewater treatment plant. Chemical oxidation could use UV and ozone or air stripping of volatiles to a carbon absorption unit. It should be noted that high heavy metal concentrations could require pretreatment to allow biological oxidation without concern for poisoning the microorganisms. Use of air stripping could be hindered by scaling tendencies and also require pretreatment. Elevated halide levels could cause corrosion problems and could influence the choice of treatment process.

Air stripping could be an attractive alternative process for removal of organics. Its use could require either disposal or regeneration of the carbon. A variation of this process could be to thermally desorb the organics off the carbon and then feed the concentrated vapors on-site to a thermal destruction unit.

Alternatively, one could treat for metals first using various techniques and then remove organics. Metal removals by precipitation or sorption remove a portion of the organics present. This could influence how residual solids are managed. Use of an ion exchange process would be hindered by high levels of organics.

If contaminated streams contain hexavalent chrome or cyanide, they would require initial redox treatment before commingling and resulting dilution. Treated effluents from redox processing can then be treated to remove the remaining chemicals of concern.

CASE STUDY

There is no single standard solution to the characterization and treatment of a hazardous waste stream. Selection of treatment options depends on the nature of the hazardous material and its treatability, the required effluent concentration levels, and economic considerations.

The following case study is for an industrial plant that stores its operating chemicals, consisting of solvents and other hydrocarbons, in aboveground and below-grade process vats and tanks. Fugitive waste streams (spills and leaks) are often the most difficult to identify and handle. The following are several approaches that can be used to collect the necessary information and several treatment options that have been used to handle the problem.

DATA COLLECTION. The design of a treatment system requires the collection of data about the material to be treated. A treatment problem typically has a beginning that may not appear to be a major problem. The following is a summary of the steps that are often followed when a fugitive waste stream is noticed. It is presented in a chronological framework.

TANK OR SUMP REMOVAL. The manager of the industrial facility has noticed that an underground tank or sump is losing product or that a process change requires the underground tank or sump to be removed or replaced. The plant manager may call either a consultant or contractor.

When the underground tank or sump is removed, it is found that contamination exists in the area. The work area may be either fenced or lined with plastic and filled with clean materials, while the samples are being analyzed by the laboratory. The decision to fence or backfill is often one of safety and the expected time before the remediation of work, if any, will be undertaken.

When the soil and water sample results come from the laboratory with concentrations in excess of the allowable limits, it is necessary to complete a site investigation to determine the lateral and vertical extent of the contamination so that remediation options can be developed.

SITE INVESTIGATION. The investigation and remediation of contaminated soil and water may be handled by different methods. Because the soil is often the source of the ongoing release of contaminants into the groundwater, the characterization and remediation of the soil contamination should be handled first.

Two basic approaches are available for the site investigation. When possible, a vapor probe study or soil gas survey is used to determine the lateral extent of the soil contamination. Using the vapor probe data, the quantity of soil to be treated or removed can be determined. If the contamination is deep or cannot be detected with the vapor probe equipment, soil and water samples can be collected by drilling and the samples analyzed in the laboratory.

The vapor probe method is inexpensive and very flexible for pursuing contamination paths, provides immediate results, and does minimum damage to the site. The typical cost for a study is $3000 to $6000/day. The method does have depth limitations and does not provide laboratory analysis. It is a good method to use to define the limits of the soil contamination because laboratory samples will be taken during the site excavation and monitoring well installations. The lateral extent of groundwater contamination is not detected using this method, but the quality of the water will be determined by the required monitoring wells around the excavation site. An alternative technique uses push probe technology and can combine the benefit of soil gas surveys and conventional drilling at reasonable costs, and there are other methods that can be considered. The choice of method depends highly on site-specific considerations such as depth and type of soil. The presence of DNAPLs can invoke still different approaches that are beyond the scope of this chapter.

Using drilling equipment and laboratory analysis of the samples can greatly increase the cost and accuracy of the site investigation. Drilling also allows for better definition of the vertical extent of the contamination. Because a limited number of borings can be completed in a given day, the equipment is not as flexible as tracking near-surface contamination plumes. Both soil and water contamination can be tracked with this method.

SOIL REMEDIATION. Soil contamination can be handled by several methods. Specific site conditions, including soil type, rainfall, and groundwater depth, will affect the selection of a treatment option. Typical options include

- In situ aeration, chemical treatment, or biological treatment;
- Excavation and transportation to a Class I or Class II disposal site, depending on the concentrations in the soil; or
- Excavation, on-site treatment by aeration, chemical, or biological processes and disposal as clean soil.

UNDERGROUND TANK PROJECT. A typical work site for an underground fuel tank is shown in Figure 9.2. An estimate was prepared for the quantity of construction work to be completed, and the costs for soil removal, treatment, and disposal were determined. These data are given in Table 9.2. The costs of consultant supervision and report writing were not included.

After the soil contamination problem is solved, several monitoring wells must be installed around the excavation site. Typically these are 5 to 10 cm (2 to 4 in.) in diameter. Typically, three wells are required to ensure that one well is installed down-gradient from the excavation site. The costs for well installations and quarterly monitoring for 1 year are given in Table 9.3.

GROUNDWATER CONTAMINATION. If the monitoring well samples are clean, the project is complete. If hydrocarbon contamination is found in the water samples, additional site investigation and groundwater remediation is required. The number of additional borings and monitoring wells required to determine the extent of the contamination cannot be determined in general terms. However, the basic approach is to drill additional monitoring wells in the down-gradient direction until the extent of the contamination plume is determined.

After the extent of groundwater contamination has been determined and the degree of contamination has been assessed, the treatment system is designed. Additional information about the site is needed to determine the water quality, flow rate, and disposal path.

The geologic data collected during the installation of the soil borings and monitoring wells yield information necessary for the design of the groundwater pumping system. Specific pumping capacity and the radius of influ-

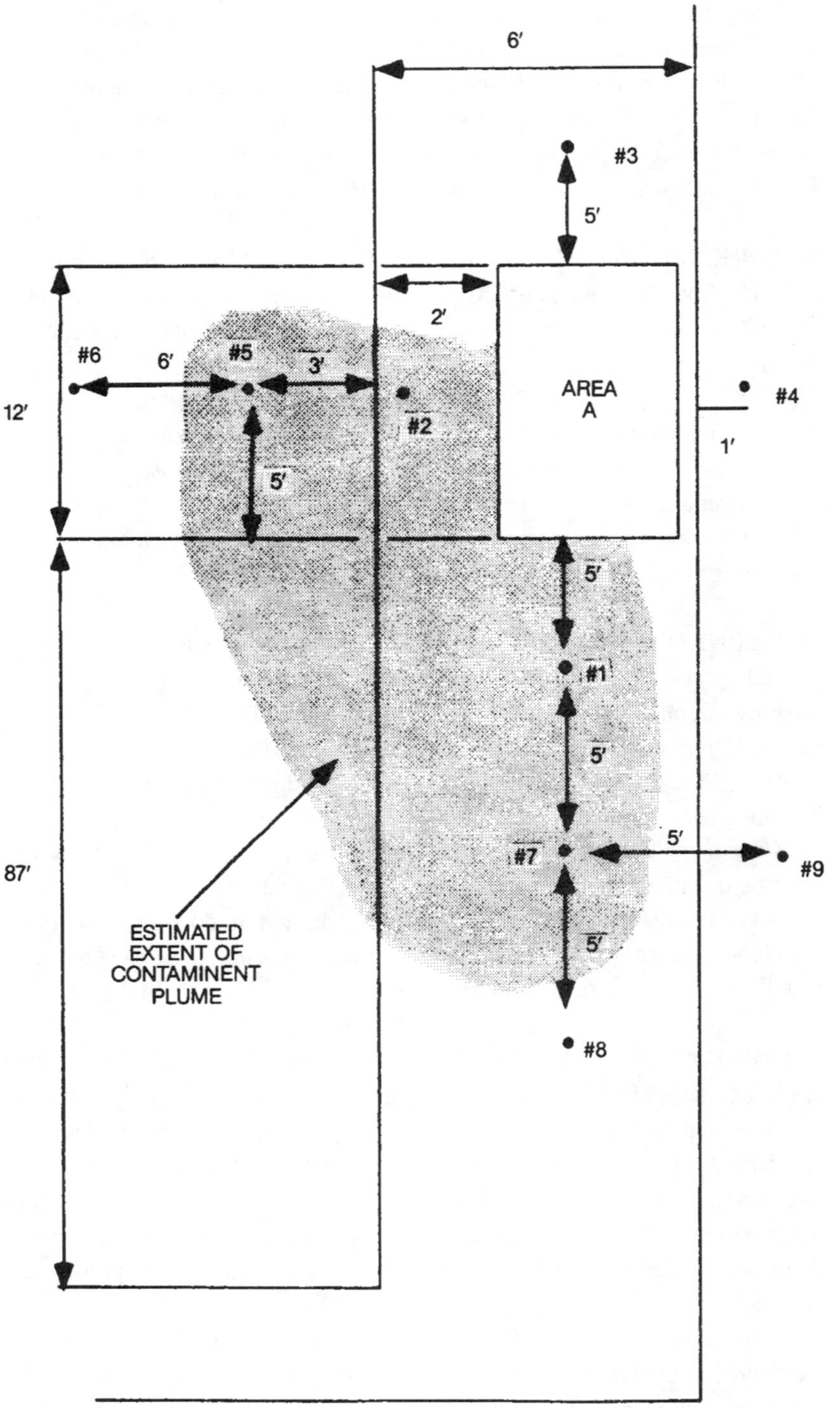

Figure 9.2 Underground storage tank site investigation (ft × 0.304 8 = m).

Table 9.2 Contractor costs for soil remediation and site work.

Description	Quantity	Units[a]	Unit price	Total
Concrete removal	232	sq ft	2.00	$464.00
Rock backfill removal	27	cu yd	30.00	810.00
Excavation soil removal	81	cu yd	30.00	2 430.00
Surface soil removal	13	cu yd	30.00	390.00
Gravel backfill placement	121	cu yd	20.00	2 420.00
Concrete sidewalk placement	174	sq ft	5.00	870.00
Concrete curb and gutter	32	ln ft	15.00	480.00
Total site work				**$7 860.00**
Soil remediation—bioremediation				
Soil treatment	94	cu yd	60.00	$5 640.00
Soil disposal	94	cu yd	30.00	2 820.00
Freight	94	cu yd	20.00	1 880.00
Total				**$10 340.00**
Hazardous waste disposal alternatives				
Subtitle C landfill	141	tons	185.00	$26 085.00
Subtitle D landfill (including freight charges)	141	tons	115.00	$16 215.00

[a]sq ft × 0.092 90 = m^2; cu yd × 0.764 6 = m^3; ln ft × 0.304 8 = m; tons × 1.016 = Mg.

ence are typically determined by the use of a pumping test. An existing monitoring well can be used for the pump test, but typically a larger and deeper well is required to provide sufficient drawdown to create an adequate radius of influence for the collection of the contaminated water. A typical drawdown curve is shown in Figure 9.3.

Table 9.3 Monitoring well installation.

Install one 5-cm (2-in.) monitoring well to 9 m (30 ft), collect soil and water samples	
Monitoring well	
Professional labor	$1 500.00
Lab analysis, three soil and one water samples. Analyze for TPH, BTEX	864.00
Drilling contractor	1 450.00
Other direct costs	125.00
Total	$3 939.00
Quarterly monitoring	
Collect quarterly water samples for 1 year from one well and analyze for TPH and BTEX. Annual cost.	
Professional labor	$800.00
Lab analysis, four water samples. Analyze for TPH and BTEX.	864.00
Other direct costs	100.00
Total	$1 764.00

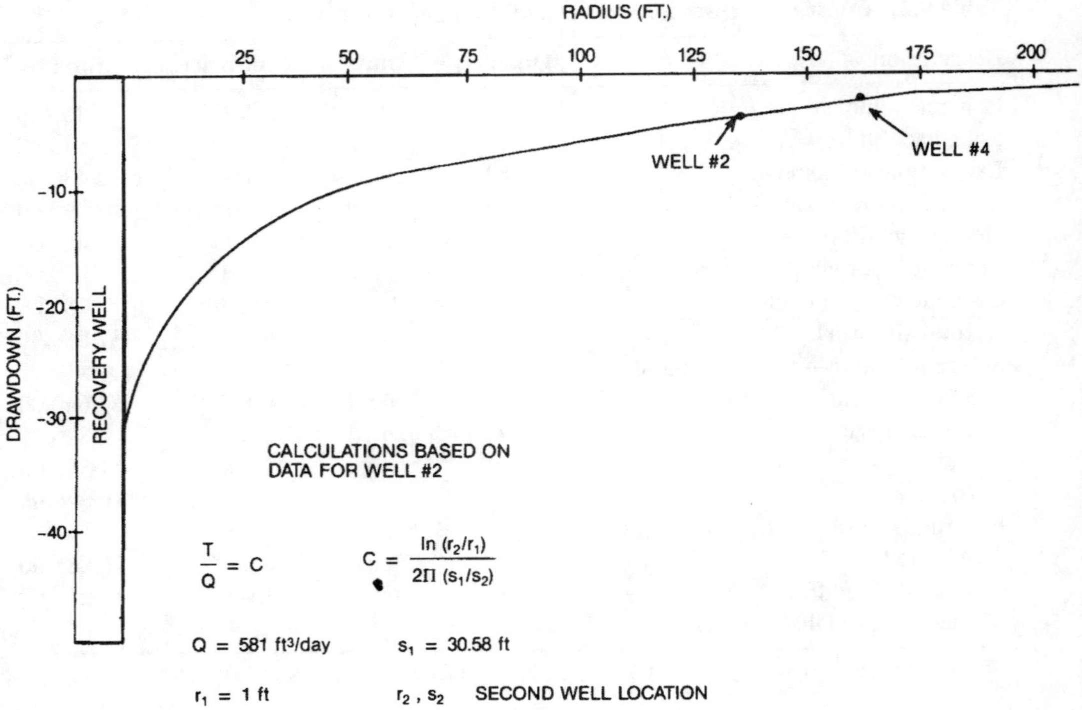

Figure 9.3 Drawdown curve (ft × 0.304 8 = m; cu ft/d × 3.277 × 10^{-4} = L/s).

After the pumping rate has been determined, the groundwater treatment system can be designed. Figure 9.4 shows a system designed to extract floating product, treat the groundwater, and reinject the treated water into an infiltration gallery up-gradient from the recovery wells. The cost of the treatment system is given in Table 9.4.

The treatment of the groundwater will continue until the necessary level of cleanup is achieved. Often the groundwater can be cleaned to an acceptable level and the equipment shut off. The monitoring wells will still be evaluated for at least 1 year, and the equipment may need to be turned on again if contamination levels start to increase.

CHLORINATED ORGANICS. Solvent contamination in groundwater can require a major cleanup effort even when the concentrations are very low. Typical treatment systems often involve air stripping and carbon adsorption. The treatment costs associated with this type of cleanup are given in Table 9.5.

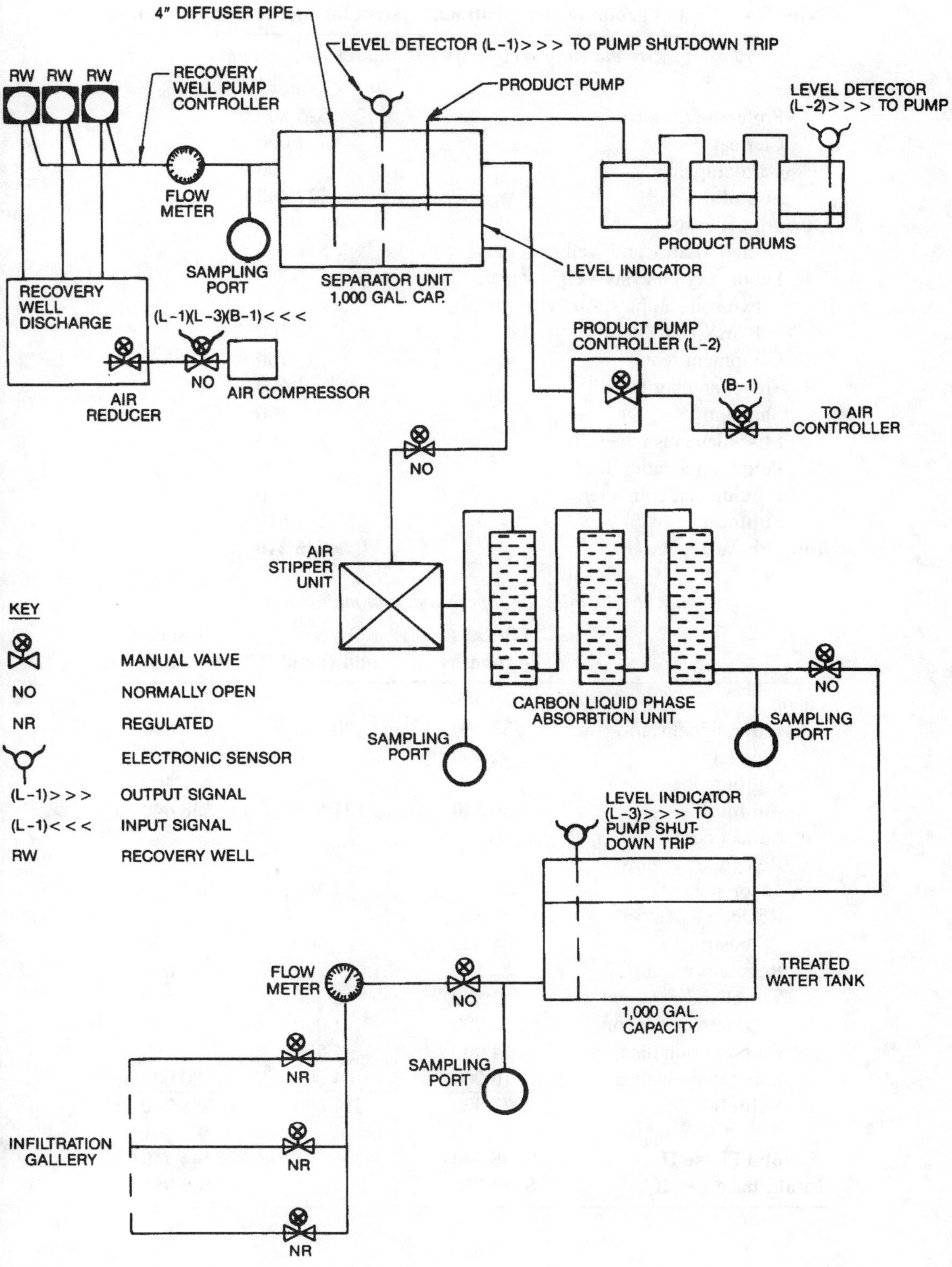

Figure 9.4 Groundwater treatment system.

Table 9.4 Cost of groundwater treatment system for fuel contamination.

Phase I Groundwater recovery wells and remediation plan

Labor	
Professional staff	$25 800
Clerical	1 000
Cartographer	600
Subtotal	**$27 400**
Other direct costs	
Driller—Labor and well materials	$12 250
Laboratory analysis—eight water, two soil samples. Analyze for BTEX and TPH.	$2 160
Equipment rental	700
Travel and per diem	1 600
Shipping	100
Miscellaneous materials	100
Permit application fees	600
Printing and computer	300
Subtotal	$17 810
Total Phase 1	**$45 210**

Phase II Installation, operation, and maintenance

	Total 24 months	**3-year additional**	**Total 36 months**
Labor			
Professional staff	$33 840	$20 000	$53 840
Clerical	1 800	1 000	2 800
Cartographer	900	500	1 400
Subtotal	**$36 540**	**$21 500**	**$58 040**
Other direct costs			
Remedial systems components	$43 000	$4 300	$47 300
Remedial area construction	4 000	400	4 400
Regulatory charges	500	500	1 000
Travel and per diem	3 500	2 000	5 500
Subcontractor labor	1 000	1 000	2 000
Carbon drum disposal	4 000	2 500	6 500
Laboratory analysis	16 000	4 000	20 000
Subtotal	**$72 000**	**$14 700**	**$86 700**
Subtotal Phase II	**$108 540**		**$144 740**
Total Phase I + II	**$153 750**		**$189 950**

Table 9.5 Cost of groundwater treatment system for chlorinated organics.

Install large diameter recovery well, develop, sample, and install well head	$10 000.00
Install all underground piping and electrica and connect to services	16 000.00
Install pumping and treatment system	50 000.00
System startup and compliance sampling	7 500.00
System operation and monitoring for 3 years	130 000.00
Total	$213 500.00

SUGGESTED READINGS

Anderson, M.J., and Whitcomb, P.J. (1996) Optimize Your Process-Optimization Efforts. *Chem. Eng. Prog.*

Freeman, H.M. (Ed.) (1989) *Standard Handbook of Hazardous Waste Treatment and Disposal.* McGraw-Hill, New York.

Treworgy, E.D. (1988) Simulating Hazardous Waste Treatment. Chem. Eng.

U.S. Environmental Protection Agency (1987) *A Compendium of Technologies Used in the Treatment of Hazardous Waste.* EPA-625/8-87-014, U.S. EPA, Washington, D.C.

Water Environment Federation (1997) *Water Environ. Res.,* **69,** 385.

Water Pollution Control Federation (1987) *Removal of Hazardous Wastes in Wastewater Facilities: Halogenated Organics.* Manual of Practice No. FD-11, Water Pollut. Control Fed., Alexandria, Va.

Appendix A Acronyms and Abbreviations

The hazardous waste treatment field uses many acronyms and abbreviations, as do many technical fields. This list contains the most frequently used acronyms and abbreviations.

Acronym	Definition
AA	Atomic Absorption
AA	Assistant Administrator (ARCS)
ABN	Acid/base neutral semivolatile compound
ACGIH	American Conference of Governmental Industrial Hygienists
ACM	Asbestos Containing Material
ACS	American Chemical Society
AEA	Atomic Energy Act
AG	Attorney General
AGI	American Geological Institute
AGT	Above-Ground Tank
AHERA	Asbestos Hazard Emergency Response Act
AIC	Acceptable Intake for Chronic Exposure
AIChE	American Institute of Chemical Engineers
AISE	Acceptable Intake for Subchronic Exposure
AISI	American Iron and Steel Institute
ALR	Action Leakage Rate
ANOVA	Analysis of Variance
ANSI	American National Standards Institute
AOC	Area of Contamination
APCD	Air Pollution Control Device

Acronym	Definition
API	American Petroleum Institute
APR	Air-Purifying Respirator
AQCR	Air Quality Control Region
AQMA	Air Quality Maintenance Area
ARAR	Applicable or Relevant and Appropriate Regulations
ARCS	Alternative Remedial Contracting Strategy
ASTM	American Society for Testing and Materials
ASTSWMO	Association of State and Territorial Solid Waste Management Officials
ATSDR	Agency for Toxic Substances and Disease Registry
AWMD	Air and Waste Management Division
AWQC	Ambient Water Quality Criteria
BAP	Benzo (a) pyrene
BART	Best Available Retrofit Technology
BAT	Best Available Technology
BDAT	Best Demonstrated Available Technology
BDL	Below Detectable Limits
BEIs	Biological Exposure Indices
BEJ	Best Engineering Judgment
BHP	Biodegradation, Hydrolysis, and Photolysis
BMP	Best Management Practice
BNA	Bureau of National Affairs
BNA	Base Neutral/Acid Semivolatile Compound
BOD	Biochemical Oxygen Demand
BRB	Bromofluorobenzene
BTEX	Benzene, Toluene, Ethylbenzene, Xylene
BTGA	Best Technology Generally Available
BTU	British Thermal Unit
CA	Cooperative Agreement (ARCS)
CA	Corrective Action
CAA	Clean Air Act
CAER	Community Awareness and Emergency Response
CAM	Contract Administration Manager (ARCS)
CAMU	Corrective Action Management Unit
CAP	Capacity Assurance Plan
CAPA	Critical Aquifer Protection Area
CASAC	Clean Air Scientific Advisory Committee
CBO	Congressional Budget Office
CBOD	Carbonaceous Biochemical Oxygen Demand
CDC	Centers for Disease Control
CEL	Continuous Exposure Limit
CEM	Continuous Emission Monitor
CEMS	Continuous Emission Monitoring System

Acronym	Definition
CEPP	Chemical Emergency Preparedness Program
CEQ	Council on Environmental Quality
CERCLA	Comprehensive Environmental Response Compensation and Liability Act
CERCLIS	Comprehensive Environmental Response Compensation and Liability Information System
CFR	Code of Federal Regulations
CGI	Combustible Gas Indicator
CGL	Comprehensive General Liability
CHEMNET	Mutual Aid Network of Chemical Shippers and Contractors
CHEMIEC	Chemical Transportation Emergency Center
CHLOREP	Mutual Aid Group of Chlorine Shippers and Carriers
CHRIS	Chemical Hazardous Response Information System
CIS	Contract Information System
Cl	Chlorine
CLP	Contract Laboratory Program
CMA	Chemical Manufacturers Association
CMI	Corrective Measures Implementation
CMS	Corrective Measures Study
CN	Cyanide
CO	Consent Order
COC	Chain of Custody
COD	Chemical Oxygen Demand
COE	Army Corps of Engineers
CPE	Chlorinated Polyethylene Polymer
CPE-A	Chlorinated Polyethylene Alloy
CPF	Carcinogenic Potency Factor
CPS	Centipoise
CPSR	Contractor Purchasing System Review
CQA	Construction Quality Assurance
CR	Polychloroprene (Neoprene)
CRC	Contamination Reduction Corridor
CRC (2nd)	Community Relations Coordinator
CRDL	Contract Required Detection Limit
CRL	Central Regional Laboratory (Region 5)
CRP	Community Relations Plan
CRZ	Contamination Reduction Zone
CSPE	Chlorosulfonated Polyethylene Polymer (Hypalon)
CWA	Clean Water Act
CZMA	Coastal Zone Management Act
DAF	Dilution/Attenuation Factor
DBCP	1,2-Dibromo-3-chloropropane

Acronym	Definition
DC	Direct Current
DCA	Deoxycorticosterone Acetate or Dichloroethylene
DCE	Dichloroethylene
DCM	Dichioromethane
DDT	Dichlorodiphenyltrichloroethane
DEC	Department of Environmental Conservation
DERA	Defense Environmental Repair Act
DERP	Defense Environmental Restoration Program
DFTPP	Decafluorotriphenyl Phosphine
DI	Distilled or Deionized
DL	Detection Limit
DMP	Data Management Plan
DMR	Discharge Monitoring Report
DO	Dissolved Oxygen
DOC	Dissolved Organic Carbon
DOD	Department of Defense
DOE	Department of Energy
DOT	Department of Transportation
DQO	Data Quality Objective
DRE	Destruction and Removal Efficiency
DWPL	Drinking Water Priority List
DWRD	Drinking Water Research Division
EA	Environmental or Endangerment Assessment
EAF	Electric Arc (or Air) Furnace
EC	Emergency Coordinator
ECAO	Environmental Criteria and Assessment Office
ED	Electrodialysis
EDl0	10% Effective Dose
EDB	Ethylenedibromide
EDD	Enforcement Decision Document
EDF	Environmental Defense Fund
EDR	Electrodialysis Reversal
EECA	Engineering Evaluation and Cost Analysis
EHS	Extremely Hazardous Substance
EIL	Environmental Impairment Liability (Insurance)
EIS	Environmental Impact Statement
EM	Electromagnetic
EMA	Emergency Management Agency
EMI	Emergency Management Institute
EMSL	Environmental Monitoring and Support Laboratory
ENR	Engineering News Record
EOC	Emergency Operating Center
EOD	Explosive Ordnance Disposal

Acronym	Definition
EOP	Emergency Operations Plan
EP	Extraction Procedure
EPA	Environmental Protection Agency
EPAAR	Environmental Protection Agency Acquisition Regulation
EPCRA	Emergency Planning and Community Right-to-Know Act
EPDM	Ethylene Propylene Diene Monomer
EPIC	Environmental Photographic Interpretation Center
EQA	Environmental Quality Act
ERA	Expedited Response Action
ERB	Environmental Response Branch
ERC	Emergency Response Coordinator
ERCS	Emergency Response Cleanup Service
ERD	Environmental Response Division
ERP	Emergency Response Plan
ERRB	Emergency and Remedial Response Branch
ERRIS	Emergency and Remedial Response Inventory System
ERT	Emergency Response Team
ESATs	Environmental Service Assistance Teams
ESBEPA	Environmental Services Branch (EPA)
ESD	Environmental Services Division
ESP	Electrostatic Precipitator
EXOH	Ethanol
EZ	Exclusion Zone
FAAS	Flame Atomic Absorption Spectroscopy
FEMA	Federal Emergency Management Agency
FFDCA	Federal Food, Drug, and Cosmetic Act
FIFRA	Federal Insecticide, Fungicide, and Rodenticide Act
FIT	Field Investigation Team
FML	Flexible Membrane Liner
FOIA	Freedom of Information Act
FP	Facilities Plan
FPR	Federally Permitted Release
FR	*Federal Register*
FRC	Functional Residual Capacity
FRP	Fiberglass Reinforced Plastic
FS	Feasibility Study
FSP	Field Sampling Plan
FTGs	Field Technical Guidelines
FWPCA	Federal Water Pollution Control Act
FWQC	Federal Water Quality Criteria

Acronym	Definition
FWS	Fish and Wildlife Service
GAC	Granular Activated Carbon
GACC	Granular Activated Carbon Column
GAO	Government Accounting Office
GC	Gas Chromatography
GC/MS	Gas Chromatography/Mass Spectrometry
GEMS	Graphical Exposure Modeling System
GIS	Geographic Information System
gpm	Gallons per Minute
GWSS	Groundwater Supply Survey
HAZMAT	Hazardous Material
HCB	Hexachlorobenzene
HCL	Hydrochloric Acid
HCS	Hazard Communication Standard
HDPE	High-Density Polyethylene
HEA	Health Effect Assessment
HEP	Habitat Evaluation Procedure
HEPA	High-Efficiency Particulate Air (filter)
HH	High Hazard
HHE	Health Hazard Evaluation
HM/HW	Hazardous Material/Hazardous Waste
HMCRI	Hazardous Materials Control Research Institute
HMIR	Hazardous Material Intelligence Report
HMIS	Hazardous Materials Information System
HMR	Hazardous Material Regulations
HMTA	Hazardous Materials Transportation Act
HOCs	Halogenated Organic Compounds
HQ	Headquarters
HRS	Hazard Ranking System
HRSD	Hazardous Response Support Division
HSCD	Hazardous Site Control Division (EPA Headquarters)
HSI	Habitat Suitability Index
HSL	Hazardous Substances List
HSP	Health and Safety Plan
HSWA	Hazardous and Solid Waste Amendments (to RCRA)
HU	Habitat Unit
HWAC	Hazardous Waste Action Coalition
HWDF	Hazardous Waste Derived Fuel
HWEB	Hazardous Waste Enforcement Branch
HWSF	Hazardous Waste Storage Facility
IAG	Interagency Agreement
ICARs	International Civil Aviation Regulations
IDD	Internal Due Diligence (Audit)

Acronym	Definition
IDLH	Immediately Dangerous to Life and Health
IEM	Inhalation Exposure Methodology
IEMS	Integrated Emergency Management System
IFR	Interim Final Rule
IM	Inquiry Memorandum (ARCS)
IPL	Interim Priorities List (now NPL)
IPM	Integrated Pesticide Management
IQRPE	Independent, Qualified, Registered, Professional Engineer
IR	Installation Restoration (USATHMA's RI/PS)
IRMS	Installation Restoration Management System
IRP	Installation Restoration Program
ISSA	Intra- or Interservice Support Agreement
IWS	Ionizing Wet Scrubber
JIT	Just In Time
JOSH	Job Occupational Safety and Health
LA	Load Allocation
LAER	Lowest Achievable Emission Rate
LCP	Laboratory Certification Program
LCRS	Leachate Collection and Removal System
LCS	Low-Cost Solution
LD5O	Quantity of a substance administered either orally or by skin contact necessary to kill 50% of exposed animals in laboratory tests within a specified time
LDPE	Low-Density Polyethylene
LDR	Land Disposal Restriction
LEL	Lower Explosive Limit
LEPC	Local Emergency Planning Committee
LFL	Lower Flammable Limit
LL	Liquid Limit
LLDPE	Linear Low-Density Polyethylene
LOD	Limit of Detection
LOQ	Limit of Quantitation
LQMP	Laboratory Quality Management Plan
LRT	Liquid Release Test
LSC	Liquid Sample Concentration
LUST	Leaking Underground Storage Tank
LVZ	Low-Velocity Zone
MACT	Maximum Achievable Control Technology
MAD	Maximum Applicable Dose
MAL	Maximum Allowable Level
MBAS	Methylene Blue Active Substance
MBE	Minority Business Enterprise
MCL	Maximum Contaminant Level

Acronym	Definition
MCLG	Maximum Contaminant Level Goal
MCP	Municipal Compliance Plan
MCRT	Mean Cell Residence Time
MDL	Minimum Detection Limit
MDL	Method Detection Limit
MDWL	Maximum Drinking Water Level
MED	Minimum Effective Dose
MEFR	Maximum Expiratory Flow Rate
MF	Membrane Filter
MF	Microfiltration
mg	Milligram
mgd	Million Gallons per Day
mg/L	Parts per Million
MIT	Mechanical Integrity Test
mL	Milliliter
MMS	Minerals Management Service
MMT	Million Metric Tons
MPN	Most Probable Number
MPRSA	Marine Protection Resource and Sanctuaries Act
MS	Mass Spectrometer
MSDS	Material Safety Data Sheet
MSHA	Mining Safety and Health Administration
MSW	Municipal Solid Waste
MSWLF	Municipal Solid Waste Landfill
MVV	Maximum Voluntary Ventilation
MWTA	Medical Waste Trading Act of 1988
NAAQS	National Ambient Air Quality Standards
NACE	National Association of Corrosion Engineers
NAMS	National Air Monitoring Station
NAS	National Academy of Sciences
NCATH	National Campaign Against Toxic Hazards
NCDC	National Climatic Data Center
NCIC	National Cartographic Information Center
NCP	National Contingency Plan
NCRIC	National Chemical Response and Information Center
NCSL	National Council of State Legislatures
ND	Not Detected
NDD	National Decision Document
NEIC	National Enforcement Investigations Center
NEPA	National Environmental Policy Act
NESHAPs	National Emission Standards for Hazardous Air Pollutants
NETC	National Emergency Training Center

Acronym	Definition
NFPA	National Fire Protection Association
NFRAP	No Further Remedial Action Planned
NGA	National Governor's Association
NGVD	National Geodetic Vertical Datum
NHMP	National Human Monitoring Program
NIAID	National Institute of Allergy and Infectious Diseases
NIH	National Institutes of Health
NIOSH	National Institute for Occupational Safety and Health
NIPDWR	National Interim Primary Drinking Water Requirements
NIPDWS	National Interim Primary Drinking Water Standards
NIRS	National Inorganics and Radionuclides Survey
NIST	National Institute of Standards and Technology
NOAA	National Oceanic and Atmospheric Administration
NOAEL	No Observed Adverse Effect Level
NOD	Notice of Deficiency
NOEL	No Observed Effect Level
NOMS	National Organics Monitoring Survey
NORS	National Organics Reconnaissance Survey
NOV	Notice of Violation
NPDES	National Pollutant Discharge Elimination System
NPDWR	National Primary Drinking Water Regulations
NPDWS	National Primary Drinking Water Standards
NPL	National Priorities List
NPO	National Program Office
NPS	National Pesticide Survey
NRC	National Response Center
NRC	Nuclear Regulatory Commission
NRDAM/CME	Natural Resource Damage Assessment for Coastal and Marine Environments
NRDC	National Resource Defense Council
NRT	National Response Team
NSF	National Sanitation Foundation
NSF	National Strike Force
NSN	National Stock Number
NSP	National Screening Program
NSPS	New Source Performance Standards
NSWMA	National Solid Waste Management Association
NTIS	National Technical Information Service
NTNCWS	Nontransient Noncommunity Water System
NTP	National Toxicology Program
NTU	Nephebretric Turbidity
NWF	National Wildlife Federation

Acronym	Definition
NWS	National Weather Service
NWWA	National Water Well Association
O&M	Operation and Maintenance
OAR	Office of Air and Radiation
OBA	Oxygen Breathing Apparatus
ODA	Ocean Dumping Act
ODW	Office of Drinking Water
OECM	Office of Enforcement and Compliance Monitoring
OERR	Office of Emergency and Remedial Response
OEW	Ordnance and Explosive Waste
OGWP	Office of Groundwater Protection
OHMTADS	Oil/Hazardous Materials Technical Assistance Data System
OLM	Organic Leachate Model
OMEP	Office of Marine and Estuarine Protection
OMPC	Office of Municipal Pollution Control
OPA	Office of Policy Analysis
OPMO	Office of Program Management and Operations
OPP	Office of Pesticide Programs
OPTS	Office of Pesticides and Toxic Substances
ORD	Office of Research and Development
ORM	Otherwise Regulated Material
ORP	Office of Radiation Programs
ORS	Ordnance Ranking System
OSC	On-Scene Coordinator
OSD	Office of the Secretary of Defense
OSHA	Occupational Safety and Health Act
OSOT	Oil Spill Operations Team
OSR	Office of Standards and Regulations
OSW	Office of Solid Waste
OSWER	Office of Solid Waste and Emergency Response
OTA	Office of Technology Assessment
OTS	Office of Toxic Substances
OURM	Operable Unit Remedial Measures
OUST	Office of Underground Storage Tanks
OVA	Organic Vapor Analyzer
OW	Office of Water
OWPE	Office of Waste Programs Enforcement
OWRS	Office of Water Regulation and Standards
PA	Preliminary Assessment
PA/SI	Preliminary Assessment/Site Inspection
PAC	Powdered Activated Carbon
PAH	Polyaromatic Hydrocarbons

Acronym	Definition
PAIR	Preliminary Assessment Information Rule
PAL	Preventive Action Limit
PAPR	Powered Air-Purifying Respirator
PCA	Tetrachloroethane
PCBs	Polychlorinated Biphenyls
PCDDs	Polychlorinated dibenzo-p-dioxins
PCDFs	Polychlorinated dibenzo furans
PCE	Tetrachloroethylene
PCP	Pentachlorophenol
PDP	Preliminary Determination Phase
PDS	Personnel Decontamination Station
PE	Performance Evaluation
PE	Polyethylene
PEIS	Programmic Environmental Impact Statement
PEL	Permissible Exposure Level
PFE	Potency Factor Estimate
PFLT	Paint Filter Liquids Test
PFS	Phased Feasibility Study
PHC	Principal Hazardous Constituents
PHRED	Public Health Risk Evaluation Database
PIAT	Public Information Assist Team
PIRP	Public Information Response Program (or Plan)
PN	Public Notification
PNA	Polynuclear Aromatics
POHCs	Principal Organic Hazardous Constituents
POLs	Petroleum, Oil, and Lubricants
POTWs	Publicly Owned Treatment Works
POU	Point-of-Use Technologies
ppb	Parts per Billion
PPE	Personal Protective (Clothing and) Equipment
ppm	Parts per Million
ppt	Parts per Trillion
PQL	Practical Quantitation Level
PRP	Potentially Responsible Party
PSD	Prevention of Significant Deterioration (Air)
PSTN	Pesticide Safety Team Network
PTA	Packed Tower Aeration
PVC	Polyvinyl Chloride
PWS	Public Water System
PWSS	Public Water System Supervision
QA	Quality Assurance
QA/QC	Quality Assurance/Quality Control
QAMS	Quality Assurance Management Staff

Acronym	Definition
QAPP	Quality Assurance Project Plan
QARMP	Quality Assurance Reference Materials Project
QC	Quality Control
QCP	Quality Control Plan
QCSR	Quality Control Summary Report
RA	Remedial Action
RACT	Reasonably Available Control Technology
RAP	Remedial Action Plan
RAP	Response Action Program
RAS	Routine Analytical Service
RBC	Rotating Biological Contactor
RCP	Regional Contingency Plan
RCRA	Resource Conservation and Recovery Act of 1978
RCS	Regulated Chemical Substance
RD	Remedial Design
RD/RA	Remedial Design/Remedial Action
RDCO	Regional Document Control Officer
RDF	Refuse Derived Fuel
REEP	Reasonable Extra Efforts Program (Air)
REL	Recommended Exposure Limit
REM	Remedial Planning
REM/FIT	Remedial Planning/Field Investigation Team
RFA	Request for Application
RFD	Reference Dose
RFI	RCRA Facility Investigation
RI	Remedial Investigation
RI/FS	Remedial Investigation/Feasibility Study
RIA	Regulatory Impact Analysis
RITTA	RCRA Integrated Training and Technical Assistance
RLL	Rapid Extremely Large Leakage
RLSC	Regional Laboratory Services Coordinator
RM	Regional Manager (ARCS)
RMCL	Recommended Maximum Concentration Limit (Water)
RMICs	Recommended Maximum Impurity Concentrations
RO	Reverse Osmosis
ROD	Record of Decision
RQ	Reportable Quantity
RRT	Regional Response Team
RSPA	Research and Special Programs Administration
RSPO	Remedial Site Project Officer
RTP	Reinforced Thermosetting Plastic
RTR	Reinforced Thermosetting Resin

Acronym	Definition
RV	Residual Volume
RWS	Rural Water Survey
SAB	Science Advisory Board (of EPA)
SAP	Sampling and Analysis Plan
SAR	Supplied-Air Respirator
SARA	Superfund Amendments and Reauthorization Act of 1986
SARMs	Standard Analytical Reference Materials
SAS	Special Analytical Services
SBIR	Small Business Innovative Research
SCAP	Superfund Comprehensive Accomplishments Plan
SCBA	Self-Contained Breathing Apparatus
SCS	Soil Conservation Service
SDI	Silt Density Index
SDI	Subchronic Daily Intake
SDL	Sample Detection Limit
SDWA	Safe Drinking Water Act
SDWAA	Safe Drinking Water Act Amendments
SERC	State Emergency Response Commission
SHERP	Safety, Health, and Emergency Response Plan
SI	Site Inspection
SI units	International System of Units
SIC	Standard Industrial Category
SIM	Selected Ion Monitoring
SIP	State Implementation Plan (Air)
SITE	Superfund Innovative Technology Evaluation
SLAMSs	State and Local Air Monitoring Stations
SLN	Special Local Needs
SM	Site Manager (ARCS)
SMCL	Secondary Maximum Containment Level
SMCRA	Surface Mining Control and Reclamation Act
SMO	Sample Management Office
SMOA	Superfund Memorandum of Agreement
SMWP	Sampling and Monitoring Work Plan
SNARL	Suggested No Adverse Response Level
SNUN	Significant New Use Notice
SNUR	Significant New Use Rule
SOC	Synthetic Organic Chemical
SOP	Standard Operating Procedure
SOQ	Statement of Qualifications
SOW	Statement of Work
SP	Sampling Plan
SPCC	Spill Prevention Control and Countermeasures

Acronym	Definition
SPE	Solid Phase Extraction
SPHEM	Superfund Public Health Evaluation Manual
SQGs	Small Quantity Generators
SS	Suspended Solids
SSA	Sole Source Aquifer
SSB	Support Services Branch
SSC	Site Safety Coordinator (SSHO, SSO, and HSO)
SSC	Scientific Support Coordinator
SSHO	Site Safety and Health Office (SSC, SSO, and HSO)
SSO	Site Safety Officer (SSC, SSHO, and HSO)
SSSP	Site-Specific Safety Plan
SSU	Saybolt Seconds Universal (viscosity measurement)
STAPPA	State and Territorial Air Pollution Program Administration
STAR	Stability Array
STEL	Short-Term Exposure Limit
STORET	Storage and Retrieval for Water Quality Data
SW	Solid Waste
SWDA	Solid Waste Disposal Act
SWMUs	Solid Waste Management Units
T-EPDM	Thermoplastic EPDM
TAG	Technical Assistance Grant
TAL	Target Analyte List
TAT	Technical Assistance Team
TBC	(Advisories) To Be Considered
TCA	Trichloromethane
TCDD	Tetrachlorodibenzo (p) dioxin, 2,3,7,8- (Dioxin)
TCE	Trichloroethylene
TCL	Target Compound List
TCLP	Toxicity Characteristic Leaching Procedure
TDD	Technical Directive Document
TDS	Total Dissolved Solids
TECP	Totally Encapsulated Chemical Protection
TEGD	Technical Enforcement Guidance Document
TEP	Typical End-Use Product
TGD	Technical Guidance Document
THM	Trihalomethane
TIC	Tentatively Identified Compound
TKN	Total Kjeldahl Nitrogen
TLD	Thermoluminescence Detector
TLV	Threshold Limit Value
TLV-C	Threshold Limit Value-Ceiling
TLV-STEL	Threshold Limit Value-Short-Term Exposure Limit

Acronym	Definition
TLV-TWA	Threshold Limit Value-Time Weighted Average
TMDL	Total Maximum Daily Load
TOC	Total Organic Carbon
TOH	Total Organic Halogen
TOX	Total Organic Halide
TPH	Total Petroleum Hydrocarbon
TPQ	Threshold Planning Quantity
TRDs	Technical Resource Documents
TRI	Toxic Release Inventory
TSCA	Toxic Substances Control Act
TSD	Toxic Substance Disposal
TSD	Treatment, Storage, and Disposal
TSDFs	Treatment, Storage, and Disposal Facilities
TSP	Total Suspended Particulate
TSS	Total Suspended Solids
TTO	Total Toxic Organics
TTUs	Transportable Treatment Units (hazardous waste)
TUAC	Total Unburned Hydrocarbons
TWA	Time Weighted Average
UCR	Unit Carcinogenic Risk
UEL	Upper Explosive Limit
UF	Ultrafiltration
UFL	Upper Flammable Limit
UHWM	Uniform Hazardous Waste Management
UIC	Underground Injection Control
UNAMAP	User's Network for Applied Modeling of Air Pollution
USA-CERL	U.S. Army Construction Engineering Research Laboratory
USALMC	U.S. Army Logistics Management Center
USATHAMA	U.S. Array Toxic and Hazardous Materials Agency
USC	U.S. Code
USCG	U.S. Coast Guard
USCS	Unified Soil Classification System
USDA	U.S. Department of Agriculture
USDI	U.S. Department of Interior
USDW	Underground Source of Drinking Water
USGS	U.S. Geological Survey
USNRC	U.S. Nuclear Regulatory Commission
USPIRG	U.S. Public Interest Research Group
UST	Underground Storage Tank
UV	Ultraviolet
VOA	Volatile Organic Acid

Acronym	Definition
VOC	Volatile Organic Compound
VOST	Volatile Organic Sampling Train
VTSR	Verified Time of Sample Receipt
WAO	Wet Air Oxidation
WAP	Waste Analysis Plan
WEF	Water Environment Federation (formerly WPCF)
WLA	Wasteload Allocation
WMDDRD	Waste Minimization Destruction/Disposal Research Division
WMP	Waste Minimization Program
WP	Work Plan
WPCF	Water Pollution Control Federation (now Water Environment Federation)
WQA	Water Quality Act of 1987
WQC	Water Quality Criteria
WRAP	Waste Reduction Assessment Program
WREAFS	Waste Reduction at Federal Sites
WRITE	Waste Reduction Innovative Technology Evaluation

Index

B

C

D

E

F

G

R

S

T

U

V

W